Global Biodiversity
Volume 3
Selected Countries in Africa

Global Biodiversity

Volume 3

Selected Countries in Africa

Edited By
T. Pullaiah, PhD

Apple Academic Press Inc.
3333 Mistwell Crescent
Oakville, ON L6L 0A2
Canada

Apple Academic Press Inc.
9 Spinnaker Way
Waretown, NJ 08758
USA

© 2019 by Apple Academic Press, Inc.

First issued in paperback 2021

Exclusive worldwide distribution by CRC Press, a member of Taylor & Francis Group

No claim to original U.S. Government works

Global Biodiversity, Volume 3: Selected Countries in Africa
ISBN 13: 978-1-77463-130-0 (pbk)
ISBN 13: 978-1-77188-722-9 (hbk)

Global Biodiversity, 4-volume set
ISBN 13: 978-1-77188-751-9 (hbk)

Library and Archives Canada Cataloguing in Publication

Global biodiversity (Oakville, Ont.) Global biodiversity / edited by T. Pullaiah, PhD.

Includes bibliographical references and indexes.
Contents: Volume 3. Selected countries in Africa.
Issued in print and electronic formats.
ISBN 978-1-77188-722-9 (v. 3 : hardcover).--ISBN 978-0-42946-980-0 (v. 3 : PDF)

1. Biodiversity--Asia. 2. Biodiversity--Europe. 3. Biodiversity--Africa. I. Pullaiah, T., editor II. Title.

QH541.15.B56G66 2018   578.7   C2018-905091-8   C2018-905092-6

CIP data on file with US Library of Congress

Apple Academic Press also publishes its books in a variety of electronic formats. Some content that appears in print may not be available in electronic format. For information about Apple Academic Press products, visit our website at **www.appleacademicpress.com** and the CRC Press website at **www.crcpress.com**

Contents

About the Editor

T. Pullaiah, PhD
Former Professor, Department of Botany,
Sri Krishnadevaraya University, Anantapur, Andhra Pradesh, India,
E-mail: pullaiah.thammineni@gmail.com

T. Pullaiah, PhD, is a former Professor at the Department of Botany at Sri Krishnadevaraya University in Andhra Pradesh, India, where he has taught for more than 35 years. He has held several positions at the university, including Dean, Faculty of Biosciences, Head of the Department of Botany, Head of the Department of Biotechnology, and Member of Academic Senate. He was President of Indian Botanical Society (2014), President of the Indian Association for Angiosperm Taxonomy (2013) and Fellow of Andhra Pradesh Akademi of Sciences. He was awarded the Panchanan Maheshwari Gold Medal, the Dr. G. Panigrahi Memorial Lecture award of the Indian Botanical Society and Prof. Y.D. Tyagi Gold Medal of the Indian Association for Angiosperm Taxonomy, and a Best Teacher Award from Government of Andhra Pradesh. Under his guidance 54 students obtained their doctoral degrees. He has authored 46 books, edited 17 books, and published over 330 research papers, including reviews and book chapters. His books include *Ethnobotany of India* (5 volumes published by Apple Academic Press), *Flora of Andhra Pradesh* (5 volumes), *Flora of Eastern Ghats* (4 volumes), *Flora of Telangana* (3 volumes), *Encyclopaedia of World Medicinal Plants* (5 volumes, 2nd edition), and *Encyclopaedia of Herbal Antioxidants* (3 volumes). He was also a member of Species Survival Commission of the International Union for Conservation of Nature (IUCN). Professor Pullaiah received his PhD from Andhra University, India, attended Moscow State University, Russia, and worked as a Postdoctoral Fellow during 1976–78.

Contributors

Lawrence Misa Aboagye
CSIR-Plant Genetic Resource Research Institute, Bunso, Ghana

Komla Elikplim Abotsi
Department of Botany and Plant Ecology, University of Lomé, Togo

Thiombiano Adjima
University Ouaga1 Pr Joseph Ki-Zerbo, 03 BP 7021 Ouagadougou 03, Burkina Faso,
E-mail: adjima_thiombiano@yahoo.fr

Komlan M. Afiademanyo
Department of Zoology, University of Lomé, Togo

Wafaa M. Amer
Department of Botany and Microbiology, Faculty of Science, Cairo University, Egypt,
E-mail: wafaa_amer@hotmail.com

Komina Amévoin
Department of Zoology, University of Lomé, Togo

Diop Sall Aminata
National Parks Directorate, Senegal, E-mail: aminat71@yahoo.fr

Benjamin Annor
CSIR-Crops Research Institute, P.O. Box 3785, Kumasi, Ghana

Judith L. Arnolds
South African National Biodiversity Institute, Kirstenbosch Research Centre, Private Bag X7,
Claremont 7735, Cape Town, RSA, E-mail: j.arnolds@sanbi.org.za

Jenneh F. Bebeley
Sierra Leone Agricultural Research Institute (SLARI), PMB 1313, Tower Hill, Freetown, Sierra Leone,
West Africa

John Bukombe
Tanzania Wildlife Research Institution (TAWIRI), Box 661 Arusha, Tanzania

Hagir Mahagoub EL-Nasry
University of Khartoum, faculty of Forestry, Khartoum North, Sudan

Khaled S. Etayeb
Department of Zoology, Faculty of Sciences, Tripoli University, Libya, E-mail: khaledetayeb@yahoo.com

Mondjonnesso Gomina
Department of Zoology, University of Lomé, Togo

Danni Guo
South African National Biodiversity Institute, Kirstenbosch Research Centre, Private Bag X7,
Claremont 7735, Cape Town, RSA, E-mail: d.guo@sanbi.org.za

Abdelnasir Ibrahim Hano
University of Khartoum, faculty of Forestry, Khartoum North, Sudan

Ouedraogo Issaka
University Ouaga1 Pr Joseph Ki-Zerbo, 03 BP 7021 Ouagadougou 03, Burkina Faso

Nacoulma Blandine Marie Ivette
University Ouaga1 Pr Joseph Ki-Zerbo, 03 BP 7021 Ouagadougou 03, Burkina Faso

Dimobe Kangbeni
University Ouaga1 Pr Joseph Ki-Zerbo, 03 BP 7021 Ouagadougou 03, Burkina Faso

Hamza Kija
Tanzania Wildlife Research Institution (TAWIRI), Box 661 Arusha, Tanzania

Kouami Kokou
Department of Botany and Plant Ecology, University of Lomé, Togo, E-mail: kokoukouami@hotmail.com

Nicephor Lesio
Tanzania Wildlife Research Institution (TAWIRI), Box 661 Arusha, Tanzania

Stephen Liseki
Tanzania Wildlife Research Institution (TAWIRI), Box 661 Arusha, Tanzania

Talaat Dafalla Abdel Magid
University of Bahri, College of Natural Resources, Khartoum North, Sudan

Mohammed H. Mahklouf
Department of Botany, Faculty of Sciences, Tripoli University, Libya, E-mail: mahklouf64@yahoo.com

Honori Maliti
Tanzania Wildlife Research Institution (TAWIRI), Box 661 Arusha, Tanzania

Gueye Mathieu
Laboratory of Botany, Department of Botany and Geology, UMI 3189, IFAN Ch. A. Diop BP 206
Dakar, Senegal, E-mail: gueye_guirane@yahoo.fr, mathieu.gueye@ucad.edu.sn

Mohamaed Menioui
Scientific Institute of Rabat, Mohammed V. University, Morocco, E-mail: mohamed.menioui@gmail.com

Jean B. Mikissa
National School of Forest and Water, Gabon, E-mail: jbmikissa@gmail.com

Ameer Awad Mohammed
Wildlife Research Centre, Khartoum North, Sudan

Emmanuel Bayani Ngoyi
General direction for Environmental, Gabon, E-mail: scoutgabon@yahoo.fr

Ally K. Nkwabi
Tanzania Wildlife Research Institution (TAWIRI), Box 661 Arusha, Tanzania,
E-mail: nkwabikiy@yahoo.com, ally.nkwabi@tawiri.or.tz

Prince E. Norman
Sierra Leone Agricultural Research Institute (SLARI), PMB 1313, Tower Hill, Freetown, Sierra Leone,
West Africa, E-mail: penorman2008@yahoo.com/p.norman@slari.gov.sl

Yvonne S. Norman
Sierra Leone Agricultural Research Institute (SLARI), PMB 1313, Tower Hill, Freetown, Sierra Leone,
West Africa

Patrick Ofori
CSIR-Soil Research Institute, Academy Post Office, Kwadaso, Kumasi, Ghana

Michael Kwabena Osei
CSIR-Crops Research Institute, P.O. Box 3785, Kumasi, Ghana, E-mail: oranigh@hotmail.com

Ouedraogo Oumarou
University Ouaga1 Pr Joseph Ki-Zerbo, 03 BP 7021 Ouagadougou 03, Burkina Faso

Flore Koumba Pambou
Institute for Research on Agriculture and Forests, Gabon, E-mail: fkoumbapambo@gmail.com

Houinsodé Segniagbeto
Department of Zoology, University of Lomé, Togo

Janatu V. Sesay
Sierra Leone Agricultural Research Institute (SLARI), PMB 1313, Tower Hill, Freetown, Sierra Leone, West Africa

Stanford Mudenda Siachoono
Department of Zoology and Aquatic Sciences, School of Natural Resources, Copperbelt University, Kitwe, Zambia, E-mail: stanford.siachoono@cbu.ac.zm

Ahmed A. H. Siddig
University of Khartoum, faculty of Forestry, Khartoum North, Sudan; Harvard Forest, Harvard University – 324 North Main St., Petersham MA 1366, USA, E-mail: asiddig@eco.umass.edu

B. A. Taïbou
Ecological Monitoring Center, Rue Leon Gontran Damascus, Fann Residence, BP 15532 Dakar, Senegal, E-mail: taibou@cse.sn

Abbreviations

AAs	authorized association
ABS	access and benefit sharing
AGEOS	The Gabonese Agency for Space Studies and Observations
AIDS	acquired immune deficiency syndrome
ANB	National Biosafety Authority
ANEAFPF	The National Agency for the Execution of Activities of the Forest-Wood Sector
ANPA	The National Agency for Fisheries and Aquaculture
ANPN	The National Agency of National Parks
AOAD	Arab Organization for Agricultural Development
APGRC	Agricultural Plant Genetic Resources and Research Centre
ARC	Agricultural Research Centre
ARI	Animal Research Institute
BNARI	Biotechnology and Nuclear Agricultural Research Institute
CAWM	College of African Wildlife Management MWEKA
CBD	Convention on Biological Diversity
CFR	Cape Floristic Kingdom
CITES	Convention on Trade in Endangered Species
CNB	National Biosafety Committee
CNDD	National Commission for Sustainable Development
CONSERE	Higher Council of Natural Resources and the Environment
CPMR	Centre for Plant Medicine Research
CR	critically endangered
CREMA	Community Resource Management Areas
CRI	Crops Research Institute
DBEB	Department of Botany and Environmental Biology
DD	data deficient
DE	Department of the Environment
EEZ	Exclusive Economic Zone
EGA	Environmental General Authority
EHF	Ebola hemorrhagic fever
EIA	environmental impact assessment
EN	endangered
ESC	Ecological Monitoring Center
EVD	Ebola virus disease

FC	Forestry Commission
FNC	Forest National Corporation
FORIG	Forestry Research Institute of Ghana
GBIF	Global Biodiversity Information Facility
GCAs	Game Controlled Areas
GCFB	Guinea-Congo Forest Biome
GHG	green house gases
GPRS	Ghana Poverty Reduction Strategy
GRs	Game Reserves
GSBAs	Globally Significant Biodiversity Areas
IAS	Invasive Alien Species
IBA	Important Bird Area
IOC	International Ornithological Committee
ITCZ	Intertropical Convergence Zone
IUCN	International Union for Conservation of Nature
IUCN	The International Union for the Conservation of Nature
KNUST	Kwame Nkrumah University of Science and Technology
LC	least concern
LDCs	least developed countries
LEAP	List of East African Plants
LKHP	Lower Kihansi Hydropower Project
MBREMP	Mnazi Bay and Ruvuma Estuary Marine Park
MEA	Multilateral Environmental Agreements
MIMP	Mafia Island Marine Park
MPAs	Marine Protected Areas
NAPCP	National Program of Action to Combat Desertification
NBSAP	National Biodiversity Strategy and Action Plan
NCA	Ngorongoro Conservation Area
NCAA	Ngorongoro Conservation Area Authority
NE	not evaluated
NFI	National Forest Inventory
NFP	National Forestry Program
NIP	National Irrigation Program
NT	near-threatened
NWFP	non wood forest products
OKNP	Outamba Kilimi National Park
PA	Protected Areas
PADP	Protected Areas Development Program

PNBC   National Park of Basse Casamance
PNDS   National Park of Saloum Delta
PNIM   National Park National of Madeleine Islands
PNLB   National Park of Langue de Barbarie
PNNK   National Park of Niokolo-Koba
PNOD   National Birds Park of Djoudj
PRSP   Poverty Reduction Strategy Paper
RFFN   Ferlo Nord Wildlife Reserve
RNICS   Nature Reserve of Community Interest of Somone
RNP   Popenguine Nature Reserve
ROK   Ornithological Reserve of Kalissaye
RPS   Rangeland Development Strategy
RSFG   Special Wildlife Reserve of Gueumbeul
SARI   Savanna Agricultural Research Institute
SNE   National Environment Strategy
TaCMP   Tanga Coelacanth Marine Park
TanBIF   Tanzania Biodiversity Information Facility
TAWA   Tanzania Wildlife Management Authority
TAWIRI   Tanzania Wildlife Research Institute
TFAP   Tanzania Forestry Action Plan
TKA   traditional knowledge associated
TWS   Tiwai Wildlife Sanctuary
UAE   Useful Agricultural Surface
UCC   University of Cape Coast
UNFCCC   United Nation Framework Convention on Climate Change
VHFs   viral hemorrhagic fevers
VU   vulnerable
WD   Wildlife Division
WMAs   Wildlife Management Areas
WWF   World Wide Fund

Preface

The term 'biodiversity' came into common usage in the conservation community after the 1986 National Forum on BioDiversity, held in Washington, DC, and publication of selected papers from that event, titled *Biodiversity*, edited by Wilson (1988). Wilson credits Walter G. Rosen for coining the term. Biodiversity and conservation came into prominence after the Earth Summit, held at Rio de Janeiro in 1992. Most of the nations passed biodiversity and conservation acts in their countries. Biodiversity is now the buzzword of everyone from parliamentarians to laymen, professors, and scientists to amateurs. There is a need to take stock on biodiversity of each nation. The present attempt is in this direction.

The main aim of the book is to provide data on biodiversity of each nation. It summarizes all the available data on plants, animals, cultivated plants, domesticated animals, their wild relatives, and microbes of different nations. Another aim of the book series is to educate people about the wealth of biodiversity of different countries. It also aims to project the gaps in knowledge and conservation. The ultimate aim of the book is for the conservation of biodiversity and its sustainable utilization.

The present series of the four edited volumes is a humble attempt to summarize the biodiversity of different nations. Volume 1 covers *Biodiversity of Selected Countries in Asia*, Volume 2 presents *Biodiversity of Selected Countries in Europe,* Volume 3 looks at *Biodiversity of Selected Countries in Africa*, and Volume 4 contains *Biodiversity of Selected Coountriess in the Americas and Australia*. In these four volumes, each chapter discusses the biodiversity of one country. Competent authors have been selected to summarize information on the various aspects of biodiversity. This includes brief details of the country, ecosystem diversity/vegetation/biomes, and species diversity, which include plants, animals and microbes. The chapters give statistical data on plants, animals, and microbes of that country, and supported by relevant tables and   A ☐
diversity with emphasis on crop plants or cultivated plants, domesticated animals, and their wild relatives. Also mentioned are the endangered plants and animals and their protected areas. The book is profusely illustrated. We hope it will be a desktop reference book for years to come.

Biodiversity of some countries could not be presented in this book. This needs explanation. I tried to contact as many specialists as possible from these countries but was unable to convince these experts to write chapter on biodiversity of their country.

The book will be useful to professors, biology teachers, researchers, scientists, students of biology, foresters, agricultural scientists, wild life managers, botanical gardens, zoos, and aquaria. Outside the ⸤A will be useful for lawmakers (parliamentarians), local administrators, nature lovers, trekkers, economists, and even sociologists.

Since it is a voluminous subject, we might have not covered the entire gamut; however, we tried to put together as much information as possible. Readers are requested to give their suggestions for improvement for future editions.

I would like to express my grateful thanks to all the authors who contributed on the biodiversity of their countries. I thank them for their cooperation and erudition.

I wish to express my appreciation and help rendered by Ms. Sandra Sickels, Rakesh Kumar, and the staff of Apple Academic Press. Their patience and perseverance has made this book a reality and is greatly appreciated.

—T. Pullaiah, PhD

CHAPTER 1

Phytodiversity of Burkina Faso

NACOULMA BLANDINE MARIE IVETTE, OUEDRAOGO ISSAKA, OUEDRAOGO OUMAROU, DIMOBE KANGBENI, and THIOMBIANO ADJIMA

University Ouaga1 Pr Joseph Ki-Zerbo, 03 BP 7021 Ouagadougou 03, Burkina Faso, E-mail: adjima_thiombiano@yahoo.fr

1.1 About Burkina Faso

Burkina Faso is a landlocked country located in the heart of West Africa with an area of around 274,200 km². Its neighboring countries are Niger in the East, Benin, Togo, Ghana, and Ivory Coast (Côte d'Ivoire) in the South and Mali in the North (Figure 1.1). The relief is relatively flat with some sandstone cliffs in the West (Banfora) and the East (Gobnangou). The Mouhoun, the Pendjari, the Comoé, the Nazinon, and the Nakambé are the main rivers encountered in the country. The soils are predominantly ferruginous and raw mineral.

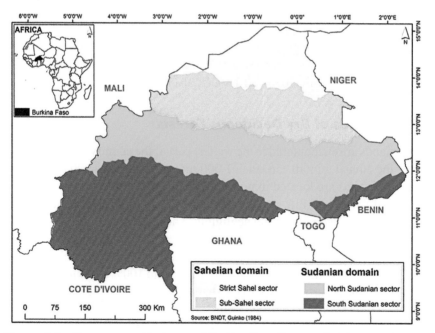

Figure 1.1 Phytogeographical map of Burkina Faso (Source: Guinko, 1984).

Based on annual average rainfall distribution Burkina Faso is divided into three ecoclimatic zones: the Sahelian zone in the northern part of the country, with annual rainfall of 300–600 mm; a transitional Sudano-Sahelian zone in the central region, with a total rainfall of 600–900 mm; and the Sudanian zone in the southern part, with annual rainfall of 900 to 1,100 mm. Therefore, the country's landscape, vegetation types and species diversity

☐   ☐

Regarding the vegetation, the country is subdivided into two phytogeographic domains (Sudanian and Sahelian domains) and each domain into two sectors (Figure 1.1). The Sudanian domain, which covers the Southern and Central parts of the country includes the South-Sudanian and North-Sudanian sectors. The northern part is covered by the Sahelian domain divided into two sectors (sub-Sahel and strict Sahel) according to the zonation done by Guinko in 1984. These vegetation types range from the shrubsteppe in the northern part, to the islands/patches of dense forests in the southern part of the country.

1.2   Status of Ecosystems in Burkina Faso

In the following paragraphs, the different vegetation types encountered in Burkina Faso from the south to the north are presented based on our own field observations and studies as well as on literature (Guinko, 1984; Ouédraogo, 2006; Ouedraogo et al., 2009; Thiombiano and Kampmann, 2010; Project BKF/015 – IFN 2 et IGB, 2015; Nacoulma, 2012; Gnoumou, 2013; Sambare, 2013; Traoré, 2013; MEEVCC, 2017).

1.2.1   Islands of Dry Deciduous Forests

These stands are encountered in the wettest part of the country (southern part) where the annual rainfall can reach 1,000 mm with more than 90 rainy days. The main stands are located in protected areas, particularly in the classified forest and partial wildlife reserve of Comoé Léraba. At the country level, the area of these islands is quite negligible. These islands are the most dense vegetation types found in Burkina Faso (Figure 1.2) with about 1,000 trees or shrubs per hectare and a basal area of 38 m²/ha. The mean total height of trees is 16 m with a rate of cover ranging from 97 to 100% (Gnoumou, 2013). The flora of these particular forests is characterized by a rich group of woody species (*Guibourtia copallifera, Gardenia nitida, Strychnos usambarensis, Alafia scandens, Landolphia hirsuta, Dalbergia hostilis, Salacia pallescens, Strophanthus sarmentosus, Ceiba pentandra* and *Mallotus oppositifolia*) as

Figure 1.2 Islands of dry deciduous forests of *Guibourtia copallifera* (Photo Credit, Thiombiano A., 2010).

well as herbaceous species (*Cissus petiolata, Commelina africana, Commelina subulata, Cyathula prostrata, Cyperus diffusus, Doryopteris kirkii, Elytraria marginata, Malaxis chevalieri, Hibiscus surattensis, Oplismenus hirtellus* and *Pouzolzia guineensis*). These forests are rich in azonal species, which are characteristics of wetter regions and are extremely vulnerable to disturbances such as bushfires and selective harvesting of wood. In unprotected areas, these forests were almost degraded by farmers in favor of farms and other plantations such as the cashew tree (*Anacardium occidentale*).

1.2.2 Woodlands

Woodlands occur mainly in the southern part of Burkina Faso with rainfall about 800 to 1,000 mm per year. According to the phytogeographical subdivision, woodlands are essentially confined to the south-Sudanian sector. In the protected areas of the Eastern region, cords of woodlands are found along the drains and valleys at river basins level (Ouédraogo, 2009). Woodlands represent the second most forested type of vegetation in the country (Figure 1.3) with sometimes densities reaching 556 trees per hectare and a basal area of 22 m²/ha (Table 1.1a). The density of juveniles is very heterogeneous. Values can be very variable from one woodland to another one

Figure 1.3 Woodland (Photo Credit, Thiombiano A., 2012).

(1,890 to 16,600 individuals/ha). The mean volume of wood is estimated at 28 m³/ha. Woodlands can store up to 52 Mg C/ha (Ouédraogo et al., 2009, Traoré, 2013; MEEVCC, 2017). The mean height of trees can reach sometimes 14 m according to Ouédraogo (2009).

The dominant and characteristic species such as *Anogeissus leiocarpa, Isoberlinia doka, Pterocarpus erinaceus, Afzelia africana, Cola cordifolia, Khaya senegalensis, Diospyros mespiliformis, Mitragyna inermis, Tamarindus indica*   the physiognomy of woodlands (Ouédraogo, 2009; Gnoumou, 2013; Traoré, 2013; Table 1.1a). The herbaceous layer of these vegetation types is characterized by *Wissadula amplissima, Setaria barbata, Achyranthes argentea, Desmodium velutinum, Hoslundia opposita, Peristrophe bicalyculata, Capparis sepiaria.*

The continuous extension of agricultural lands coupled with the uncontrolled and fraudulent exploitation of Woodlands' species (e.g., *Pterocarpus erinaceus*) lead to the   of these vegetation types in protected areas and sacred woods.

By comparing data from Ouadba (1997) cited in the country monograph of the biological diversity in 1999 to recent data from the land occupancy database (Project BKF/015 – IFN 2 et IGB, 2015), the area covered by woodlands in Burkina Faso drastically decreased from 287,000 ha to 7,658 ha (MEECV, 2017) indicating a loss of 97.33% of their total area within 15 years. These vegetation types are present in the form of relics and are generally found in the Boucle du Mouhoun, Cascades, South-Center, Eastern, Hauts-Bassins, and Southwestern regions.

1.2.3 Savannas

The savanna in its multiple faces is the most common vegetation type found in Burkina Faso with an area of 11,392,033 ha representing 41.7% of the country's total area (MEEVCC, 2017). It includes a tree and/or shrub layer scattered in a continuous grass cover. The tree species' rate of cover in savannas ranges from 2 to 50%. In fact, the savanna contains a rich floristic procession forming a continuum from the North-Sudanian to the South-Sudanian sector. There is a high diversity of savannas due to the diversification of habitats varying from lowlands to hills and other armored hillocks.

Overall, there are three types of savannas that can be distinguished in Burkina Faso. They are tree savannas, shrub savannas, and grass savannas. Another type namely savanna woodland, which describes a type of vegetation at the transition between savannas and forests might be observed. However, this savanna woodland is often merged into the large set of tree savannas.

1.2.3.1 Savanna Woodlands

In Burkina Faso, savanna woodlands are encountered in the North and South Sudanian sectors with a more pronounced presence in the western part than in the Eastern part of the country. Trees' rate of cover in this vegetation type reaches 45% with a mean height of 8 m for native trees species (Figure 1.4). The density of trees is about 468 individuals /ha with a basal area ranging from 7 to 13 m²/ha (Table 1.1a). Juveniles are proliferating in this savanna with densities varying from 15,512 to 28,424 juveniles per hectare (Traoré, 2013). Considering the flora, these savannas are distinct in the tree layer by species such as *Isoberlinia doka, Daniellia oliveri, Pterocarpus erinaceus, Vitellaria paradoxa, Burkea africana, Terminalia avicennioides* and also by *Acacia gerrardii* more specifically in the National Park of Arly located in the Eastern region of the country. The herbaceous vegetation of savanna woodlands is dominated by *Andropogon gayanus, Hyparrhenia involucrata* and *Andropogon pseudapricus* (Ouédraogo, 2009; Nacoulma, 2012).

Figure 1.4   Savanna woodland of *Terminalia avicenniodes* (Photo Credit, Ouedraogo O., 2009).

Statistics data on the evolutionary trend of savanna woodland are almost nonexistent. However, they could present regressive tendencies similar to the woodlands especially given that they colonize rich soils, exposing these vegetation types to the farming activities.

1.2.3.2   Tree Savannas

These vegetation types are found in the North and South Sudanian sectors. The grass cover is continuous with variable height and coverage depending on the location in the protected or nonprotected environment (Figure 1.5). The herbaceous layer reaches 2 m with 80% of coverage in a well-protected environment without grazing (case of protected areas in Eastern region). However, in the communal areas, the herbaceous layer hardly reaches 45 cm due to the intensity of grazing. The woody species that characterize tree savannas are *Vitellaria paradoxa, Lannea acida, Terminalia avicennioides, Daniellia oliveri, Anogeissus leiocarpa, Burkea africana* and *Pterocarpus erinaceus* (Traoré, 2013). *Andropogon gayanus, Andropogon pseudapricus, Hyparrhenia involucrata, Spermacoce stachydea* are the prominent herbaceous species found in these savannas (Ouédraogo, 2009). The density of the trees is estimated at 645 individuals/ha and a basal area of 11 m²/ha with a volume of wood of 30 m³/ha (Table 1.1a). This vegetation type has the highest abundance of juveniles (31,412 individuals/ha). The mean carbon sequestered by woody species in this savanna is estimated at 68.65 t/ha (MEEVCC, 2017).

Tree savannas represent about 16% (4,291,000 ha) of the total area of the country (Ouadba, 1997). In 2012, they accounted for 4.8% (1,316,101 ha) of the total area (MEEVCC, 2017), revealing a critical loss of 69% (between 1997 and 2012) of the original area.

1.2.3.3   Shrub and Grass Savannas

Shrub and grass savannas occur in the same phytogeographical zone on 10,075,932 ha representing 36.9% of the country total area (MEEVCC, 2017) and the essential of Burkina Faso vegetation. In terms of flora, shrub (Figure 1.6) and grass (Figure 1.7) savannas are characterized by the following woody species: *Acacia seyal, A. dudgeoni, Detarium microcarpum, Balanites aegyptiaca, Vitellaria paradoxa, Combretum glutinosum, C. nigricans* (Table 1.1a). The herbaceous layer consists mainly of *Andropogon pseudapricus, Loudetia togoensis,* and *Pennisetum pedicellatum*. The density of shrubs and some trees reach 667 individuals/ha with a small basal area of 5 m² /ha (Ouédraogo, 2009; Traoré, 2013; Table 1.1a). The density in juvenile individuals is, however, remarkable (21 331 individuals /ha). As for the volume of wood, it remains low at about 20 m³/ha. Moreover, the amount of carbon sequestered is about 40 t/ha (MEEVCC, 2017).

A variant of these savannas, which are in fact grass savannas (Figure 1.7), occupies lowlands and large alluvial plains with a woody cover rate less than 2%, a low density of shrubs (ca. 92 individuals/ha) and basal area of 3 m² /ha (Ouédraogo et al., 2008). Current climatic data, land use patterns, geographic distribution of species, and   composition indicate a trend toward the sahelization of Sudanian domain. The areas remain virtually unchanged with a decrease of 1% (−109,068 ha). Even though the difference is not observed at the physiognomy level within shrub

1.2.4   Steppes

The steppes are vegetation types that are specific to the Sahelian phytogeographic domain, recorded both in sub-Sahelian and Sahelian strict sectors. They are easily identifiable on the field by a scattered tree and/or shrub layer with a discontinuous grass cover whose mean height hardly exceeds 40 cm.

Figure 1.5   Tree savanna of *Sterculia setigera* (Photo Credit, Thiombiano A., 2010).

Figure 1.6   Shrub savanna (in the foreground) (Photo Credit, Ouedraogo O., 2009).

Figure 1.7 Swampy shrub savanna with grassy tendency of alluvial plains (Photo Credit, Ouedraogo O., 2009).

This vegetation type covers an area of 4,676,134 ha or 17.2% of the total area of Burkina Faso (MEEVCC, 2017). Steppes are recognizable by a very remarkable discontinuity between the green islands. These vegetation types consist of woody and herbaceous species of high fodder quality supporting the livestock breeding, one of the main activity of rural populations in the Sahel. In this area, two types of steppes were recorded: the trees steppe and shrubs or grassy steppes.

1.2.4.1   Trees Steppe

Tree steppes are sparse vegetation including generally small trees (Figure 1.8) with a low rate of woody cover. They are more common in the sub-Sahelian sector and cover an area of 178,189 ha, or 0.7% of the total area of the country (MEEVCC, 2017). The characteristic species of tree steppes are: *Pterocarpus lucens, Combretum micranthum, Grewia flavescens, Commiphora africana, Acacia erythrocalyx, Acacia tortilis,* and *Dalbergia melanoxylon.* In the herbaceous layer, there are *Schoenefeldia gracilis, Pennisetum pedicellatum, Senna obtusifolia, Zornia glochidiata, Aristida adscensionis, Aristida kerstingii, Eragrostis tremula,* and *Digitaria ciliaris* (Tindano et al., 2015; Zerbo et al.,

Figure 1.8   Tree steppe (Photo Credit, Thiombiano A., 2012).

2016). The tree density is about 434 individuals/ha and a basal area of 4 m²/ ha with a volume of wood equals to 12 m³/ha (Table 1.1a). The carbon content sequestered by these vegetation types is 22 t per hectare. Juvenile recruitment reaches 5,144 individuals per hectare (MEEVCC, 2017).

1.2.4.2   Shrub and Grassy Steppes

This type of vegetation is most prevalent in the strict Sahel over an area of 4,497,945 ha or 16.5% of the country's total area. It consists of shrubs scattered in the discontinuous grass cover (Figure 1.9). The rate of cover is low and varies between 10 and 30%. The height of herbaceous layer exceeds 60 cm and is composed mainly of annual species such as *Schoenefeldia gracilis, Zornia glochidiata, Cenchrus biflorus, Aristida adscensionis, Aristida kerstingii, Eragrostis tremula, Panicum laetum,* etc. The woody species characterizing shrub and grassy steppes are generally *Acacia seyal, Acacia tortilis* var. *raddiana, Balanites aegyptiaca,* and *Acacia laeta* (Table 1.1a).

The density of shrub steppes in the Sahel is 305 individuals/ha with a low basal area (3 m²/ha) (see Table 1.1a). Regeneration is also low with 790 juveniles/ha. The volume of wood is estimated at 5 m³/ha. The shrub and

Figure 1.9 Shrub and grassy steppe (Photo Credit, Thiombiano A., 2005).

grassy steppes sequester only 9 t of carbon per hectare (MEEVCC, 2017). The steppe landscape consists only of a relic of woody and herbaceous species that have been able to adapt to the high pressure of grazing and the aridity of the climate. Glacis and wetlands of the steppes are increasingly invaded by *Senna obtusifolia*, a species with low forage value. The occurrence and invasion of this species might be an indicator of the stage of steppes' physical degradation.

Referring to the statistics of the 1999 monograph and data from the land occupancy database in 2012, only steppes among the different types of vegetation in Burkina Faso experienced a   increase in their area of about 290% (+3,476,134 ha). Thus, much of the savannas have been converted into steppes, which indicates a high state of degradation of the country's vegetation.

1.2.5 Tiger Bush

The tiger bush forms a band of vegetation confined to the north of Gorom-Gorom. This type of vegetation is characterized by a succession of bare soil and covered areas with wood cover rates that can vary from 55 to 90% (Figure 1.10). Established in the lowlands with sandy-clay soil, these tiger bush

Figure 1.10   Tiger bush (Photo Credit, Ouedraogo A., 2005).

are characterized by species such as *Acacia ehrenbergiana, Ziziphus mauritiana, Bauhinia rufescens, Calotropis procera, Piliostigma reticulatum, Dichrostachys cinerea, Guiera senegalensis, Boscia angustifolia, Boscia senegalensis, Maerua crassifolia, Commiphora africana* (Ouédraogo, 2006 Table 1.1a). Tiger bush is relatively unaffected by the human pressure in the strict Sahel, but is severely affected by drought, which causes severe mortality of trees and shrubs (Ouédraogo, 2006).

1.2.6   Riparian Forests

The riparian forests refer to the vegetation type developing along the rivers (Sambaré, 2013). Based on the structural and ecological characteristics, the riparian forests are divided into two main types: gallery forests and riparian strips.

The gallery forests consist of a strip of vegetation strictly linked to the river, and constitutes the interface between the river and the surrounding vegetation (Figure 1.11). They are found mainly along the permanent and semipermanent rivers of the Sudanian domain such as Mouhoun, Nazinon, Comoé, and Pendjari. The area covered by gallery forests is quite small and estimated at 121,151 ha which represents 0.4% of the country total area (MEEVCC, 2017). The gallery forests of the North-Sudanian sector are dominated by *Pterocarpus santalinoides, Mitragyna inermis, Diospyros mespiliformis, Daniellia oliveri,* and *Anogeissus leiocarpa*. In the South-Sudanian sector *Berlinia grandiflora, Vitex chrysocarpa, Syzygium guineense, Dialium guineense, Cola laurifolia, Morelia senegalensis, Diospyros mespiliformis,*

Figure 1.11   Gallery forest along Pendjari river (Photo Credit, Ouedraogo O., 2009).

and *Carapa procera* are prominent (Sambaré, 2013). The density of gallery forests is 438 individuals/ha and the basal area is 22 m²/ha. These types of forests can sequester 50 t of carbon per hectare. There are nearly 29,108 juveniles/ha in gallery forests of Burkina Faso (MEEVCC, 2017). The riparian strips consist of a thin band of vegetation more or less narrow, located along the drainage axis. They are often encountered along the permanent and semi-permanent rivers of the Sahelian domain such as Sirba and Gourol. The thin vegetated band is often dominated by *Anogeissus leiocarpa*, *Acacia seyal* and *Mitragyna inermis* (Table 1.1b). The dominant species in riparian strips are: *Diospyros mespiliformis*, *Piliostigma reticulatum*, *Anogeissus leiocarpa*, *Balanites aegyptiaca* and *Acacia raddiana* in the strict Sahel sector. In the Sub-Sahel sector, *Acacia seyal*, *Piliostigma reticulatum*, *Balanites aegyptiaca*, *Mitragyna inermis* and *Acacia sieberiana* are the dominant woody species found in riparian strips. The riparian strips of the Sahel have a mean density of 233 individuals/ha with a basal area of 8 m²/ha (Traoré, 2013). Concerning the regeneration, there are nearly 7,404 juveniles per hectare.

At the level of riparian forests, the loss of biodiversity is especially accelerated by the construction of the facilities such as bridges and hydro-electric and hydro-agricultural developments that cause changes in hydrological regimes. Thus, the riparian forests have lost over the last two decades more than the half (55.12%) of their area since 1997 (Ouadba, 1997; MEEVCC, 2017).

Table 1.1a Vegetation Types of Burkina Faso and Their Characteristics

Vegetation types	Phytogeographic sectors	Rainfall (mm)	Area according to the land occupancy database (ha)	Density (trees and shrubs/ha)	Basal area (m²/ha)	Characteristic plant species
Island of dry deciduous forest	South-Sudanian	1000	–	1,000	38	*Guibourtia copallifera, Gardenia nitida, Dalbergia hostilis*
Woodland	South & North-Sudanian	800–1000	7,658	556	22	*Anogeissus leiocarpa, Isoberlinia doka, Pterocarpus erinaceus*
Savanna woodland	South & North-Sudanian	700–1000	–	468	7–13	*Vitellaria paradoxa, Burkea africana, Terminalia avicennioides*
Tree savanna	South & North-Sudanian	700–1000	1,316,101	645	11	*Vitellaria paradoxa, Lannea acida, Daniellia oliveri*
Shrub & grass savannas	South-Sudanian, North-Sudanian & Sub-Sahel	500–800	10,075,932	667	5	*Acacia seyal, Acacia dudgeoni, Combretum glutinosum, Andropogon pseudapricus, Loudetia togoensis*
Tree steppes	Sahel	300–400	178,189	434	4	*Acacia tortilis, Dalbergia melanoxylon, Balanites aegyptiaca*
Shrub & grassy steppes	Sahel	300–400	4,497,945	305	3	*Schoenefeldia gracilis, Zornia glochidiata, Cenchrus biflorus*
Tiger bush	Sahel	300–400	–			*Bauhinia rufescens, Guiera senegalensis, Boscia angustifolia, Boscia senegalensis*

Source: Ouédraogo (2006, 2009); Thiombiano and Kampmann (2010).Project BKF/015 – IFN 2 et IGB, (2015); Gnoumou (2013); Traoré (2013); MEECV (2017).

Table 1.1b Vegetation Types of Burkina Faso and Their Characteristics

Vegetation types	Area (ha)	Tree density (ind./ha) (These values given by the expert are very low!)	Basal area (m²/ha)	Characteristic plant species
Riparian strip	–	233	8	*Anogeissus leiocarpa, Acacia seyal, Mitragyna inermis*
Gallery forest	121,151	438	22	*Pterocarpus santalinoides, Mitragyna inermis, Diospyros mespiliformis, Daniellia oliveri, Berlinia grandiflora, Syzygium guineense, Carapa procera*

Source: Traoré (2013).

1.3   Status of Phytodiversity in Burkina Faso

The state of knowledge about plant species in Burkina Faso reveals relatively advanced investigations for vascular plants grouped in the Spermatophyta (subphylum of Angiospermae and Gymnospermae). However, the data on the lower plants (Bryophyta, Lycopodiophyta and Equisetophyta) are limited and almost nonexistent.

1.3.1   Pteridophyta (Ferns)

The current botanical inventories reveal 26 species of ferns and relatives recorded in Burkina Faso (Thiombiano et al., 2012). They are distributed over 14 genera and 12 families (Table 1.2).

1.3.2   Gymnospermae

The flora of Burkina Faso does not contain spontaneous species of Gymnosperms. Existing Gymnosperm species are all introduced and used as ornamental plants. Therefore, documentation on these species is very

Table 1.2 Number of Fern Species Per Taxonomic Group

Branch	Class	Order	Family	Genera	Species
Pteridophyta	Pteridopsida	Pteridales	Adiantaceae	*Adiantum*	2
				Doryopteris	1
				Pityrogramma	1
	Filicopsida	Hydropteridales	Azollaceae	*Azolla*	1
		Isoetales	Isoetaceae	*Isoetes*	3
		Polypodiales	Lomariopsidaceae	*Bolbitis*	1
			Oleandraceae	*Nephrolepis*	2
			Osmundaceae	*Osmunda*	1
			Parkeriaceae	*Ceratopteris*	1
			Thelypteridaceae	*Thelypteris*	2
	Lycopodiopsida	Lycopodiales	Lycopodiaceae	*Lycopodiella*	1
		Selaginellales	Selaginellaceae	*Selaginella*	1
	Equisetopsida	Salviniales	Marsileaceae	*Marsilea*	6
		Ophioglossales	Ophioglossaceae	*Ophioglossum*	3
Total	4	8	12	14	26

Source: Thiombiano et al. (2012).

limited. Table 1.3 shows a total of 6 woody species of Gymnosperms belonging to 5 genera, 2 families, 2 orders and 2 classes (Soma 2012; Thiombiano et al., 2012). Among these species, 5 were assessed by the IUCN Red List and ranked under the status of Least Concern (LC). Only *Araucaria excelsa* (Lamb.) R. Br. has not been evaluated. However, it appears in the IUCN catalog (The IUCN Red List of Threatened Species, 2017).

1.3.3   Angiospermae (Dicotyledonous and Monocotyledonous)

The flora of Burkina Faso is dominated in term of species diversity and richness by the subphylum of Angiospermae including Monocotyledonous and Dicotyledonous classes. This total flora of the country is evaluated to 2,067 species, which include 124 cultivated species (Thiombiano et al., 2012). The 1943 noncultivated species of vascular plants (herbaceous and woody) belong to 738 genera and 133 families (Zizka et al., 2015). Indeed, according to Thiombiano et al. (2012), 531 species compose the woody flora representing 25.69% of the country's local flora. The remaining 74.31% represents the herbaceous flora mainly dominated by Poaceae, Legumes, and Cyperaceae. Poaceae one of the main symbol of savanna ecosystems reveals 10 tribes, 85 genera, and 316 species.

Table 1.3   List of Woody Gymnosperm Species in Burkina Faso

Scientific name	Family	Order	Class
Abies alba Miller	Pinaceae	Pinales	Pinopsida
Abies grandis (Douglas ex D. Don) Lindley	Pinaceae	Pinales	Pinopsida
Araucaria excelsa (Lamb.) R. Br.	Pinaceae	Pinales	Pinopsida
Cupressus sempervirens L.	Pinaceae	Pinales	Pinopsida
Cycas revoluta Thunb. [cult.]	Cycadaceae	Cycadales	Cycadopsida
Encephalartos transvenosus Stapf & Burtt Davy	Cycadaceae	Cycadales	Cycadopsida

[cult.] = cultivated.

Source: Soma (2012); Thiombiano et al. (2012).

1.4   Hotspots of Burkina Faso

Based on different scenarios (Figure 1.12) recent investigations highlight
two hotspots of plant diversity (in terms of plant species richness) were identified in Burkina Faso: the Southwest around the sandstone massif of Mount
Ténakourou in the Kénédougou province and the Southeast around the W
National Park (Schmidt et al., 2017).

These investigations also   ☐A
try as hotspots for threatened and endangered species (Figure 1.13A–D;
Schmidt et al., 2017).

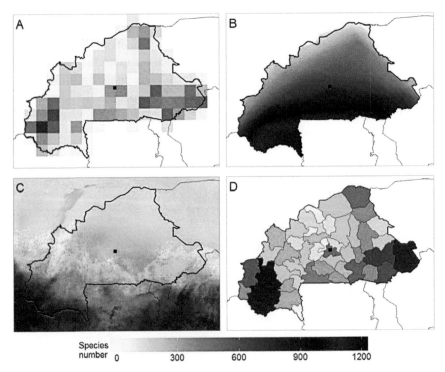

Figure 1.12   Vascular plant species richness in Burkina Faso [Note the diversity
hotspots in the southeast and southwest of the country. A: Species
number based on occurrence records; B: Based on overlaid Extent of
Occurrence; C: Based on the modeled distribution of 1,107 species;
D: Per political province based on occurrence records (Schmidt et al.,
2017 Zootaxa)].

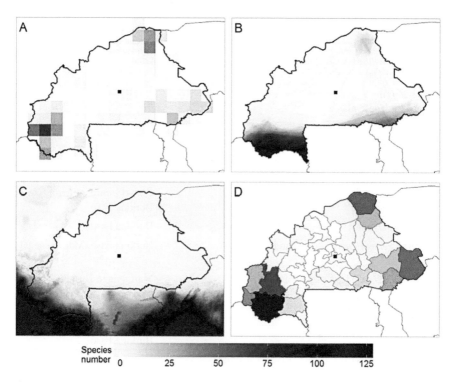

Figure 1.13 Potentially threatened and near-threatened vascular plant species in Burkina Faso [The colors indicate the number of species classified as potentially Critically Endangered (CR), Endangered (EN), Vulnerable (VU) and Near-Threatened (NT). A: Species number based on occurrence records; B: based on overlaid Extent of Occurrence; C: based on the modeled distribution of 1,107 species; D: per political province based on occurrence records (Schmidt et al., 2017, Zootaxa)].

1.5 Endemic Species

The flora of Burkina Faso dominated by therophytes and phanerophytes has a wide distribution in Africa and is representative of West African flora. This might be explained by the general monotonous relief of the country even though, mountains have been recognized as the endemic center of biodiversity. Consequently, there is a low presence of endemic species in Burkina Faso, which is a general characteristic of West African savannas (Zizka et al., 2015). However, the country has 7 endemic species that are present in West Africa (Table 1.4; Thiombiano et al., 2012; Sakandé et al., 2016). The Figures 1.14–1.15 and 1.16–1.17, respectively, illustrate some species

Table 1.4 Endemic Species of West Africa Present in Burkina Faso

Species	Family	Areas of endemism
Batopedina tenuis (A.Chev. ex Hutch. & Dalziel) Verdc. Endemic to the West African Sudanese plateaus	Rubiaceae	Ghana and Burkina Faso.
Borassus akeassii Bayton, Ouédr. & Guinko	Arecaceae	From Senegal to Burkina Faso
Brachystelma simplex Schltr.		
Subsp. *banforae*	Apocynaceae	Côte d'Ivoire, Nigeria, Burkina Faso
Isoetes jaegeri Pitot	Isoetaceae	Burkina Faso
Pandanus senegalensis Huynh	Pandanaceae	Burkina Faso, Mali, Senegal
Panicum nigerense Hitchc.	Poaceae	Mali, Niger, Burkina Faso, Nigeria
Polycarpaea billei J.-P. Lebrun	Caryophyllaceae	Senegal, Mali, Côte d'Ivoire, Ghana, Togo, Burkina Faso

Source: Thiombiano et al. (2012); Sakandé et al. (2016).

Figure 1.14 *Batopedina tenuis* (A.Chev. ex Hutch. & Dalziel) Verdc (Photo Credit, Thiombiano A., 2010).

Figure 1.15 *Borassus akeassii* Bayton, Ouédr. & Guinko (Photo Credit, Thiombiano A., 2010).

Figure 1.16 *Pandanus senegalensis* Huynh en Inflorescence (Photo Credit, Thiombiano A., 2010).

Figure 1.17 *Pandanus senegalensis* Huynh in fructification (Photo Credit, Thiombiano A., 2010).

Batopedina tenuis; *Borassus akeassii* and *Pandanus senegalensis* (Figure 1.18). Among these species, only *Isoetes jaegeri* Pitot is endemic to Burkina Faso. This fern was collected from the Banfora Cliffs (Toussiana) in the Comoé's Province (Zizka et al., 2015).

1.6 Threats

The main threats of phytodiversity in Burkina Faso are rank from natural to anthropogenic constraints as well as combined factors. Indeed, the decline in rainfall, drought and soil depletion and climate change has been reported as the main threats (Mbayngone and Thiombiano, 2011; Gaisberger et al., 2017). The rainfall shortage leads to mortality of nonresistant species like *Pterocarpus lucens* and *Dalbergia melanoxylon* (Hahn-Hadjali and Thiombiano, 2000). For these species, large cemeteries of stand dead wood might be observed in the northern part of the country as a result of consecutive periods of drought.

At the species level, climate change, the reduction or even disappearance of certain habitats and human pressure are the main causes of threat that impact negatively many species in the country different climate zones (Thiombiano and Kampmann, 2010). Some plant species are heavily threatened either because the climatic conditions do not allow their regeneration, or due to the overexploitation (Hahn-Hadjali and Thiombiano, 2000; Thiombiano and Kampmann, 2010; Gaisberger et al., 2017). Specially, the widely collection of vital parts such as the roots (*Securidaca longipedunculata, Sarcocephalus latifolius, Parinari curatellifolia*), the   (*Bombax costatum, Annona senegalensis*), the seeds (*Vitellaria paradoxa, Parkia biglobosa* and *Acacia macrostachya*), the shoots (*Borassus aethiopum*) and leaves (*Adansonia digitata, Afzelia africana*) might be detrimental to targeted species. Other human factors   the survival of plant species are the uncontrolled bush   which, whatever their regime that lead to selection in the   by eliminating the most sensitive species over the years (Thiombiano and Kampmann, 2010; Gaisberger et al., 2017).

Considering the combined threats in the country, a number of food tree species (16 species) are highly threatened over large areas of their distribution (45 to 78% of the area); on average 60.5% of the distribution area of these tree species is highly threatened (Gaisberger et al., 2017). The Figure 1.18 presents the threat level of these food tree species calculated from a

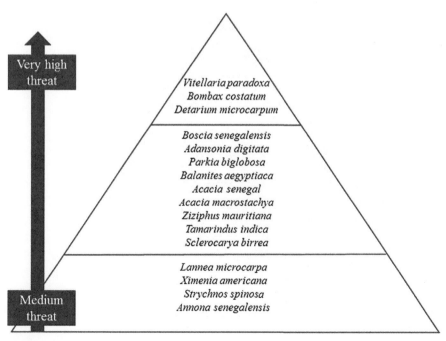

Figure 1.18   Threat level of 16 food tree species Reprinted from (Gaisberger et al., 2017 https://creativecommons.org/licenses/by/4.0/).

combination of several factors (overexploitation, overgrazing, bush   ☐
cotton cultivation, and climate change).

Among the 20 most species-rich plant families in Burkina Faso, Apocynaceae, Lamiaceae, Euphorbiaceae, Phyllanthaceae, Malvaceae had the
highest proportion of species   ☐A
(Schmidt et al., 2017). While considering the life forms, these authors found
that Geophytes and Hemicryptophytes had the highest proportion of threatened or near threatened species. However, the magnitude of threat is related
to their respective habitats. Thus, of all habitats, forests and gallery forests
have the highest proportion of potentially threatened or near-threatened species (Figure 1.19; Zizka et al., 2015; Schmidt et al., 2017).

Indeed, the abuse and uncontrolled use of chemical products including
systemic pesticides in cotton cultivation in Burkina Faso is another devastating phenomenon that seriously threatens habitats and their biological

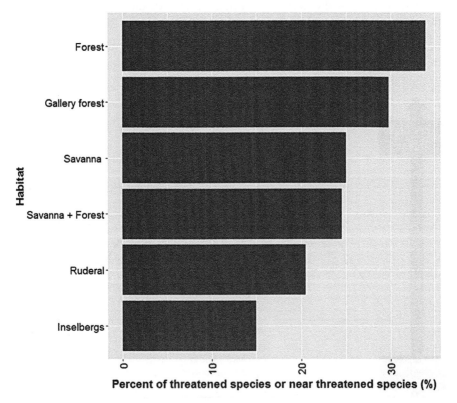

Figure 1.19 Percent of threatened or near-threatened species in Burkina Faso
according to their habitats (Schmidt et al., 2017, Zootaxa).

diversity. Illiteracy, poverty and the vulnerability of rural populations lead to unsavory practices in the use of chemical products. Commonly farmers use pesticides intended for cotton production or other prohibited products in food crop   cash crops and even market gardening near watercourses, which compromises undeniably the viability of aquatic biota and ecosystems. More than 60% of the country (representing up to 25% of the southern part) is under chemicals threat.

In recent years, Burkina Faso has experienced a mining boom, where the country is ranked fourth in Africa for gold production, and it has the third highest exploration activity in the continent; production is expected to continue its rapid growth (Gaisberger et al., 2017). However, the ranging is not without consequences on ecosystems and biodiversity. The   of mining activities includes habitat destruction in the mining sites and habitat degradation in the surrounding areas (Gaisberger et al., 2017). Mining is both a direct driver of deforestation through the physical encroachment of quarries on forest or vegetated lands, and also an indirect factor of forest resource degradation (Gaisberger et al., 2017) through the openings of tracks and the installations of facilities mostly spontaneous in the case of artisanal mining sites. The use of persistent hazardous chemical products (mercury, cyanide, etc.) renders the soils of exploited sites unproductive for many years after closure of the mine (MEEVCC, 2017). Similarly, the different watersheds adjoining these mining sites are strongly affected. Thus, animal mortality near to their watering has been noticed through some gold mining sites. Around such sites, soil contamination caused death and a virtual absence of plant regeneration. Actually, with the expansion of mining sites throughout the country, some protected areas are under threat.

1.7 Conservation Status of Woody Species in Burkina Faso

The International Union for the Conservation of Nature (IUCN) has set up criteria for classifying species into nine categories. However, like many other countries, several taxa in Burkina Faso have not been assessed by the IUCN Red List. The status of each of the 531 woody species recorded in the country was verified by introducing their scientific name (the Latin binomial) on the IUCN site (http://www.iucnredlist.org/search). Of these 531 species in, only 32 species (6.03%) were assessed by the IUCN red list against 491 species (92.47%), which were not evaluated. Only one species (0.19%) had no sufficient data to be evaluated (Table 1.5). However, 8 species (1.51%)

do not even appear in the IUCN catalog. Those are *Abrus melanospermus* Hassk, *Acacia hockii* De Wild, *Euphorbia poissoni* Pax, *Ficus scott-elliotii* Mildbr. & Burret, *Manilkara multinervis* (Baker) Dubard, *Philenoptera cyanescens* (Schumach. & Thonn.) Roberty, *Pouteria alnifolia* (Baker) Roberty, *Pseudocedrela kotschyi* (Schweinf.) Harms (The IUCN Red List of Threatened Species, 2017).

In addition, from one locality to another, species have not the same status. For instance, a species may be threatened in a neighboring country and not in Burkina Faso. Likewise, a species may be   A ☐ the IUCN Red List and not threatened in a given country (Neuenschwander et al., 2011). Table 1.6 shows the respective conservation status of the 32 woody species in Burkina Faso, assessed by IUCN.

Based on the occurrence number of each species, Schmidt et al. (2017) ranked 350 species (both woody and herbaceous), representing 18% of the national   as potentially threatened (CR = 20 species, EN = 150 species, VU = 93 species) or potentially near-threatened (NT = 87 species) in accordance with the nine IUCN categories. Of these species, 117 are woody and subwoody (CR = 06 species, EN = 42 species, VU = 32 and NT = 37) (Table 1.7). For these authors, potentially threatened and near-threatened species are highly concentrated in the southern part of the country.

Finally, the previous   A ☐ according to the ecological zones. This   ☐A ogy of the species, the human pressure and the structures of populations in the different plant communities (Thiombiano and Kampmann, 2010).

Table 1.5 Proportion of Woody Species in Burkina Faso Assessed By IUCN

Status according to IUCN	Absolute values	Relative values (%)
Evaluated	32	6.03
☐ ☐	1	0.19
Endangered (EN)	4	0.75
Least Concern (LC)	17	3.20
Minor risk/Least Concern (LC)	1	0.19
Minor risk /Near-threatened (NT)	1	0.19
Vulnerable (VU)	8	1.51
Not evaluated (NE)	491	92.47
Not listed in the IUCN catalog	8	1.51

Table 1.6 Conservation Status of 32 Woody Species in Burkina Faso According to IUCN

Species	IUCN status
Acacia ehrenbergiana Hayne	Least Concern (LC)
Afzelia africana Sm. ex Pers.	Vulnerable (VU)
Albizia ferruginea (Guill. & Perr.) Benth.	Vulnerable (VU)
Bauhinia purpurea L.	Least Concern (LC)
Borassus aethiopum Mart.	Least Concern (LC)
Borassus akeassii Bayton, Ouédr. & Guinko	Least Concern (LC)
Calamus deerratus G.Mann & H. Wendl.	Least Concern (LC)
Cissus doeringii Gilg & M.B randt	Least Concern (LC)
Cissus rufescens Guill. & Perr.	Least Concern (LC)
Dalbergia melanoxylon Guill. & Perr.	Minor risk/Near-threatened (NT)
Desmodium barbatum (L.) Benth.	Least Concern (LC)
Detarium microcarpum Guill. & Perr.	Least Concern (LC)
Dichrostachys cinerea (L.) Wight & Arn.	Least Concern (LC)
Diospyros ferrea ssp. *ebenus* (Thwaites) Bakh.	Endangered (EN)
Elaeis guineensis Jacq.	Least Concern (LC)
Erythrina senegalensis A. DC.	Least Concern (LC)
Isoberlinia doka Craib & Stapf	Least Concern (LC)
Khaya grandifoliola C.DC.	Vulnerable (VU)
Khaya senegalensis (Desr.) A.Juss.	Vulnerable (VU)
Leptoderris brachyptera (Benth.) Dunn	Least Concern (LC)
Mimosa pudica L.	Least Concern (LC)
Pavetta lasioclada (K. Krause) Mildbr. ex Bremek.	Vulnerable (VU)
Philenoptera laxiflora (Guill. & Perr.) Roberty	Least Concern (LC)
Pteleopsis habeensis Aubrév. ex Keay	Endangered (EN)
Pterocarpus lucens Lepr. ex Guill. & Perr.	Least Concern (LC)
Pterocarpus santalinoides DC.	Minor risk/Least Concern (LC)
Raphia sudanica A. Chev.	□ □
Terminalia ivorensis A. Chev.	Vulnerable (VU)
Vepris heterophylla (Engl.) Letouzey	Endangered (EN)
Vitellaria paradoxa C.F.Gaertn.	Vulnerable (VU)
Warneckea fascicularis (Planch. ex Benth.) Jacq.-Fél.	Endangered (EN)
Xylopia elliotii Engl. & Diels	Vulnerable (VU)

Table 1.7 Conservation Status of Woody Species in Burkina Faso According to Schmidt et al. (2017)

Critically Endangered (CR)	Endangered (EN)	Vulnerable (VU)	Near-threatened (NT)
Albizia ferruginea (Guill. & Perr.) Benth.	Acacia gerrardii Benth.	Aedesia glabra (Klatt) O.Hoffm.	Adenodolichos paniculatus (Hua) Hutch. & Dalziel
Lannea fruticosa (Hochst. ex A.Rich.) Engl.	Annona glauca Schumach. & Thonn.	Cassipourea congoensis DC.	Adenostemma caffrum DC.
Pandanus brevifrugalis Huynh	Anthocleista procera Lepr. ex A.Chev.	Dalbergia hostilis Benth.	Afraegle paniculata (Schumach.) Engl.
Pavetta oblongifolia (Hiern) Bremek.	Blumea crispata (Vahl) Merxm.	Dombeya buettneri K. Schum.	Antidesma rufescens Tul.
Requienia obcordata (Lam. ex Poir.) DC.	Cadaba glandulosa Forssk.	Guibourtia copallifera Benn.	Carapa procera DC.
Tarenna pavettoides (Harv.) Sim	Calamus deerratus G.Mann & H.Wendl.	Haematostaphis barteri Hook.f.	Clerodendrum polycephalum Baker
	Canarium schweinfurthii Engl.	Hibiscus micranthus L.f.	Combretum racemosum P.Beauv.
	Christiana africana DC.	Ipomoea verbascoidea Choisy	Diospyros abyssinica (Hiern) F.White
	Cissus rubiginosa (Welw. ex Baker) Planch.	Khaya grandifoliola C.DC.	Eriosema molle Hutch. ex Milne-Redh.
	Clappertonia ficifolia (Willd.) Decne.	Kinghamia macrocephala (Oliv. & Hiern) C.Jeffrey	Erythroxylum emarginatum Thonn.
	Coffea ebracteolata (Hiern) Brenan	Leptadenia pyrotechnica (Forssk.) Decne.	Ficus polita Vahl
	Cordia sinensis Lam.	Markhamia tomentosa (Benth.) K.Schum. ex Engl.	Flabellaria paniculata Cav.

Critically Endangered (CR)	Endangered (EN)	Vulnerable (VU)	Near-threatened (NT)
	Croton pseudopulchellus Pax	*Merremia pterygocaulos* (Choisy) Hallier f.	*Grewia carpinifolia* Juss.
	Croton scarciesii Scott-Elliot	*Multidentia pobeguinii* (Hutch. & Dalziel) Bridson	*Harungana madagascariensis* Lam. ex Poir.
	Dalechampia scandens L.	*Newbouldia laevis* (P.Beauv.) Seem. ex Bureau	*Hibiscus sterculiifolius* (Guill. & Perr.) Steud.
	Drypetes floribunda (Müll. Arg.) Hutch.	*Periploca nigrescens* Afzel.	*Melastomastrum capitatum* (Vahl) A.Fern. & R.Fern.
	Embelia guineensis Baker	*Psophocarpus palustris* Desv.	*Mussaenda elegans* Schumach. & Thonn.
	Euphorbia paganorum A.Chev.	*Psydrax subcordata* (DC.) Bridson	*Nesphostylis holosericea* (Baker) Verdc.
	Euphorbia poissoni Pax	*Rhus longipes* Engl.	*Ochna afzelii* R.Br. ex Oliv.
	Ficus craterostoma Warb. ex Mildbr. & Burret	*Rutidea parviflora* DC.	*Ochna rhizomatosa* (Tiegh.) Keay
	Gardenia nitida Hook.	*Sabicea brevipes* Wernham	*Pavetta lasioclada* (K.Krause) Mildbr. ex Bremek.
	Hibiscus diversifolius Jacq.	*Solanum aculeatissimum* Jacq.	*Rhynchosia hirta* (Andrews) Meikle & Verdc.

Table 1.7 (Continued)

Critically Endangered (CR)	Endangered (EN)	Vulnerable (VU)	Near-threatened (NT)
	Hibiscus longisepalus Hochr.	*Tarenna thomasii* Hutch. & Dalziel	*Rhynchosia pycnostachya* (DC.) Meikle
	Hibiscus physaloides Guill. & Perr.	*Tragia tenuifolia* Benth.	*Sarcocephalus pobeguinii* Pobeg.
	Hymenocardia heudelotii Müll. Arg.	*Vangueriella spinosa* (Schumach. & Thonn.) Verdc.	*Synsepalum pobeguinianum* (Pierre ex Lecomte) Aké Assi & L.Gaut.
	Indigofera terminalis Baker	*Vernonia adoensis* Sch.Bip. ex Walp.	*Triclisia subcordata* Oliv.
	Lannea egregia Engl. & K.Krause	*Voacanga thouarsii* Roem. & Schult.	*Uapaca heudelotii* Baill.
	Psophocarpus monophyllus Harms	*Xylopia aethiopica* (Dunal) A.Rich.	*Vigna gracilis* (Guill. & Perr.) Hook.f.
	Ruellia praetermissa Schweinf. ex Lindau	*Cayratia delicatula* (Willems) Desc.	*Vigna luteola* (Jacq.) Benth.
	Sabicea venosa Benth.	*Jasminum obtusifolium* Baker	*Waltheria lanceolata* R.Br. ex Mast.
	Salacia pallescens Oliv.	*Keetia leucantha* (K.Krause) Bridson	*Zanha golungensis* Hiern
	Salacia stuhlmanniana Loes.	*Uncaria africana* G.Don	*Adenia cissampeloides* (Planch. ex Hook.) Harms

Critically Endangered (CR)	Endangered (EN)	Vulnerable (VU)	Near-threatened (NT)
	Strychnos usambarensis Gilg		*Ceropegia sankuruensis* Schltr.
	Terminalia ivorensis A.Chev.		*Dalbergia bignonae* Berhaut
	Vigna kirkii (Baker) J.B.Gillett		*Indigofera macrophylla* Schumach. & Thonn.
	Zanthoxylum leprieurii Guill. & Perr.		*Ipomoea mauritiana* Jacq.
	Cissus diffusiflora (Baker) Planch.		*Mikania chevalieri* (C.D.Adams) W.C.Holmes & McDaniel
	Cyphostemma lageniflorum (Gilg & M.Brandt) Desc.		
	Pentatropis nivalis (J.F.Gmel.) D.V.Field & J.R.I.Wood		
	Triclisia patens Oliv.		
	Cayratia debilis (Baker) Suess.		
	Cissus polyantha Gilg & M.Brandt		

Source: Schmidt et al. (2017).

1.8   Conclusion

The total flora of Burkina Faso is evaluated to 1943 non-cultivated species of vascular plants belonging to 738 genera and 133 families with 74.31% herbaceous species and 25.69% of woody species. Twenty-six species of ferns were recorded. No local Gymnosperm species has been recorded. The country holds two hotspots of plant diversity: the Southwest around the sandstone massif of Mount Ténakourou in the Kénédougou province and the Southeast around the W National Park. The vegetation types consisted of islands of dry deciduous forests to steppes, and tiger bush *via* different types of savannas. Only one fern species *Isoetes jaegeri* Pitot is endemic to Burkina Faso. The phytodiversity, in general, is under multiscale-threat including habitat, family, life-form and species level. The identified main threats are rank from climate to anthropogenic pressure.

Keywords

- ecosystems
- endemic species
- phytodiversity
- riparian forests

References

Gaisberger, H., Kindt, R., Loo, J., Schmidt, M., Bognounou, F., Da, S. S., et al., (2017). Spatially explicit multi-threat assessment of food tree species in Burkina Faso: A fine-scale approach. *PLoS One, 12*(9), e0184457.https://doi.org/10.1371/journal.pone.0184457.

Gnoumou, A., (2013). *Diversity and spatio-temporal dynamics of the vegetation of the classified forest and the partial wildlife reserve of Comoé Léraba (South-West of Burkina Faso).* PhD thesis, University of Ouagadougou.

Guinko, S., (1984). *The vegetation of the Upper Volta.* Volume 1 – Doctoral Thesis. University of Bordeaux III.

Hahn-Hadjali, K., & Thiombiano, A., (2000). Perception of Endangered Species in Gourmantché environments (Eastern Burkina Faso). *Berichte des Sonderforschungsbereichs, 268*, Band 14, Frankfurt a. M. pp. 285–297.

Mbayngone, E., & Thiombiano A., (2011). Degradation of protected areas through the exploitation of plant resources: Case of the Pama Partial Wildlife Reserve, Burkina Faso (West Africa). *Fruits, 66*, 187–202. doi: 10.1051/fruits/2011027.

MEEVCC, (2017). *Second National Forest Inventory of Burkina Faso.* Draft report (Version 3).

Nacoulma, B. M. I., (2012). *Dynamics and conservation strategies of vegetation and phytodiversity of the ecological complex W National Park in Burkina Faso.* PhD thesis, University of Ouagadougou.

Neuenschwander, P., Sinsin, B., & Goergen, G., (2011). Protection de la nature en Afrique de l'Ouest: Une liste rouge pour le Bénin. *Nature Conservation in West Africa: Red List for Benin.* International Institute of Tropical Agriculture, Ibadan, Nigeria.

Ouadba, J. M., (1997). *Development of a national monograph on biological diversity: Collection of biological data, ecological considerations.*

Ouédraogo, A., (2006). *Diversity and dynamics of woody vegetation in Eastern Burkina Faso.* PhD thesis, University of Ouagadougou.

Ouédraogo, O., (2009). *Phytosociology, dynamics and productivity of the vegetation of the Arly National Park (South East of Burkina Faso).* PhD thesis, University of Ouagadougou.

Ouédraogo, O., Thiombiano, A., Hahn-Hadjali, K., & Guinko, S., (2009). Diversity and dynamics of the juvenile woody vegetation of the Arly National Park (Burkina Faso). *Candollea, 64*(2), 257–278.

Ouédraogo, O., Thiombiano, A., Hahn-Hadjali, K., & Guinko, S., (2008). Diversity and structure of the woody plant communities in the National Park of Arly. *Flora et Vegetation Sudano-Sambesica, 11*, 5–16.

Projet BKF/015-IFN 2 et IGB, (2015). User Guide for the 2012 Land Cover Database (BDOT) of Burkina Faso.

Sakandé, M., Sanogo, S., & Beentje, H., (2016). *Tree identification guide of Mali.* Royal Botanic Gardens, Kew, Royaume uni.

Sambaré, O., (2013). *Diversity Phytosociology and structure of riparian forests of Burkina Faso (West Africa).* PhD thesis, University of Ouagadougou.

Schmidt, M., Zizka, A., Traoré, S., Ataholo, M., Chatelain, C., Daget, P., et al., (2017). Diversity distribution and preliminary conservation status of the flora of Burkina Faso. *Phytotaxa, 304,* 1–215. doi: org/10.11646/phytotaxa.304.1.

Soma, S., (2012). *Production of plants, uses and perception of local species by the nursery gardeners of the city of Ouagadougou.* MSc thesis. University of Ouagadougou.

The IUCN Red List of Threatened Species (2017). Version 2017–1. <www.iucnredlist.org>.

Thiombiano, A., & Kampmann, D., (2010). *Biodiversity Atlas of West Africa, Volume II: Burkina Faso.* Ouagadougou & Frankfurt/Main.

Thiombiano, A., Schmidt, M., Dressler, S., Ouédraogo, A., Hahn, K., & Zizka, G., (2012). Checklist of the Vascular Plants of Burkina Faso. *Boissiera, 65*, 1–391.

Tindano, E., Ganaba, S., Sambare, O., & Thiombiano, A., (2015). Sahelian inselberg vegetation in Burkina Faso. *Bois et Forêts des Tropiques, 325*(3), 21–33.

Traoré, L., (2013). *Influence of climate and protection on woody vegetation in western Burkina Faso.* PhD thesis, University of Ouagadougou.

Zerbo, I., Bernardh-Römermann, M., Ouédraogo, O., Hahn, K., & Thiombiano, A., (2016). Effects of climate and land use on herbaceous species richness and vegetation composition in West African savanna ecosystems. *J. Botany.* http://dx.doi.org/10.1155/2016/95236851.

Zizka, A., Thiombiano, A., Dressler, S., Nacoulma, M. I. B., Ouédraogo, A., Ouédraogo, I., et al., (2015). The vascular plant diversity of Burkina Faso (West Africa): A quantitative analysis and implications for conservation. *Candollea, 70*, 9–20. doi: http://dx.doi.org/10.15553/c2015v701a2.

CHAPTER 2

Biodiversity in Egypt

WAFAA M. AMER

Department of Botany and Microbiology, Faculty of Science, Cairo University, Egypt,
E-mail: wafaa_amer@hotmail.com

2.1 Introduction

Egypt is located between 22°–32 N°; it is a low-lying country on the N.E. Africa with total area of 1.01 million km². According to El-Hadidi and Hosni (2000), Egypt is divided into four main geomorphologic units and 16 subunits namely: (I) Western desert (6 subunits), (II) Nile land (2 subunits), (III) Eastern desert (3 subunits) and (IV) Sinai Peninsula (5 subunits) (Plate 2.1A).

Egypt placed in the subtropical dry belt, is the most arid country in N. Africa. The coldest months of the year are between December and February, 4°C is the lowest temperature (except the western desert oases reaches –5°C). While the hottest are between June and August, 42°C is the maximum temperature (except the western desert oases with maximum 52°C).

The rainy season mostly between November and March, based on the climatic provinces as outlined in Figure 2.1B, the Semi-arid belt or the Mediterranean belt (I.a), receives 100–220 mm/year, the Arid belt II.1 and II.2 receive 50–30 and 50–10 mm/year, respectively. While, the hyper-arid belt III.1, III.2 and III.3 receive 50–0.0, 20–10 and 50–10 mm/year, respectively, with low number of species and a low number of endemics (Ayyad and Ghabbour, 1986).

2.2 Biodiversity in Egypt: An Overview

Flora of Egypt consists of 2076 species, in addition to 151 infra-specific taxa from 725 genera representing 120 families. Gymnosperms are represented by 6 species, monocots by 430 species and the rest (1640 species) belong to dicots. Egyptian flora has a special interest due to its unique mixture of native African and Asiatic species. The plant species are distributed not equally into four phytogeographic regions of Egypt namely: Eastern desert, Western desert (including Oases and depressions), Nile land, and Sinai peninsula (Amer, 2008). The Sinai Peninsula and Mediterranean stripe showed the higher species diversity (Ghabbour and Mikhail,

1998). While Nubian desert, depressions and Uweinate mountain out of the Western desert showed the lowest species diversity (Boulos and Barakat, 1998). Egyptian flora contains a low number of endemic species amounted 61 species (Boulos, 1995) and no endemic families (Wickens, 1976), most of the endemic species are grouped to the following families: Lamiaceae, Liliaceae, Scrophulariaceae, and Asteraceae. The Sinai Peninsula containing 33 endemic species 24 of them recorded in South Sinai, Sinai endemic species comprising 60.7% of the total endemic species in Egypt (Hegazy and Amer, 2001). The largest families in flora of Egypt according to Boulos (2009) are: Poaceae (241 species), Fabaceae (233 species), Asteraceae (230 species), Brassicaceae (101 species), Caryopyllaceae (85 species), Chenopodiaceae (75 species), and Scrophulariaceae (63 species). More than 470 bird species (6 endemic) have been recorded in Egypt and among them 150 are resident. The Egyptian wetlands in Red Sea, Mediterranean Sea and Lake Nasser are important habitats for the bird migrations in which during autumn a huge number of white storks, common cranes and white pelicans are traced annually passing through this area. Terrestrial mammals (95 species), amphibians (9 species), reptiles (97 species) are represented by few number of species (Table 2.1), this is mainly attributed to the low rainfall, aridity and the sparse vegetation.

The coastal and marine environment is distinguished by   □ - tats and threatened species especially the turtles (5 species), sharks (c. 5 species), sea cucumber, bivalves, 209 species of coral reefs, mangrove trees and many visiting birds among them: white eyed gulls, sooty falcons, and ospreys. In addition to, 800 species of seaweeds and seagrasses, c. 800 species of mollusks, 1000 species of crustacea and 350 species of echinodermata occur in the marine environment. Special remarks for Dugong (mermaid) species, a total of 50 Dugong (mermaid) were recorded. This is one of the most threatened species in the world, and reaches more than 2 meters in length. It is usually found in seagrass beds. Later, the two characteristic freshwater reptiles the soft-shelled turtle (*Trionyx triunguis*) and Nile crocodile (*Crocodylus niloticus*), were traced in Nile water especially in Lake Nasser.

2.3 Ecosystem Diversity in Different Geomorphic Units

The biodiversity of Egypt can be summarized as general features of the four main geomorphologic units (Plate 2.1A), namely: (1) Western desert, (2) Nile land, (3) Eastern desert, and (4) Sinai Peninsula.

Table 2.1   Species Diversity in Egypt (*= Low Estimate)

Category	Taxonomic group	No. of species	No. of endemics
I. Flora	– Bryophyta	337	–
	– Pteridophyta	16	–
	– Gymnospermae	6	–
	– Angiospermae	2,072	62
II. Fauna:	– Insecta	15,000	–
1. Terrestrial fauna	– Arachnida	1517	–
2. Marine fauna & Flora	– Amphibia	9	1
	– Reptilia	97	6
3. Fresh water fauna	– Aves	470	6
	– Mammalia	95	6
	– Invertebrata	1,740*	–
	– Fish (Chondrichthyes & Osteichthyes)	669*	–
		5	–
	– Reptilia (Turtles)	17	–
	– Mammalia	209	–
	– Coral	800	–
	– Molluscs	c. 1000	–
	– Crustaceans	350	–
	– Echinodermata	800	–
	– Seaweeds of Seagrasses	124	–
	– Invertebrata	70	–
	– Fish (Osteichthyes)	2	–
	– Reptilia		

2.3.1   Western Desert Unit

It extends from the Mediterranean coast in the North to the Egyptian Suda-nese border in the South, and from the Nile valley in the East to the Egyp-tian-Libyan border in the West. It covers 2/3 of the Egyptian land (c. 681,000 km^2). It consists of six geomorphologic subunits and they are discussed in the following subsections.

2.3.1.1   Mediterranean Coastal Plain Subunit

It is the northern part of the western desert extends from Alexandria west-wards to Sallum, it varies in width between 5–25 km from the Mediterranean

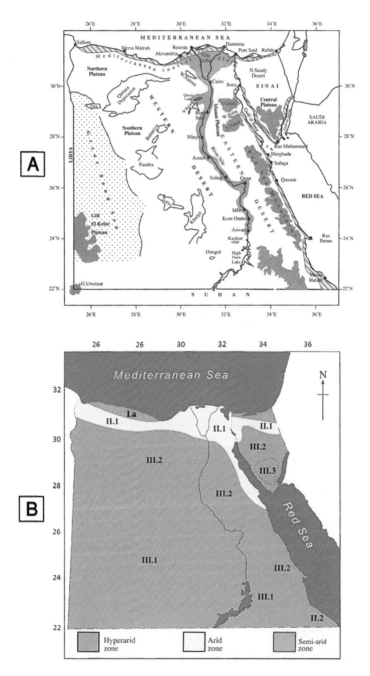

Figure 2.1 Maps of Egypt (A): The main geographic units, (B): The climatic provinces.

Sea. This subunit is the richest part of the western desert in biodiversity. Its high diversity owing to its relatively high rainfall (Zahran and Willis 2009). About 1,060 plant species were recorded (El-Hadidi and Hosni, 2000). Therophytes (67%) are the dominant species, which flourished during the rainy season (winter) followed by geophytes (11%), halophytes and helophytes (11%), while the lowest is the stem succulents (0.1%). The calcareous sand dunes next to the seashore characterized by psammophytes, among them: *Ammophila arenaria, Ononis vaginalis, Reseda alba, Orlaya maritima, Lotus polyphyllos, Silene succulenta, Crucianella maritima, Echium sericeum*, and others. Followed by the outer calcareous rocky ridge dominated by *Thymelaea hirsuta, Gymnocarpos decander, Reaumuria hirtella*. While the inner calcareous rocky ridge is characterized by Chasmophytics, namely: *Herniaria hemistemon, Carduus getulus, Plantago notata, Teucrium polium, Thymus capitatus, Echinops spinosissimus, Peganum harmala* and *Asphodelus microcarpus*. The earlier two rocky ridges enclosed a salt marsh dominated by halophytic species, among them: *Juncus rigidus, J. acutus, Halocnemum strobilaceum, Limoniastrum monopetalum, Zygophyllum album, Atriplex halimus*, and others. Southern border of this subunit is the cultivated plateau and the characteristic species are: *Centaurea glomerata, Chrysanthemum coronarium, Papaver rhoeas, Senecio desfontainei, Matthiola longipetala, Convolvulus althoides, Launaea nudicaulis, Achillea santalina, Adonis dentata* and *Hordeum leporinum*. Due to the relatively high rainfall which induced a notable vegetation diversity, accordingly animal diversity, among the traced mammal species is the endangered rodents namely: *Alloctaga tetradactyla* and *Jaculus orientalis* known as *four-toed jerboa* and *Greater Egyptian jerboa*, respectively. In addition to, the common *Vulpes vulpes* (Red fox), *Hemiechinus auritus* (Long-eared hedgehog), and *Psammomys obesus* (Fat sand rat) are also found. The migratory birds cross this area to the south in spring season; millions of *Calandrella cinerea* (Short-toed larks) was estimated 10,000 bird/mile to west Alexandria. While, *Cursorius cursor* (Cream colored courser), *Chersophilus duponti* (Dupont's lark) and *Oenanthe moesta* (Red-rumped wheatear) are the characteristic resident birds. The most common terrestrial gastropod mollusk is the desert snail (*Eremina desertorum*).

2.3.1.2 Northern Plateau Subunit

It is a Hamada Plateau followed the Mediterranean coastal plain. It is covered by stones and gravel extends from the Libyan borders and eastwards to the Qattara Depression. The plateau dissected by shallow ephemeral wadis.

Extensive areas of this plateau have been reclaimed for cultivation. The habitats with deeper sand deposits support the growth of xeric perennial species (c. 72 species are perennials); among them: *Thymalaea hirsuta, Anabasis articulata, Helianthemum stipulatum, H. lippii, Echiochilon fruticosum, Artemisia monosperma, Deverra tortuosa* and *Calligonum polygonoides*. Among the common annual species are: *Polycarpaea repens, Neurada procumbens, Trigonella maritima, Anthemis microsperma, Senecio desfontainei* and *Achillea santolina*.

The harsh desert conditions in this plateau with its sparse vegetation and high wind regimes sustain very limited animal species. Among the recorded mammals, the endangered *Gazella leptoceros* (Slender-horned gazelle), the desert water-independent Fennec Fox (*Vulpes zerda*). From birds: *Falco biamicus* (Lanner falcon), and from reptiles: *Acanthodoctylus scutellatus* (Nidua lizard), *Psammophis aegyptius* (Saharan sand – snake) and *Cerastes vipera* (Sand-dwelling viper).

2.3.1.3   South Plateau Subunit

It extends from south Qattara Depression to the central part of the western desert. It is characterized by the presence of a series of large depressions interrupted by mobile sand areas. This area is characterized by occidental and contracted type (Bornkamm and Kehl 1990). Among the recorded 53 plant species are: *Cornulaca monacantha, Fagonia arabica* and *Suaeda vermiculata* among the plant species; while the fauna are similar to that of the northern plateau.

2.3.1.4   Great Sand Sea Subunit

This Great Sand Sea sweeps down the western half of the desert to join the Great Selima sand in Sudan. It is dominated by mobile sand dunes which was formed as a result of Aeolian geomorphological processes. The sand mobility and scarcity of rainfall render the species existence difficult. Among the traced grass species is the *Stipagrostis pulmosa* and from reptiles *Sphenops sepsoides* (Audouin's sand skink) and *Psammophis aegyptius* (Saharan sand snake) and *Jaculus jaculus* (Lesser Egyptian jerboa) from mammals.

2.3.1.5   Depressions Subunit

A series of large depressions formed by wind erosion where the areas of weaker rocks in the Nubian sandstone have been preferentially eroded

(El-Hadidi and Hosni 2000). These depressions from North to South are: Wadi El-Natrun, Qattara, Moghra, Qara, Siwa, Bahariya, Farafra, Dakhla, Kharga, Kurkur and Dungul. Qattara is the largest (19,500 km^2) depression in the western desert (50–134 m b.s.l.) and the smallest are Moghra and Qara. According to Amer (2008), 9 species were confined to Farafra Oasis: *Bromus diandrus, Polycarpon succulentum, Cutandia memphitica, Vicia lutea, Polycarpon tetraphyllum, Plantago amplexicaulis, Oligomeris linifolia, Nymphaea nouchali* and *Avena sterilis.* Other 45 species were confined to Bahariya Oasis (e.g., *Erucaria hispanica, Heliotropium lasiocarpum, Bromus diandrus, Allium ampeloprasum, Brassica juncea, Lotus halophilus, Chenopodium glaucum, Homognaphalium pulvinatum* and *Geranium dissectum*), and 6 species were confined to Dakhla Oasis (e.g., *Capparis aegyptia, Asphodelus fistulosus, Althaea ludwigii, Aerva javanica, Senna italica* and *Abutilon pannosum*). Seven species showed consistency to Kharga Oasis (e.g., *Erodium cicutarium, Sesbania sesban, Silene villosa, Conyza aegyptiaca, Rhynchosia minima, Stipagrostis scoparia* and *Kickxia elatine*), and 54 species were confined to Faiyum region (e.g., *Digitaria ciliaris, Sonchus asper, Cuscuta planiflora, Physalis angulata, Geropogon hybridus, Orobanche aegyptiaca, Ammi visnaga, Raphanus raphanistrum* and *Matricaria recutita*). The present study exhibited 39 species which were confined to Wadi El-Natrun and included among others *Cornulaca monacantha, Atriplex halimus, Artemisia monosperma, Moltkiopsis ciliata, Agathophora alopecuroides, Farsetia aegyptia, Limbarda crithmoides, Deverra tortuosa, Convolvulus lanatus* and *Typha elephantina.* While, the common halophytic species dominating Qattara depression are: *Zygophyllum album, Nitraria retusa, Salsola tetranda, Cressa cretica, Juncus rigidus, J. actus* and *Halocnemum strobilaceum.*

The notable vegetation diversity in these depressions (except Qattara) is due to the water availability as springs in Kharga, Dakhla and Farafra or Nile water seepage in Kurkur. The combination of local ground water and the sedimentary soils makes these depressions suitable for agriculture. According to Bornkamm (1986), the characteristic features of some depressions are: (1) Large part of Qattara depression is covered by salt crust. (2) Kurkur and Dungul depressions are dominates by Dom-palm (*Hyphaene thebaica*) and date-palm (*Phoenix dactylifera*), in addition to the fan-palm (*Medemia argun*) which is   A⎯   *Gossypium herbaceum* var. *africana* is a relict species   A⎯ (Hamed et al., 2015).

Among the common birds of these depressions are: *Cercotrichas galactotes* (Rufous bush robin), *Cursorius cursor* (Cream colored courser) and *Streptopelia turtur* (Turtle dove). Fewer reptiles were recorded in these depressions, among them, *Acanthodactylus scutellatus* (Fringe – toed lizard) and *Varanus griseus* (Desert monitor) are important. It is notable to mention that *Tarentola mindiae* (Mindy's Gecko) is a characteristic gecko inhabiting *Acacia* trees in Qattara depression. Twelve mammal species were recorded and among them *Dipodillus henleyi* (Henley's dipodil), *Arvicanthus niloticus* (Nile-rat), and *Felis chaus* (Swamp cat) are the important.

2.3.1.6 Mountainous Area Subunit

This area is situated in the extreme southwest, it divides to Gilf El Kebir (700 m a.s.l.) plateau (sandstone) and southern Gebel Uweinat (600–1900 m a.s.l.) on Egyptian border with Libya and Sudan. The scarcity of rainfall and the extreme aridity, support about 72 of xeric species in Gebel Uweinat-El Gilf Elkebir district, out of them Uweinate mountain possesses 66 species (Boulos and Barakat 1998). Among these species are: *Panicum turgidum, Zilla spinosa, Trichodesma africanum, Schouwia purpurea, Acacia tortilis* (Figure 2.3B), *A. raddiana, Crotalaria thebaica, Fagonia arabica,* and the characteristic species to this area is the endangered *Maerua crassifolia* tree. Fauna is little known in this area due to its extreme aridity. *Tarentola annularis* (Egyptian gecko) is the only reptile recorded and the insectivorous birds *Oenanthe leucopyga* (White-crowned black wheatear) and *Ammomanes deserti* (Desert lark). From the mammals the highly endangered *Ammotragus lervia* (Barbary sheep; Figure 2.3E), were occasionally traced in this mountainous area.

2.3.2 *Nile Land Unit*

It is the most fertile land in Egypt and extends from the Sudanese border in the South to the Mediterranean Sea in the North. This is the cultivated land, containing 712 plant species, the majority are weeds of the farmland, in addition to this number 42 species are hydrophytes. It is divided into two subunits.

2.3.2.1 The Nile Valley Subunit

It extends from the Sudanese border in the South to the Nile Barrages in the North. South of the Aswan High Dam the upper part of the valley is now flooded forming lake Nasser, while the remaining part (c. 900 km long) flows in an elongated S-shaped pattern to reach Cairo. The valley soil has been formed by the deposition of sand silts and clays when the Nile flooded and this soil nature is responsible for its high fertility.

2.3.2.2 The Nile Delta Subunit

This area extends northwards (c. 160 km), from Cairo to northwestwards to Alexandria and North eastwards to Port Said. The delta is dominated by alluvial sediments, forming the fertile clay soil suitable for crop cultivation (Figure 2.2D).

The   and fauna of these two subunits are nearly similar, while the general feature is as follows: The aquatic vegetation of the River Nile system within the Egyptian borders (including Nile valley, Nile Delta and Fayium depression),   □And
19 emergent species (Zahran and Willis 2003). The submerged communities are dominated by: *Myriophyllum spicatum, Najas horrida, N. marina, Ceratophyllum demersum, Potamogeton crispus, P. pectinatus, Ludwigia stolonifera, Nymphaea caerulea, N. lotus* and *Potamogeton nodusus*. The   communities are dominated by: *Eichhornia crassipes, Azolla filiculoides, Lemna gibba* and *Pistia stratiotes*. The emergent communities are dominated by: *Cyperus alopecuroides, C. articulatus, C. difformis, Echinochloa stagnina, Eleocharis capitata, Paspalum distichum, Persicaria salicifolia, P. senegalensis, Phragmites australis* and *Typha domingensis*, this ecosystem is outlined in Figure 2.2C.

The plant diversity in Nile system shows that Nile delta is the richest, while Nile in Fayium depression has the lowest diversity. Among the characteristic hydrophytes of the Nile Delta: *Elodea canadensis, Pistia stratiotes, Lemna minor* and *Ranunculus rionii*. While, the characteristic to Nile valley are: *Potamogeton perfoliatus, P. trichoides, Najas horrida* and *Vallisneria spiralis*.

The Common Nile algae are Chlorophyta dominated by: *Ankistrodesmus convolutes, Clodophora glomerata, Coelastrum microsporum, Oedogonium* sp. and *Pediastrum simplex*; and the Bacillariophyta species are *Cocconeis pediculus, Gomphonema angustatum* and *Synedra ulna*. While the cyanophyta are dominated by *Anabaena flas-aquae, Boriza* sp., *Nostoc microscopium*, and *Oscillatoria formosa* (Zahran and Willis 2009).

The common terrestrial annual weed species in the Nile valley and Delta are: *Melilotus indicus, Vicia montana, Trifolium resupinatum, Lotus arabicus, Polypogon monspeliensis, Echinochloa colona, Emex spinosa, Sisym-*

Figure 2.2   General feature of different Egyptian habitats. (A) Desert in rainy season,
(B) Mediterranean wetland, (C) Saluga Nile island, (D) Cultivated land
"barley field," (E) Mangrove in S. red sea coast "*rhizophora mucronata*,"
(F) Halophytic vegetation.

brium irio, Chenopodium murale and *Rumex dentatus*. While, the cultivated
canal bank, shade or wind break trees are: *Eucalyptus citriodora, E. rostrata,
Salix benegalensis, Morus alba, Dalbergia sissoo, Phoenix dactylifera,
Casuarina equisetifolia, Tamarix nilotica* and *Acacia nilotica*.

The most common wild bird species of this subunit are: *Flaco tinnunculus* (Kestrel), *Upupa epops epops* (Hoopoe), *Corvus ruficollis* (Raven), *Bubulcus ibis* (Cattle egret), *Montacilla flava pygmaea* (Egyptian wagtail), *Anthus pratensis* (Meddow pipit), *Anthene noctus saharae* (Little owl), *Bubobubo ascalaptus* (Pharaoh owl), *Gallinula chloropus* (Figure 2.3G), and *Anthya nyroca* (Wild duck) both are affected negatively by hunters. During 2010, 150,000 birds belonging to 56 species of migratory birds were recorded. Birds have increased in Lake Nasser such as *Aopochen aegyptiacus* and more than 200,000 birds. The notable invasive species to the Nile water are: *Eichhornia crassipes* and *Procambarus clarkia* (EE AA, Annual Report, 2010).

Reptiles and amphibians of this subunit: This group is not well represented in whole Egypt, 18 reptiles are traced which include *Acanthodactylus boskianus* (Egyptian leopard lizard), *Naja haje* (Egyptian Cobra; Figure 2.3J), *Chalcides ocellatus, Psammophis sibilans* and *Tarentola annularis* (Common agama). The common amphibians are: *Bufo regularis* (Toad) and *Ptychadera mascareniensis* (Mascarene frog). The characteristic species of Lake Nasser is the *Crocodylus niloticus* (Nile crocodile; Figure 2.3I). Mammal species are *Rousettus aegyptiacus, Felischaus, Rattus rattus, R. norvegicus,* and *Canis aureus. Argiope trifasciata* is the common spider out of the 18 spiders traced in this unit. Among the common ⬜A are *Bagrus bajad* (Forskals ⬜O*reochromis niloticus niloticus* (Nile Tilapia), *Sarotherodon galilaeus* (Tilapia), and *Tilapia zillii* (Tilapia).

2.3.3 Eastern Desert Unit

It extends from the Nile land eastward to the Gulf of Suez and Red Sea, embracing ca. 226,000 km² (22% of the total area of Egypt). The Eastern Desert consists of a backbone of high rugged mountains running parallel (Red Sea mountains igneous and limestone) to the Red Sea Coast. About 576 plant species were recorded in the eastern desert. However, 53 species are the dominant species. The sector between Cairo-Suez road in the north to Idfu-Mersa Alam road in the south showed the presence of 433 species (Hassan, 1987). Red Sea mountain groups maintain a higher number of species than open deserts such as Samiuki mountain group containing 125 species and Nugrus group contain 92 species (Kassas and Zahran, 1971). Shayeb El-Banat group comprises 183 species (Hegazy and Amer, 2001).

West of the Red Sea mountains lie two broad plateau the northern is limestone while the southern is Nubian Sand stone, both are dissected by dense networks of wadis. The water resources in Eastern desert based mainly on

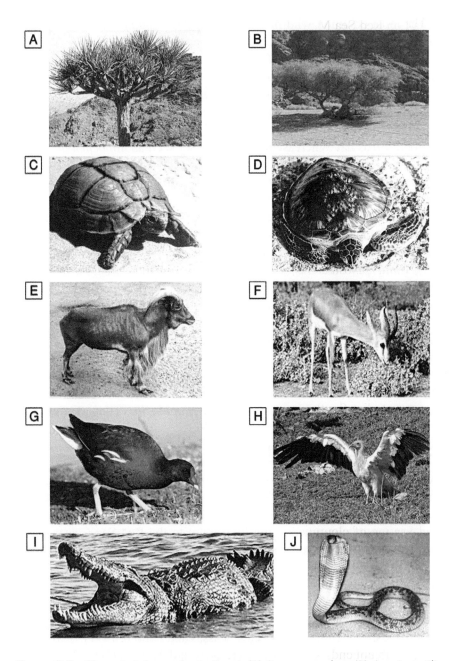

Figure 2.3 Characteristic species in Egypt. (A) *Dracena ombet*, (B) *Acacia tortilis*,
(C) *Testudo kleinmanni*, (D) *Chelonia mydas*, (E) *Ammotragus lervia*,
(F) *Gazella Dorcas*, (G) *Gallinula chloropus*, (H) *Torgas tracheliotus*, (I)
Crocodylus niloticus, (J) *Naja haje*.

rainfall on Red Sea Mountains which collected in the eastern desert wadis, then follow westward downstream to the Nile. Eastern desert unit consists of three subunits as given in the following subsections.

2.3.3.1 The Galala Desert Subunit

This subunit acquired its name from the mountain blocks of north and south Galala separated by Wadi Araba drainage system. In addition to the Northern plateau of the Esatern Desert and the Mazza plateau (the inner limestone plateau, which separated eastwards from Red Sea by littoral salt marshes and runs westward to Nile Valley. Gravel desert is sterile in general except in places where adequate depths of sand enabling the growth of annuals and some xeric woody species. Among these species are: *Acacia tortilis* (Figure 2.3A), *Fagonia arabica, Launaea spinosa, Hammada elegans, Panicum turgidum, Zilla spinosa* and *Pulicaria crispa*. While, the wadis of this subunit is dominated by *Scrophularia deserti, Matthiola longipetala, Iphiona mucronata, Diplotaxis harra, Astragalus sieberei, Helianthemum kahiricum, Cleome arabica, Lavandula stricta, Trichodesma africanum,* and *Crotalaria aegyptiaca*, the general feature of this subunit is shown in Figure 2.2A.

2.3.3.2 The Arabian Desert Subunit

This subunit lies South of the Galala Desert subunit and bounded along its entire eastern length by the southern part of the Gulf of Suez and the Red Sea proper. It comprises of a narrow coastal desert, igneous mountain chain and the inner sand stone plateau, which runs westward to delimit the Nile Valley. The igneous mountain chain ends with Gebel Elba group, the extensive granite block at the Egyptian-Sudanese borders.

The water resources availability in this subunit from the orographic condensation of clouds at high altitudes supports the growth of 158 species including 46 ephemerals. Among the characteristic plant species to this subunit are *Galium setaceum, Lappula spinocarpos, Arnebia hispidissima, Plantago afra, Iflago spicata, Commelina forskohlii, Solenostemma argel,* and the frutescent vegetation patches of *Moringa peregrina* trees at the foot of mountain, individuals of *Balanites aegyptiaca* trees and *Ochradenus baccatus* shrubs.

This subunit ends southward with Gebel Elba group, the richest vegetation among the other Red Sea mountains. About 427 species (El-Hadidi and Hosni 2000), were recorded from Gebel Elba group. The desert vegetation of this group is characterized by dense woody species as *Acacia tortilis* subsp. *raddiana, Balanites aegyptiaca, Ochradenus baccatus.* The herbaceous species are *Aerva lanata, Fagonia isotricha, Morettia philaeana, Senna holosericea, Cucumis prophetarum* and *Euphorbia granulata.* The mountain vegetation is characterized by *Euphorbia cuneata, Commiphora opobalsamum, Cadaba rotundifolia, Seddera arabica, Maytenus senegalensis* in addition to the threatened *Dracena ombet* (Figure 2.3A) trees and few Pteridophytes namely *Ophioglossum polyphyllum, Chelianthes coriacea* and *Anogramma leptophylla* are   A] region.

The characteristic birds to mountains are: *Neophron percnopterus* (Pharaoh's chicken) and *Pterocles coronatus* (Crowned sandgrouse). While, reptiles are: *Ptyodactylus hasselquistii* (Fan-footed gecko), *Agama spinosa* (Spiny agama), and the common *Cerastes cerastes* (Horned viper). The characteristic mammals are the endangered *Capra nubiana* (Nubian ibex) and *Gazella dorcas* (Dorcas gazelle; Figure 2.3F).

2.3.3.3   The Red Sea Coastal Plains Subunit

It is the border strip of the Eastern Desert coasts of the Red Sea proper and Gulf of Suez. Its width varies from few meters to more than five kilometers, it may reach 15 km at Gebel Elba district in the Southern corner. The Egyptian sector of Elba mountain group contains 285 species, Elba sector occupies the second position in plant diversity after Sinai. Its flora contains 52 families (Hegazy and Amer, 2001). This subunit extends to include the Red Sea islands.

The halophytic species dominating the Red Sea salt marshes are characterized by: *Halocnemum strobilaceum, Arthrocnemum macrostachyum, Zygophyllum album, Nitraria retusa, Tamarix nilotica, Juncus rigidus* and *Limonium pruinosum.* Southern littoral plain is characterized by mangrove vegetation dominated by *Avicennia marina* and the less codominant *Rhizophora mucronata* (Figure 2.2F). A number of birds breed in mangrove habitat which includes *Ardeola striata* (Striated Heron), *Platylaea leucorodia* (Spoonbill), *Egretta gularis* (Reef heron) and *Pandion haliaetus* (Osprey), while the resident bird is *Larus hemprichii* (Sooty gull). Some crustaceans (shrimps and crabs) are adapted to this littoral area among them *Cassiopea* sp. and *Ocypode* sp. Generally, the mangrove habitats are rich in biodiversity including the following species: 40 insects, 36 algal, 65 crusta-

Red Sea in Egyptian borders includes 40 islands, 22 out of them are protected, all of them are uninhabited. The vegetation of these islands is generally low and sparse consisting of few halophytic species which include *Avicennia marina* and the land Hermit crabs (*Coenobita* sp.), two species of lizards (*Hemidactylus turcicus* and *Mesalina guttulata*; known as Turkish Gecko and small spotted lizard; respectively). Marine turtles *Eretmochelys imbricata* (Hawksbill) and *Chelonia mydas* (Green turtle; Figure 2.3D), nesting the beaches of the southern islands. Among the recorded birds are *Sula leucogaster* (Brown boody), *Falcon concolor* (Sooty Falcon), *Torgas tracheliotus* (Egyptian vultule; Figure 2.3H) and *Larus leucophthalmus* (White-eyed gull), 30% of world *Larus* populations breeds on the northern Islands. Wadi El-Gemal islands harbor the highest community of *Falcon concolor* (Sooty falcon), the monitoring program of this endangered bird counted 340 bird, and 94 nests in 2007. While, the hawksbill turtles richest site is the "El-Giftoun Island" where 255 nests were monitored in 2007 (EE AA, Annual Report, 2007).

2.3.4 *Sinai Peninsula Unit*

Sinai is a triangular area c. 61000 km² of the Egyptian area. It is situated at the head of Red Sea bordered eastward by Gulf of Agaba and from the west by Gulf of Suez. It is the most geologically complex area in Egypt, and possesses a special ecological interest because of its variable environment and distinct flora. The remarkable note in the flora of Sinai is the visit of Delile (1809–1812) in 1778 under the command of Napolean I and Fresenius (1834), (Zahran and Willis 2009). The Flora of Sinai has the highest species diversity and plant cover in Egypt. However, the species richness is notable in north Sinai decreased southward to be minimum in the middle sector of Sinai and increased again in the mountainous region of south Sinai. Flora of Sinai comprises nearly 50% of the total species in the flora of Egypt and 65% (41 species out of 63) of the endemic species (Boulos, 2005). The number of species recorded in Sinai are 984 species (El-Hadidi, 1994/1995). Among these, 171 are in the Mediterranean sector, 203 in Gulf of Suez, and 394 in south Sinai (Hegazy and Amer, 2001). This unit is divided into five subunits as given in the following subsections.

2.3.4.1 Southern Mountains Subunit

This area is the southern half of Sinai and comprises of Precambrian igneous and metamorphic rocks similar to those of the eastern desert mountains. These mountains (nine peaks) including the highest mountain (St Katherine Mountain) in Egypt reached 2642 m a.s.l. This subunit receives high quantities of rainfall than the other parts in Sinai. It Contains 540 plant species, out of them 24 are endemics. The wadis are dominated by: *Hammada elegans, Retama raetam, Zilla spinosa, Deverra tortuosa,* and *Iphiona mucronata.* The high latitude species dominated by *Origanum syriacum, Nepeta septemcrenata, Euphorbia santa-catharine, Thymus bovei, Teucrium decaisnei, Phlomis aurea, Ephedra alata, Ballota undulata,* and *Juniperus phoenicea.* The characteristic fauna are: the endangered lizard *Uromastyx ornata* (Ornate Spiny-tailed lizard), the endemic *Coluber sinai* (Sinai banded snake) and *Telescopus hoogstraali* (Hoogstraal's cat snake) and the dangerous snake *Echis coloratus.* Birds include *Torgas tracheliotus* (Egyptian vultule; Plate 2.3H), *Carpodacus synoicus* (Sinai rose finch), and the rare *Strix butleri* (Hume's tawng owl).

2.3.4.2   Central Plateau Subunit

This plateau declines gradually northward from 1600 to 750 m a.s.l. over distance 110 km to El Igma and El Tiharea. This plateau is dissected by Wadi El Arish drainage system, while small tributaries run westward to Gulf of Suez and eastward to the Jordan Rift valley. This subunit has a larger dry period and the annual rainfall of 20–100 mm, it decreased to 12mm/year in the Gulf of Suez.

The Flora of Wadi El Arish dominated by *Cleome arabica, Aizoon canariense, Cucumis prophetarum, Pergularia tomentosa, Schismus barbatus, Salvia aegyptiaca, Reseda decursiva, Citrullus colocynthis* and *Anthemis melampodia.* The reptiles characteristic to this plateau are: *Laudakia stellio* (bright bued starred Agma), *Pseudocerasters persicus* (False horned viper) and the golden eagle (*Aquila chrysaetos*). From mammals the Indian crested porcupine (*Hystrix indica*) and the vespertilionid bat (*Barbastella barbastellus*).

2.3.4.3   Northern Sandy Desert Subunit

This subunit lies northern to the central plateau and much lower (c. 200–300 m a.s.l.), excluding Gebel Halal-Gebel Yallag group (c. 800–1100 m a.s.l.) The plain is dominated by xeric species like *Tamarix nilotica, Retama raetam, Hammada elegans, Zygophyllum occineum, Fagonia arabica, Farsetia aegyptiaca, Aerva javanica, Zygophyllum album, Zilla spinosa, Gymnocarpos decarder, Hyoscyamus muticus, Crotalaria aegyptiaca, Launaea spinosa,* and *Blepharis ciliaris.*

2.3.4.4 Mediterranean Coastal Belt Subunit

It is the narrow coastal plain consists of saline coastal mudflats and lagoons; the largest is the Bardawil. The region extends Port Said in the west to Rafah at Egypt-Palestinian Authority boundaries in the east. About 625 species are recorded in this Mediterranean subunit, some of these species were recorded along the Mediterranean stripe in whole Egyptian coast. According to Gibali (1988), the number of recorded species in the eastern sector (Sinaitic sector) are 171 species. This belt bordered from south by mobile sand dunes. This coastal belt lies under the influence of the Mediterranean Sea with a relatively shorter dry period and annual rainfall from 100–200 mm/year, supports vegetation reached 50% on sand dunes. The dominant species are: *Ammophila arenaria, Eremobium aegyptiacum, Lotus arabicus, Moltkiopsis ciliata, Cyperus capitatus, Eichinops spinosissimus, Pancratium maritimum* and *Artemisia monosperma*. Halophytic species dominating the coastal wetlands include *Halocnemum strobilaceum, Arthrocnemum glaucum, Suaeda vera, Cressa cretica, Juncus rigidus, Zygophyllum album, Limoniastrum monopetalum* and *Mesembryanthemum crystalinum*. The general feature of the coastal wetland is outlined in Figure 2.2B.

Among the resident birds are: *Dendrocopus syriacus* (Syrian Woodpecker), *Serinus serinus* (Serin) and *Muscicapa striata* (Spotted and the *Cursorius cursor* (Cream-colored courser). The coastal belt hosting the endangered Egyptian tortoise (*Testudo kleinmanni*; Figure 2.3C). From reptiles *Acanthodactylus scutellatus* (Nidua lizard), *Varanus griseus* (Desert monitor) and *Chamaeleo chamaeleon* (Common chameleon) and *Coluber jugularis* (Syrian black snake) have been recorded while mammals include *Lepus capensis* (Cape Hare), *Hystrix indica* (Indian crested porcupine), *Felis margarita* (Sand cat) and *Vulpes zerda* (Fennec fox).

2.3.4.5 Coasts of Red Sea Gulfs Subunit

It is a narrow coastal plain or rocky slopes bordering the coasts of Suez and Aqaba Gulfs. The littoral vegetation of the gulfs salt marshes is dominated by halophytes like *Halocnemum strobilaceum, Arthrocnemum glaucum, Nitraria retusa, Tamarix nilotica* and *Limonium pruinosum*. These gulfs are terminated by Ras Mohammed, which are considered as the most northern mangrove (*Avicennia marina*) habitats in the world. Some water springs support the xeric species which include *Diplotaxis acris, Gymnocarpos decander, Hyoscyamus muticus, Pulicaria undulata, Zilla spinosa,* and *Zygophyllum coccineum*.

The coastal wadis of the Red Sea gulfs are dominated by plant species namely: *Artemisia judaica, Acacia raddiana, Hammada elegans, Cleome droserifolia, Iphiona mucronata, Fagonia mollis, Pancratium turgidum, Ochradenus baccatus, Leptadenia pyrotechnica, Moringa peregrina, Crotalaria aegyptiaca, Salvadora persica,* and *Launea spinosa.*

2.4 Diversity of Wild Medicinal Plants

There is no complete inventory of medicinal plants in Egypt. It is clear that most medicinal plants are collected from wild populations and seriously threatened them and the disappearance of the medicinal plants from their natural habitats has been unseen consequences (Batanouny, 1999). Around 200 wild species were known in Egypt for its medicinal values (Amer, 2008). Among the species listed in pharmacopeia are *Ammi majus, A. visnaga, Citrullus colocynthis, Datura stramonium, Glycyrrhiza glabra, Hyoscyamus muticus, Plantago agra, P. ovata, Senna alexandrina,* and *Urginia maritima.*

2.5 Diversity of Cultivated Plants and Domestic Animals

The cultivated plants in Egypt amounted as 29 vegetable crops (121 grown varieties, in 1000 accessions); 22 fruit crops (297 grown varieties, in 280 accession); 7 medicinal and aromatic plants in 44 landraces. From the 19,265 field crop accessions in Egypt, 51.7% only have been deposited in the National Gene Bank of Egypt (GRPI Report 2004–2005).

The domestic animals in Egypt are limited due to lack of enough natural vegetation and the shortage in the cultivated land area. Accordingly, the ownership pattern in Egypt is fragmented managed by small holders. The national counts are 3.3 millions of buffalo's heads, 3.4 millions of cow's heads, the cows distributed in 6 local breeds and 7 imported breeds (GRPI Report 2004–2005).

2.6 Biodiversity Conservation Effort in Egypt

2.6.1 *In Situ* Conservation in Protected Areas (PAs)

To 2016, 30 protected areas were established comprising c. 15% of the Egyptian land. Some of these protected areas will be mentioned to reflect its biodiversity.

2.6.1.1   Ashtoom El Gameel

Declared as a protected area by Prime Minister's decree no. 459 of 1985, it occupies the northeastern corner of Lake Manzala close to Port Said, and covers an area of about 35 km^2 It regarded as an internationally important wetland owing to the settlement of a large number of birds in winter. The most common plant species are of salt tolerant and halophytes, which include *Phragmites australis, Netraria retusa, Artemisia monosperma, Juncus rigidus* and *Suaeda aegpytiaca* and hydrophytes like *Potamogeton* sp., *Ceratophyllum* sp. and *Najas* sp. The area is an internationally important wintering place for numerous migratory birds like *Phalacrocorax carbo, Egretta alba, Ardea cinerea, Tadorna tadorna, Anas crecca, A. clypeata, Circus aeruginosus, Fulica atra, Recurvirostra avosetta, Charadrius alexandrinus, Calidris alba, C. minuta, C. alpina, Tringa totanus, Larus genei, Chlidonias hybridus, Alcedo atthis, Motacillidaes,* and *Luscinia svecica.* Numerous fishes are caught in the lake and at the entrance, freshwater fishes like *Tilapia* sp., Siluridae, Cyprinidae and Serranidae and Euryhaline or haline fishes like Mugilidae, Anguillidae, Serranidae, Sparidae, Clupeidae, and Pleuronectidae.

2.6.1.2   Elba Protectorate

Declared as a protected area by the Prime Minister's decree no. 450 of 1986, it covers an area of 35,600 km^2. It includes four distinct types of ecosystem: the mangrove forests of the Red Sea coast and its numerous islands, the Doaib region, the Gebel Elba region and the Abraq region. The Gebel Elba region has large mangrove communities along the Red Sea coast, which are the most important breeding sites for marine birds. Plant diversity in Elba is also remarkable; records include 458 species of flowering plants and ferns and the vegetation is particularly flourishing after incidents of rainfall. Among the interesting plant species are the endemic *Biscutella elbensis, Dracaena ombet,* and *Anogramma leptophylla.* At lower altitudes, in mountain wadis and foothills, there is dense parkland dominated by *Acacia tortilis, Delonix elata, Aerva persica, Euphorbia cuneata* and the mangrove trees (*Avicennia marina* and *Rhizophora mucronata;* Figure 2.2A) in coastal salt-marshes. Gebel Elba supports a rich faunal diversity in Egypt. Forty species of birds, relict Afro-tropical bird species namely *Struthio camelus* (Ostrich) and *Torgos tracheliotus* (Lappet face vulture), both disappeared from most of their former North Africa/Middle-eastern range. Twenty-three species of mammals including the endangered *Dugong dugon* (Sea cow), while other groups of animals include *Ictonyx striatus* (Zoril), *Proteles cristatus* (Aardwolf),

Griffon Vulture (*Gyps fulvus*), *Ammotragus lervia* (Barbary sheep) and *Panthera pardus pardus* (Leopard). Thirty species of reptiles and only one amphibian species have been recorded from this protectorate.

2.6.1.3   El-Omayed Biosphere Reserve

Declared as a protected area by the Prime Minister's decree no. 3216 of 1996, it covers an area c. 700 km² and is located some 83 km² to the west of Alexandria nearly 15 km² south of the Mediterranean shore. It incorporates a variety of habitat types, animal and plant communities, traditional Bedouin settlements, and patterns of land use. 288 plant species have been recorded from this Biosphere reserve which include *Ammophila arenaria, Euphorbia paralias, Pancratium maritimum* (dominating the coastal calcareous sand dunes). The inland rocky ridges hosting *Thymellaea hirsuta, Gymnocarpus decandrum, Plantago albicans* and *Asphodelus microcarpa*. Halophytic species like *Salicornia fruticosa, Cressa cretica, Atriplex halimus*, etc. dominated the marshy depressions. The calcareous soil is a part from hammada desert and its characteristic species are *Artemisia monosperma, Hammada elegans, H. scorpia, Anabasis articulata, Achillea santolina, Alhagi graecorum, Arisarum vulgare, Artemisia herba-alba, Foeniculum vulgare, Lycium shawii, Suaeda pruinosa* and *Salsola tetrandra*. Around 70 bird species, 30 species of reptiles and amphibians, 600 insects species including c.100 species of below ground insects have been recorded in this reserve. The most important species of mammals include *Dorcas gazelle*, the Eastern Mediterranean endemic mole rat, Gerbils, Fennec, Red fox and the North African endemic and threatened rat.

2.6.1.4   Lake Qaroun Protectorate

Declared as a protected area by the Prime Minister's decree no. 943 of 1989, it covers an area of 250 km². It receives the agricultural drainage of Faiyum Governorate, and some ground water from a few natural springs in its bottom. Lake Qaroun is regarded as an internationally important wetland, for resident and winter migratory birds. 88 species of birds have been spotted here. Among the internationally important populations are *Podiceps nigricollis* (Black-necked grebe), *Anas clypeata* (Shoveler), *Larus genei* (Slender-billed gull) and *Sterna albiforns* (Little tern). About 100 plant species are identified which include *Zygophyllum coccineum, Tamarix nilotica, Cynanchum acutum, Phragmites australis, Juncus rigidus, Suaeda aegyptiaca* and *Salicornia fruticosa*. The lake houses more than 10 kinds of fish, including garfish, eel, cod, muskellunge, halibut, striped bass and shrimps. From mammals five species, including Egyptian hyena, red fox, beaver, kudu and gnu have been recorded. Moreover, the reserve houses rare kinds of reptiles including the Egyptian Cobra, red-spotted and coral snake.

2.6.1.5 Nabq Multiple Use Area

Declared as a protected area by the Prime Minister's decree no.1511 of 1992, it covers an area c. 600 km² of the S. Sinai Peninsula. This area incorporates a variety of ecosystems ranging from the marine to the mountainous and harbors rich populations of corals, other marine animals and sea grasses. It provides food and shelter for numerous resident and migratory birds. The shores at Nabq provide the northernmost limit of the mangrove trees *(Avicennia marina)* in the Red Sea region. Heron, Spoonbill and Osprey birds have sustainable breeding populations in and around the mangrove trees. The plant cover consists of 164 species, including the common xeric species *Zygophyllum album, Cleome droserifolia, Halocnemum strobilaceum, Salicornia fruticosa, Avicennia marina, Anabasis articulata* and the notable largest single stands of Arak bushes (*Salvadora persica*) in the Middle East.

2.6.1.6 Ras Mohammed National Park

Declared as a protectorate by the Prime Minister's decree no. 1068 of 1983, it occupies the southern tip of the Sinai Peninsula at the southern end of the Suez and Aqaba Gulfs. The total area of this national park is about 750 km². Ras Mohammed National Park is remarkably rich in biodiversity as has been shown by the numerous baseline studies of its fauna and flora. Plant species includes *Halophila ovalis* and *Halophila sipulacea* from the sea grasses, and a wide variety of halophytes inhabiting the coastal hypersaline mudflats and mangroves *(Avicennia marina)* and the common xeric *Acacia raddiana* trees. Ras Mohammed is important as a site for the coral reef; it hosts c. 200 species of corals including 125 species of soft corals. 1000 fish species, 25 species of sea urchins, more than a 100 species of molluscs and 150 species of crustaceans have been recorded. The area is an internationally important site for soaring birds migration. A great number of the world populations of *Ciconia ciconia*, white stork passing through it. Marine water hosts also the threatened Green Turtle (*Chelonia mydas*) and *Eretmochelys imbricata* (Hawksbill Turtle), while the terrestrial threatened mammal species include Dorcas Gazelle (*Gazella dorcas)*, and Nubian Ibex Capra (*Ibex nubiana)* were also traced.

2.6.1.7 Zaraneek Protected Area

It was declared a protectorate by Prime Minister's decree no. 1429 in 1985. It is a shallow water body in the northern coastal part of Sinai Peninsula, it covers an area of 595 km². This protectorate is an internationally important site for resident and migratory birds and has recently been recognized as a Ramsar site. It is also among the important locations for fishing and quail hunting, which attract tourism and traditional Bedouin settlements. Thousand of migratory birds pass through Zaranik Protected Area during autumn which include *Phoenicopterus ruber* (Greater flamingos), *Pelecanus onocrotalus* (White pelican), *Ixobrychus minutus* (Little bittern), *Nycticorax nycticorox* (Night herns), *Anas querquedula* (Gargany), and others. While, the most common plants species are *Panicum turgidum, Stipagrostis scoparia, Artemisia monosperma, Thymelaea hirsuta* and *Zygophyllum album.*

2.6.2 *Ex Situ* Conservation

The potential plant genetic resources especially the threatened and economic species were conserved *ex-situ* in botanical and ornamental gardens, zoos as well as gene banks, and herbaria.

2.6.2.1 Herbaria

To the date of this work, 27 herbaria were recorded according their establishment date and there are about 6 small herbaria; we have no information about their specimen number. Among the important herbaria are: Cairo Flora and Taxonomy Research Dept., Horticulture Institute, Agricultural Research Centre; Botany Dept. Faculty of Science, Cairo University; Botany Dept. Faculty of Science, Alexandria University; Botany Dept. Faculty of Science, Ain Shams University; Orman Garden and Botany Dept. Faculty of Science, Assiyut.

2.6.2.2 Ornamental Gardens

The diversity of cultivated ornamental species of old and newly introduced perennial plants grouped under 110 families (Khaliefa and Loutfy, 2006). Hamdy et al. (2007), studied the six major historical botanic gardens in Cairo namely Zohriya, Aquarium, Ezbekiya, The Zoo, Orman and Horreya. These gardens were established by the Khedive Ismail (1863–1879). The flora of these gardens consists of 962 species of vascular plants that belong to 490 genera and 125 families. The most species-rich garden is the Orman (835), followed by Zohriya (358) and the Zoo (325), while the lowest number is found in the Horreya garden (62). Generally, the most species-rich families are Leguminosae (86 species), Cactaceae (74), Palmae (56) and Euphorbiaceae (45), whereas Myrtaceae, Crassulaceae, Apocynaceae and Acanthaceae comprise the lowest number of species (21, 21, 18 and 17 species, respectively). The genera most rich in species is *Ficus* (35 species), followed by *Euphorbia* (24), *Agave* (22) *Opuntia* (21), *Aloe* (16), and *Clerodendrum* (9). However, certain genera showed/ limited occurrence in a single garden, particularly in the Orman, examples of these genera include *Yucca, Sansevieria, Opuntia, Kalanchoe, Ferocactus* and *Epiphyllum*.

2.6.2.3   Gene Banks

During the past 8 decades, a number of institutions and individuals collected plant germplasm based on their interest or research needs with the absence of a national program for genetic resources conservation. In 2005, a national program for *ex-situ* conservation in Egypt has been developed. A National Plant Genetic Resources Unit "The Gene bank of Egypt" is being established. Now there are some gene banks and propagation centers, among of them:

1. **El Sheikh Zuwayed bank:** A Plant Genetic Resources Station at El Sheikh Zuwayed at the North Sinai has been established in 1997. Field collections of fruit species are maintained at this station (18 acres). The bank containing two cold stores one running at $-20°C$ and the other 50 m^{3} running at $-4°C$. Among the wild plants stored in the gene bank are *Moringa peregrina, Capparis sinaica* and *C. spinosa.* Species grown and distributed among the bedouins as small plants planted in their fields are *Thymus vulgaris, Balanites aegyptiaca, Urginea maritima, Salvia officinalis* and *Origanum syriacum.*

2. **El-Hammam breeding center:** It is a newly established center Near El-Omyed protected area; it is a breeding center used for growing the wild desert economic species under conditions similar to its natural habitats. This project granted by the Egyptian Academy of Scientific Research and Technology, and supervised by Dr. Prof. Batanowny, from the Botany Dept. Faculty of Science, Cairo University. Some species were succeeding to grow include *Thymus vulgaris, Balanites aegyptiaca, Urginea maritima, Capparis sinaica,* and *Ziziphus spina-christi.*

2.7   Threats Affecting Biodiversity

Threats affecting biodiversity in different ecosystems in Egypt are mainly overuse, pollution, unsustainable use of marine and land resources and invasive species. The percentages of the key pressures on biodiversity in protected areas are outlined in Figure 2.4, among these are:

- Unsustainable exploitation of marine resources and unplanned rapid economic development.
- Deterioration of   Ⓐ breeding and nursery areas in many areas. Some of the coastal lakes are polluted from the social pressures.
- Wetlands degradation due to the excessive expansion in scooping coastal lakes for implementing development projects and intrude of sewage drainage from agricultural and cities.
- Climate change, inducing shortage in rainfall and sandstorms.
- Overgrazing induce erosion of the vegetation cover.
- Excessive use of chemical fertilizers and pesticides in agricultural land, which led to the disappearance of most of the wildlife species.
- Invasive species, especially palm weevil, grasses and various agricul-

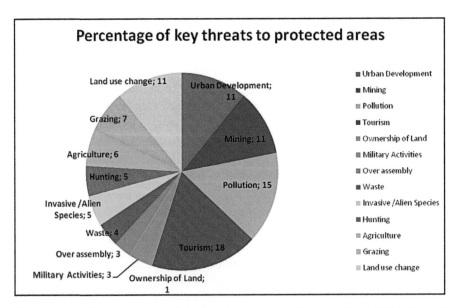

Figure 2.4   The percentage of the key threats to the Egyptian land (EEAA Annual Report 2010).

2.7.1 Threatened Species in the Egyptian Flora

The rate of species extinction in Egypt is around 3.38%. The harsh climatic conditions of this area are the main limiting factors for species extinction. Most of the species with high potentiality are under pressure and overuse and its low productivity due to scanty rainfall. Some examples of the threatened species according to Amer (2008) are discussed in the following subsections.

2.7.1.1 The Threatened Wild Crop Relatives

The threatened wild crop relatives according to Amer (2013) the following genera were nominated: *Hordeum* (3 species), *Trifolium* (18 spp.), *Medicago* (18 spp.), *Vicia* (15 spp.), *Sorghum* (4 spp.), *Melilotus* (7 spp.), *Trigonella* (11 spp.), *Gossypium* (1 spp.), *Cucumis* (4 spp.), *Linum* (3 spp.), and *Solanum* (10 spp.).

2.7.1.2 Severe Threatened Annual Species of Medicinal Uses

Severe threatened annual species of medicinal uses are: *Anastaticha hierochuntica, Bidens schimperi, Centaurea ammocyanus, Cucumis ficifolius, C. pustulatus, Hordeum marinum, H. murinum, H. spontanum, Hypecoum aequilobum, Merremia aegyptia, Papaver dubium, P. decaisnei, Phlomis aurea,* and *Pulicaria sicula* (Amer, 2008).

2.7.1.3 Highly Threatened Woody Perennials

Highly threatened woody perennials are *Balanites aegyptiaca, Commiphora opobalsamum, C. quadricincta, Dracena ombet, Gossypium herbaceum* (Amer, 1999), *Mimosa pigra, Moringa peregrina, Rhamnus lycioides, Salvadora persica, Tephrosia uniflora,* and *Ziziphus spina-christi* (El-Hadidi and Fahmy, 1992).

2.7.2 Invasive Species

Invasive species are those spread on a wide ecological range threaten the biological diversity. Among 211 species recorded in the Egyptian list of alien invasive species, 21 were recorded in the worst world list. To date, no reliable data or information is available about most of these species, their spread and negative impacts on the Egyptian environment and economy.

Among the species recorded as invasive in the Egyptian environment and mentioned in the worst global list are: Hydrophytes: Water hyacinth (*Eichhornia crassipes*), Caulerpa algae *Caulerpa taxifolia;* from terrestrial plant species: Black mimosa (*Mimosa pigra*); Honey mesquite (*Prosopis glandulosa*); Lantana (*Lantana camara*) and Horse tamarind – Leucaena (*Leucaena leucocephala*); from Crustaceans: Green crab (*Carcinus maenas*); Khapra beetle (*Trogoder magranarium*) and Sweet potato whitefly (*Bemisia tabaci*); from fishes common carp (*Cyprinus carpio*), Nile perch (*Lates niloticus*) and Western mosquito Fish (*Gambusia affinis*);Mammals: Black rat – Ship rat (*Rattus rattus*); Amphibians: Cane toad (*Bufo marinus*); Viruses: Bunchy top virus (Banana bunchy top virus), Rinderpest virus, and others (EE AA Annual Report 2010).

Keywords

- Egyptian habitats
- gene banks
- threats
- wild medicinal plants

References

Amer, M. W., (1999). Egyptian cotton: Relict *Gossypium herbaceum* L. in Egypt. *Bulletin of Faculty of Science*, Assiut University, *28*(2-D), 161–172.

Amer, M. W., (2008). Egyptian flora: Status and future prospective. *Taeckholmia, 28*, 29–47.

Amer, M. W., (2013). Diversity of crop wild relatives in Egyptian flora. *Crop Wild Relatives*, *9*, 42–44.

Ayyad, M. A., & Ghabbour, S. I., (1986). Hot deserts of Egypt and the Sudan. In: Evenari, M., et al., (eds.). *Ecosystems of the World, 12B, Hot Desert and Arid Shrublands*. Amsterdam: Elsevier.

Batanouny, K. H., (1999). *Wild Medicinal Plants in Egypt*. Academy of Scientific Research and Technology Egypt and IUCN, Switzerland.

Bornkamm, R., & Kah, L., (1990). The plant communities in the western desert of Egypt. *Phytocoenologia, 19(2), 149–231*.

Bornkamm, R., (1986). Flora and vegetation of some small oasis in south Egypt. *Phytocoenologia, 14, 275–284*.

Boulos, L., & Barakat, H., (1998). Some aspects of the plant life in the western desert of Egypt. *J. Union Arab Biologists Cairo, 5*(B), 79–94.

Boulos, L., (2005). *Flora of Egypt, Alismataceae-Orchidaceae*, vol. 4. Al Hadara Publishing, Cairo, Egypt.

Boulos, L., (2009). *Flora of Egypt Checklist Revised Annotated Edition*. Al Hadara Publishing, Cairo, Egypt.

EEAA: Egyptian Environmental Affairs Agency, (2007), Annual Biodiversity Report, Nature Conservation Sector.

EEAA: Egyptian Environmental Affairs Agency, (2010), Annual Report, Egypt State of Environment.

El-Hadidi, M. N., & Fahmy, A. G., (1992). *The Plant Red Data Book of Egypt, I: Woody Perennials.* The Palm Press, Cairo University Herbarium.

El-Hadidi, M. N., & Hosni, A. H., (2000). *Flora Aegyptiaca. Part 1,* vol. I, Palm Press, Cairo University Herbarium.

El-Hadidi, M. N., (1994/1995). Materials for excursion flora of Egypt (EFE). *Taeckholmia, 15,* 1–233.

Ghabbour, S. I., & Mikhail, W. A., (1998). Does phyto-diversity coincide with zoo-diversity in Egypt. *J. Union Arab Biologists Cairo, 5*(B), 59–78.

Gibali, M. A., (1988). *Studies on the Flora of Northern Sinai.* MSc Thesis, Cairo University.

GRPI: Genetic Resources Policy Initiative, (2004–2005). Egypt Phase I.

Hamdy, R. S., Abd El-Ghani, M. M., Youssef, T. L., & El-Sayed, M., (2007). The floristic composition of some historical botanical gardens in the metropolitan of Cairo, Egypt. *African J. Agric. Res., 2*(11), 610–648.

Hamed, A., Kord, M., & Amer, W., (2015). *Wild Cotton in Egypt: Physiological and Molecular Investigations of Wild Cotton.* Lambert Academic Publishing, Germany.

Hassan, L. M., (1987). *Studies on the Flora of Eastern Desert, Egypt.* PhD thesis, Cairo University.

Hegazy, A. K., & Amer, W. M., (2001). Altitudinal and latitudinal diversity of the flora on the Eastern and Western sides of the Red Sea. *Symposium on Natural Resources and Their Conservation in Egypt and Africa.* Institute of African Research and Studies, Cairo University.

Kassas, M. A., & Zahran, M. A., (1971). Plant life on the coastal mountains of the Red Sea. Egypt. *Indian Bot. Soc. Golden Jubilee, 50A,* 571–589.

Khaliefa, F. S., & Loutfy, M. H., (2006). *Ornamental Cultivated Plant Collection.* Ministry of Agriculture Cairo.

Wickens, G. E., (1976). *The Flora of Jebel Marra and its Geographical Affinities.* Majesty's Stationary Office, London.

Zahran, M. A., & Willis, A. J., (2003). *Plant Life in the River Nile in Egypt.* MARS publication House, Cairo.

Zahran, M. A., & Willis, A. J., (2009). *The Vegetation of Egypt,* 2nd edition. Springer Science and Business Media B.V.

CHAPTER 3

Biodiversity in Gabon: An Overview

**JEAN B. MIKISSA,[1] FLORE KOUMBA PAMBOU,[2] and
EMMANUEL BAYANI NGOYI[3]**

[1]National School of Forest and Water, Gabon, E-mail: jbmikissa@gmail.com
[2]Institute for Research on Agriculture and Forests, Gabon, E-mail: fkoumbapambo@gmail.com
[3]General direction for Environmental, Gabon, E-mail: scoutgabon@yahoo.fr

3.1 Introduction

Gabon, located in Central Africa, crossed by Equator, between latitudes 2° 30' N and 3° 55' S, is a forest country par excellence. It has an important forest area of nearly 23 million hectares (87% of the territory) of which 40% remain primary. The flora and fauna are all the more remarkable, as the equatorial forest is still well preserved. This forest ecosystem meets a mosaic of natural environments (savannas, estuaries and lagoons, mangroves, mountains), indicators of composite biological diversity. Despite some measures, including those related to conservation and sustainable use, there is a gradual loss of biodiversity and degradation of ecosystem services. By ratifying the Convention on Biological Diversity (CBD) on 14 March 1997, Gabon has therefore embarked on the implementation of a rational policy to combat the erosion of biological diversity. It is in this context that we take stock of the trends in biodiversity, the threats to biodiversity and the measures taken to manage biodiversity.

3.2 Major Ecosystems in Gabon

Gabon has an important forest area representing almost 85% of the territory (FAO, 2015) (Mayaux et al., 2004) (Figure 3.1). The Gabonese forest has six types of vegetation: flooded and swampy forests, coastal basin forests, mountain forests, forests of the plateaus of the interior, forests in the northeastern plateaus, degraded or secondary forests. The nonforest area constitutes 15% of the territory. Indeed, the savannas appear scattered throughout the territory. They are located in Ogooué-Maritime, Nyanga, Ngounié, and especially in Haut-Ogooué where they are best represented (Figure 3.1).

The Gabonese territory has an area of 26,766,700 ha of which 22,751,695 ha represent the forest area. In this forest there are two domains:

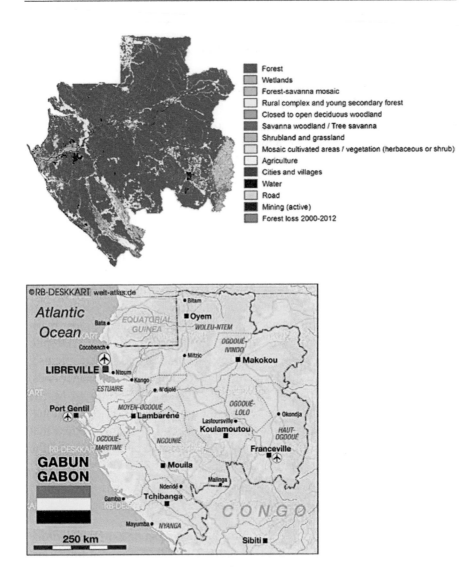

Figure 3.1 Occupation of the national territory.

Permanent Forest Estate subdivided into   A ☐   -
ing all protected areas (National Parks, Reserves, Hunting Areas and Historic
Sites) with a surface area of 4 million hectares or more than 18%, in productive public forests or forests with an area of 14,500,000 hectares, i.e., 62%.

The Domaine Forestier Rural with an area of 4,500,000 hectares or 20%.
The mangroves cover an area of 3950 km². The most important are located

in Mondah Bay (35,000 ha), the Komo estuary (85,000 ha) and the Ogooué delta (80,000 ha). Smaller areas exist in the Fernand-Vaz, Iguela, Ngové, Ndogo and Banio lagoons.

3.3 Plant Diversity in Gabon

In addition to algae, lichens and bryophytes, the former estimates of the number of plant species in Gabon were 6,000 to 8,000 (Breteler, 1989, 1990; Morat and Lowry, 1997). Sosef et al. (2006) estimate that this number ranges from 7,000 to 7,500. This indicates that it is difficult to estimate the total number of plant species present in Gabon. It should be noted that 1900 species were described in the Flora of Gabon in 1988 with an endemism rate of 20% (Table 3.1). Recently, the checklist of vascular plants, based on data from more than 65,000 herbarium specimens, identified 4,170 species of vascular plants, 82 of which were introduced (Sosef et al., 2006, Table 3.2 and Figure 3.2). Among the plant species known to date, a group of herbaceous plants living on land in the forest and on the rock walls and trunks of trees, *Begonia,* has been studied in detail. Thus, of the 121 African *Begonia*

Table 3.1 Some Endemic Plants in Gabon

Family	Species
Acanthaceae	*Pseudocalyx macrophyllus*
Apocynaceae	*Baissea longipetiolata*
Balsaminaceae	*Impatiens floretii*
Combretaceae	*Combretum exilii*
Connaraceae	*Cannarus gabonensis*
	Cnestis uncata
Euphorbiaceae	*Croton loukandesis*
	Pycnocoma tholonii
Leguminosae/ Caesalpinioideae	*Anthonota ferruginea*
	Dalbergia librevillensis
	Loesenera gabonensis
Melastomataceae	*Dicellandra descoingsii*
Rubiaceae	*Chassalia tchibangensis*
	Tarenna ogouensis
Zingiberaceae	*Aframomum inversiflorum*
	Costus fissiligulatus

Table 3.2 Number of Vascular Plants in Gabon

	Lycopsida (Lycophytes)	*Pteropsida* (Pteridophytes)	*Pinopsida* (Gymnospermes)	*Magnoliopsida* (Angiospermes)	**Total**
Family	2	23	2	159	186
Genus	4	54	2	1,237	1,297
Species	17	158	3	4,532	**4,710**

Source: Sosef et al., (2006).

species, 50 are found in Gabon, 16 of which are endemic (Sosef et al., 2006). From these data on the *Begonia* the Dutch botanist Sosef admitted three shelters in Gabon: the Crystal Mountains, the Massif du Chaillu and the Doudou Mountains. These three regions constitute the centers of endemicity and correspond to the presumed refuges of the Pleistocene. It is also in these regions that the flora is the richest in genera and species of the entire Guineo-Congolese forest massif and especially the centers of diversity and speciation for *Begonia* (Sosef et al., 2006).

Gabon's forest biomass is home to more than 400 woody species that can be harvested only 13% are known to users and 3% are exploited (Figure 3.3), with okoume being the foremost. More than a third of the Gabonese forest remains primary, with a deforestation rate of less than 1% per year (Les Forêts du Bassin du Congo – Etat des Forêts, 2006).

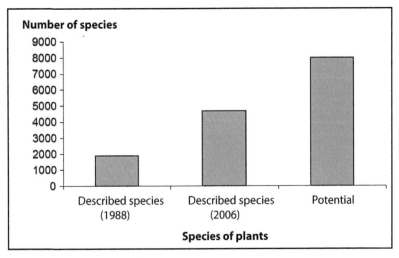

Figure 3.2 Species wealth of Gabon plants.

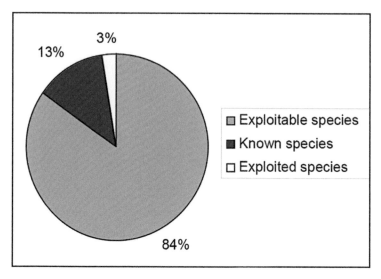

Figure 3.3 Gabon forest biomass.

3.4 Fauna Diversity in Gabon

3.4.1 Terrestrial Fauna

3.4.1.1 Mammals

The mammalian fauna comprises about 190 species including 19 species of primates, including mandrillas, colobus and the endemic Cercopithecus solatus of the forests of central Gabon, nearly 20 species of carnivores (canines, herpestidae, felidae, hyanids, mustelids and viverridae) and 13 species of artiodactyla (suidae, tragulidae and bovidae).

Gabon is also a sanctuary for some large forest mammals: the gorilla, the chimpanzee and the elephant (Christy et al., 2008). Indeed, the censuses carried out in Gabon by Tutin and Fernandez (1984) showed that the populations of these three species were around 35,000 gorillas, 64,000 chimpanzees and 74,000 elephants (Figure 3.4), respectively 30%, 35% and 11% of the world population of these animals (Christy et al., 2008). The census, done by Barnes et al. (1995) on the forest elephants, places them at 61,000 individuals. To date, Blake et al. (2007), in their inventory, estimated the forest elephants population size at 22,000 individuals in 7,592-km^2 at the north national park of Minkebe (2.9 elephants km^{-2}).

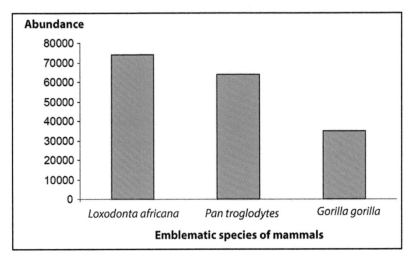

Figure 3.4 Abundance of Gabon's mammalian flagship species.

The number of species encountered in grassland savannas, densely shrubby, arborescent or wooded is relatively low compared to that of forests. The majority of mammalian species live in forests, making up the most mammalian ecosystem in Gabon (Figure 3.5). The mammalian fauna of the savannahs also contains rare species. There are, like large mammals, buffaloes, elephants, harnessed guib, and so on. On the other hand, the lion and the lycaon that former occupied these areas have disappeared now.

3.4.1.2   Birds

The avifauna is also very diverse. Indeed, resident or transient birds were estimated at over 600 species (Christy et al., 2001). The recent synoptic list

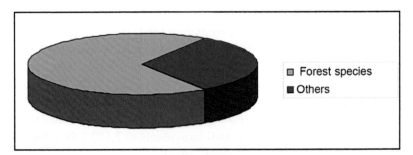

Figure 3.5 Diversity of terrestrial mammals in Gabon.

of birds in Gabon lists 749 species, including 11 threatened and 8 vulnerable. There are more than 300 sedentary species in forests or terrestrial wetland ecosystems (Vande Weghe, 2006, 2007, 2011a, 2011b). Here again, the forest ecosystem remains the richest in bird species (Figure 3.6).

3.4.1.3   Reptiles

The list of reptiles of Gabon, whose presence is duly   □ A prises 121 species including 13 species of chelonians or turtles, 3 crocodilian or crocodile species, 3 amphisbenian or amphisbene species, 70 species of ophidians or snakes and 32 species of lacertilians or lizards (Pauwells and Vande Weghe, 2008) (Figure 3.7).

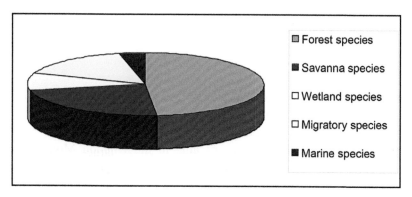

Figure 3.6   Diversity of birds in Gabon.

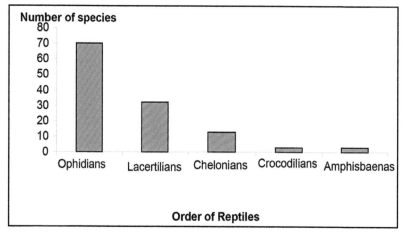

Figure 3.7   Specific richness of the different orders of Reptiles of Gabon.

3.4.1.4 Amphibians

The first systematic inventories of the Batrachian fauna in central Gabon, carried out by Blanc and Fretey (2004) in 41 sites in the Lopé Wildlife Reserve and the Bee Forest, identified 6 families, 20 genera and 43 species of Amphibians Anoures. Later, another inventory was undertaken by Burger et al. (2006) in the Gamba Protected Areas Complex, which resulted in the identification of specific wealth at four study sites. Thus, the Moukalaba-Doudou National Park was the richest site with 70 species followed by Rabi Toucan 49 species, then the National Park of Loango with 37 species, finally the poorest zone was Gamba with 20 species (Figure 3.8). In three years, the number of amphibian species has increased from 72 to 98 (Burger et al., 2006). Among the endemic species of amphibians are the *Werneria iboundji* toad, which are found only at Mount Iboundji in the Massif du Chaillu, or the leptopelis crystallinoron sticky frog, endemic to the Crystal Mountains.

3.4.1.5 Insects

The Gabonese fauna also includes a large number of invertebrates. The first insects capture campaign in 2001 identified 142,425 insect species in 22 morpho-species (Basset et al., 2004).

Another preliminary inventory of ants in the Doudou Mountains by Fisher (2004) found 310 species in this group. Placing Gabon as the richest species in this group in front of other African countries such as Tanzania

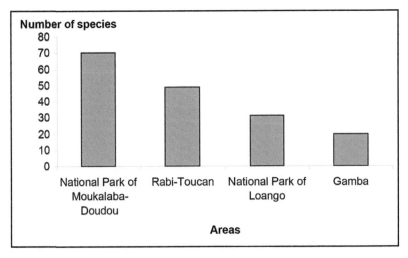

Figure 3.8 Specific wealth of Amphibians in different areas of Gabon.

with 237 species (Robertson, 1999), Madagascar 215 species (Fisher, 1998), and Ghana 176 species (Belshaw and Bolton, 1994) (Figure 3.9).

3.4.2 Coastal and Marine Fauna

3.4.2.1 Mammals

Cetacean fauna comprises 15 species: 6 species of baleen whales and 9 species of odontocetes (sperm whales and dolphins) (Van de Weghe, 2007). The endemic dolphin (*Sousa teuszii*) and the large dolphin (*Tursiops truncatus*) are considered the species most threatened by coastal development, coastal fishing and oil exploitation.

3.4.2.2 Fish

The fish biodiversity of the Gabonese fresh waters shelters 380 species of fish (Mve Beh, 2011, Pers. Comm.). Of the species identified, 4 appear to be endemic and include a new genus of *Ivindomyrus* found in the Ivindo River. According to the site http://www.fishbaseforafrica.org/, which constitutes the fish database, accessed on 20/10/2015, the nonexhaustive inventory of the fish fauna of Gabonese waters lists almost 791 species of which 510 species in marine waters and 297 species in fresh waters and 44 endemic species.

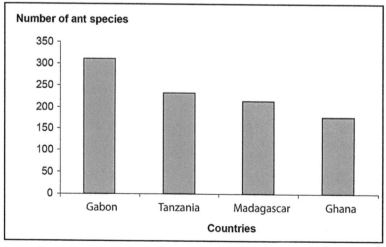

Figure 3.9 Specific richness of ants in different African countries.

3.5 Genetic and Crop Diversity in Gabon

Gabon is a weakly agricultural country where cropland accounts for only 5% of the total area. Nevertheless, the few studies show that agricultural biodiversity is moderately high. It is based on a cash crop (coffee plantations, cocoa, oil palm, sugar cane) and subsistence agriculture, with food speculation of current consumption (banana, cassava, taro, yam, peanut, vegetables, etc.).

3.6 Threatened Biodiversity in Gabon

Gabon has made many efforts to protect threatened fauna by publishing the list of 27 full-protected species (Bayani, 2017).

There are no fully or partially protected plant species except, of course, those in protected areas, which are in fact placed under a full protection regime. However, 5 forest species, Afo (*Poga oleosa*), Andok (*Irvingia gabonensis*), Douka or Makore (*Tieghemella africana*), Moabi (*Baillonnella toxisperma*) and Ozigo (*Dacryodes buttneri*), are defended, forbidden slaughter,   A ☐
January 2009 (Decree No. 0137/PR/MEFEPA of 4 February 2009). Nevertheless, in the exercise of customary rights, fruit, bark, latex or resin can be harvested (Decree No. 692/PR/MEFEPEPN of 24 August 2004).

To date, Gabon has not made any efforts on the ex situ conservation of its biological resources. However, steps have been taken to preserve the site of the Sibang arboretum, inherited from the colonial area.

3.7 Biodiversity Conservation Initiatives

Having found that protected areas were exclusively reserves, Gabon, through the national biodiversity strategy and action plan adopted by the Council of Ministers on 27 July 2000, resolved to create 13 National Parks. The political decision was announced at the World Summit on Sustainable Development (Johannesburg, 2002); the realization of this decision was made through the adoption of a set of decrees creating 13 National Parks (Figure 3.10).

The Lopé National Park is listed as a UNESCO World Heritage Site under the name "Lopé-Okanda Ecosystem and Relational Cultural Landscape." Other sites,   A ☐
Heritage List, are listed on the UNESCO Gabon Tentative List, namely: the pygmy cultural ecosystem and cultural landscape of the Minkébé massif

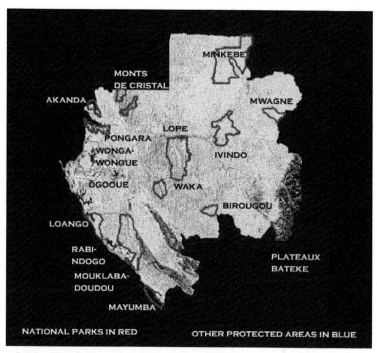

Figure 3.10 Distribution of Gabon National Parks.

(2003), Ivindo (2005), the caves of Lastourville (2005), the Batéké Plateau National Park (2005), the Moukalaba-Doudou National Park (2005) and the Monts Birougou National Park (2005). The Ipassa-Makokou Integral Nature Reserve (Northern Part of the Ivindo National Park) was recognized in 1983 by UNESCO as a Biosphere Reserve under the UNESCO Man and the Biosphere Program (UNESCO Program-MAB, The Man and the Biosphere). Gabon also registered 9 sites on the Ramsar Convention registry (Table 3.3).

3.8 Challenges and Major Threats to Biodiversity in Gabon

Biodiversity supports the functioning of ecosystems and provides ecosystem services that ensure food security, human health and the development of a country. It is also an essential component of the belief, vision and identity system of many peoples. In recent years, there has been an accelerated extinction of animal and plant species whose impact is the reduction of the

Table 3.3 Gabon Ramsar Sites

Sites	Localization	Surface (ha)
Akanda (Park National)	Estuaire	54,000
Petit Loango (Réserve de faune)	Ogooué-Maritime	480,000
Pongara (Park National)	Estuaire	92,969
Rapides de Mboungou Badouma et de Doumé	Haut-Ogooué Ogooué-Lolo	59,500
Setté Cama (Réserve de faune et Domaine de chasse)	Ogooué-Maritime	220,000
Bas Ogooué	Moyen Ogooué Ogooué Maritime	862,700
Monts Birougou (Park National	Ngounié Ogooué-lolo	536,800
Chutes et Rapides sur Ivindo	Ogooué Ivindo	132,500
Wonga-Wongue (Réserve présidentielle)	Ogooué-Maritime	380,000

biological and genetic heritage of the biosphere. This regressive dynamic is partly due to the deep disturbances and even to the advanced destruction of certain ecosystems, which result either from natural factors or from anthropogenic activities.

3.8.1 Natural Causes

Natural factors of biodiversity loss are primarily infectious diseases and invasive alien species.

3.8.2 Infectious Diseases

Diseases are threats to wildlife. Gabon's four epidemics of Ebola fever, including two in the Minkebe National Park, have contributed significantly to the decline of monkey populations in Gabon (Walsh et al., 2003). As a result, the Western Plains gorilla as a critically endangered species on the IUCN Red List of Threatened Species was classified as early as 2007.

3.8.3 Invasive Alien Species

Exotic species such as the *Stachytarpheta cayennensis* or the *Chromolaena odorata* have invaded native ecosystems and have devastating effects on the environment. Originating from an area from Florida to northern Argentina, *C. odorata* is a shrub that is considered one of the world's most invasive alien invasive species. In Gabon, this weed is bordered by the main roads along the railway. The local populations call it "Comilog" or "Maghudunu," names that express its dispersal, invasive action and impact. It is a major threat to biodiversity (competition with indigenous vegetation, impact on the reproduction of certain reptiles) and agricultural areas. Similarly, the little fire ant *Wasmannia auropunctata*, it also classified as one of the hundred invasive alien invasive species, is the typical example of an invasive animal species that causes disturbance to indigenous fauna of Gabon. Introduced to fight against certain parasites of various cultivated plants including cacao, this ant originating in South and Central America spreads gradually throughout the country. The regions, which have been infested for more than 10 years, have lost approximately 95% of their native ants (Walker, 2006, Ndoutoume-Ndong and Mikissa, 2007, Mikissa et al., 2008, 2013). *Wasmannia auropunctata* also attacks wildlife, from invertebrates to vertebrates, causing corneal damage in some wild and domestic animals (Walsh et al., 2004).

3.8.4 Anthroponic Causes

In Gabon, the main driving force responsible for the loss of biological diversity and the degradation of ecosystem services remains the development choices that result in the overexploitation of natural resources and thus affecting biological diversity. The pressure on natural resources, caused by this driving force, is exerted, among other things, by exploitation of fishery resources, agriculture, mining (oil and mining), logging and urbanization, uncontrolled and the development of infrastructure. Industrial fishing, illegally practiced by trawlers, is a major threat to marine biodiversity. In fact, they fish regularly within the prohibited coastal zone within 6 nautical miles of the beach. On some occasions, explosives are used to destroy clumps of rocks. At the coastal and continental levels, the main threats to the conservation and sustainable use of fisheries resources have been identified, including (i) the use of prohibited and nonselective fishing gear (netting, use of fish products) and (ii) weak enforcement of fisheries resources exploitation and protection regulations. The coastal zone encompasses a wide variety of habitats and ecosystems. The process of urbanization and current coastal development are increasing the pressure on these ecosystems and habitats. This results in the destruction of mangroves due to the construction of houses and the cutting of mangroves to smoke the fish. In addition, the installation of fishers in these ecosystems results in intensive fishing in spawning grounds and nurseries.

The extractive industry (mining and oil) in Gabon has a considerable impact on ecosystems. As regards mining activities, the threat to biological resources stems from the prospecting of substances, the opening up and exploitation of quarries and mines. All of these activities are responsible for the destruction of vegetation cover and wildlife habitat. For example, gold mining destroys freshwater aquatic ecosystems and reduces water quality. More than 10,000 gold miners are active in the Minkébé National Park. With regard to petroleum activities, seismic surveys generate powerful and harmful sound waves for wildlife, particularly at sea, where they affect demersal fauna and cetaceans, which are particularly sensitive to seismic tests. The impact is all the more remarkable when the surveys are carried out during periods of migration or reproduction of   Exploration drilling also contributes to the destruction of natural habitats such as hatcheries (mangroves, seagrass beds) or spawning grounds.

Hydrocarbon deposits sporadically weaken Gabon's coastal ecosystem, threatening marine organisms, invertebrates (plankton, molluscs, crustaceans),   turtles and marine mammals. During the egg-laying period, marine pollution is a serious problem for the protection and preservation of these endangered species. Petroleum activity affects the quality of water and the integrity of aquatic ecosystems. In the Rabi-Toucan region, blockage of drainage resulting from poor management of water   creates   A swamps in forests. Areas with   ☐ A sion and siltation, leading to serious consequences on freshwater ecosystems downstream.

The pressure on agricultural biodiversity lies particularly in the large-scale conversion of land as part of the establishment of agro-industrial plantations of oil palm, rubber, cocoa and coffee. This conversion is a source of loss of biological diversity and impoverishment of ecosystem services. Indeed, the plantations established are generally   and the introduced species transform the habitats.

On the other hand, following cultivation practices (slash-and-burn agriculture), soils can degrade and be exposed to erosion, leading to their progressive impoverishment.

Similarly, the modern agriculture to which our country is going to tend, stimulated by maximizing yields, can gradually lead to the homogenization of crops and livestock, as well as the introduction of invasive alien species and increased vulnerability of cultures with regard to pathogens.

Although the total forest cover of Gabon (almost 87% of the territory) has remained stable over the last decades, logging, notably through the felling of trees and the opening of roads and forest tracks, exerts the most sig-
        ⬜A
essential for the well-being of the populations.

Selective or nonselective logging has damaging effects on forest structure, both canopy and understory. It is estimated that logging results in approximately 10% loss of canopy, but up to 50% of this canopy can be affected because many trees are often damaged or destroyed during approach and slaughter of a single tree (Collomb et al., 2000). For example, the harvest rates of okoumé, the most exploited gasoline, accounting for more than 85% of total production, are of the order of 1.5 ft/hectare, resulting in direct and indirect damage to ecosystem with the destruction of 10–20% of the canopy (White, 1994).

Forestry affects the composition of wildlife. Indeed, the destruction or scarcity of fruit trees has   ⬜A
mals (Vande Weghe, 2011b). Logging is also a real threat to the survival of wildlife, as bushmeat hunters use old and newly developed trails and roads to reach isolated areas. Thus, poaching is increasingly penetrating into the forests, with an impact on biodiversity that differs according to the species under consideration. Smaller species, which are more generalized and rapidly reproductive, are much more resistant to sustained hunting than large, slow-growing and low-density species. And, the scarcity of large species leads to greater hunting pressure on smaller species. The decline and/or disappearance of hunted species have certain repercussions on the forest community, such as the loss of pollinators, seed disseminators, prey or predators. These losses can lead to a change in plant composition, a change in natural ecosystem balances, or even a likely reduction in overall biodiversity.

3.9 Concluding Remarks

The signing of the Convention on Biological Diversity by Gabon in 1992 and its ratification in 1997, as well as the elaboration and implementation of the strategies and action plans, represent to this day the perfect translation of the national policy on of "Gabon Vert," which holds the lead in the policy of the Government.

In addition to the actions mentioned above, other measures relating to the biodiversity have been taken by Gabon, including the creation of:

- The National Agency of National Parks (ANPN);
- The National Agency for Fisheries and Aquaculture (ANPA);
- The Gabonese Agency for Space Studies and Observations (AGEOS);
- The National Agency for the Execution of Activities of the Forest-Wood Sector (ANEAFPF).

The limits on the implementation of these activities and measures remain linked to ⬜A capacity building, both institutional and systemic, for optimal management of biodiversity in Gabon.

Keywords

- challenges
- fauna diversity
- threatened biodiversity
- threats

References

Barnes, R. F. D., Blom, A., Alers, M. P. T., & Barnes, K. L., (1995). An estimate of the numbers of forest elephants in Gabon. *J. Tropical Ecol.*, *11*(1), 27–37.

Basset, Y., Mavoungou, J. F., Mikissa, J. B., Missa, O., Miller, S. E., Kitching, R. L., & Alonso, A., (2004). Discriminatory power of different arthropod data sets for the biological monitoring of anthropogenic disturbance in tropical forests. *Biodiversity and Conservation*, *13*, 709–732.

Belshaw, R., & Bolton, B., (1994). A survey of the leaf litter ant fauna in Ghana, West Africa (Hymenoptera: Formicidae). *J. Hymenoptera Res.*, *3*, 5–16.

Blake, S., Strindberg, S., Boudjan, P., Makombo, C., Bila-Isia, I., Ilambu, O., et al., (2007). Forest elephant crisis in the Congo Basin. *PLoS Biol.*, *5*(4), e111.https://doi.org/10.1371/journal.pbio.0050111.

Blanc, C. P., & Fretey, T., (2004). Répartition écologique des amphibiens dans la réserve de faune de la lopé et la station biologique de la makandé (Gabon). *Bull. Soc. Zoo. Fr.*, *129*(3), 297–315.

Breteler, F. J., (1989). Gabon. In: Campbell, D. G., & Hammond, H. D., (eds.). *Floristic Inventory in Tropical Countries*. New York Botanical Garden, New York, pp. 198–202.

Breteler, F. J., (1990). *Gabon's Evergreen Forest: The Present Status and its Future*. Mitteilungen aus dem Institut für Allgemeine Botanik in Hamburg, *23a*, 219–224.

Burger, M., Pauwels, O. S. G., Branch, W. R., Tobi, E., Yoga, J. E., & Mikolo, E. M., (2006). Inventaire des amphibiens du Complexe de Gamba, Gabon. In: Alonso, A., Lee, M. E., Campbell, P., Pauwells, O. S. G., & Dallmeier, F., (eds.). Gamba, Gabon: Biodiversité d'une forêt Équatoriale Africaine. *Bull. Biol. Soc. Washington*, *12*, 79–90.

Christy, P., (2001). Gabon. In: Evans, L. D. C., & Evans, M. I., (eds.). *Importance Birds Areas in Africa and its Associated Islands: Priority Sites for Conservation, Fishpool.* Pisces Publications et BirdLife International, Newbury et Cambridge, Royaume-Uni., pp. 349–356.

Christy, P., Lahm, S. A., Pauwels, O. S. G., & Vande Weghe, J. P., (2008). *Check-List des Amphibians, Reptiles,* oiseaux et mammifères des Parcs Nationaux du Gabon. Smithsonian Institution.

Collomb, J. G., Mikissa, J. B., Minnemeyer, S., Mundunga, S., Nzao Nzao, H., Madouma, J., et al., (2000). *Un Premier Regard sur L'exploitation Forestière au Gabon.* World Resources Institute, Washington DC, USA.

Fisher, B. L., (1998). Ant diversity patterns along an elevational gradient in the Réserve Spéciale d'Anjanaharibe-Sud and on the western Masoala Peninsula, Madagascar. *Fieldiana: Zoology, 90,* 39–67.

Fisher, B. L., (2004). *Monts Doudou, Gabon: A Floral and Faunal Inventory With Reference to Elevational Variation.* California Academy of Sciences Memoirs, *28,* 269–286.

Mayaux, P., Bartholome, E., Fritz, S., & Belward, A., (2004). A new land - cover map of Africa for the year 2000. *J. Biogeography, 33*(6), 861–877.

Mikissa, J. B., Delabie, J. H. C., Mercier, J. L., & Fresneau, D., (2008). Preliminary assessment on the interactions of *Wasmannia auropunctata* in Native Ant Communities (Hymenoptera: Formicidae) of a Mosaic Gallery Forest/Savannah in Lope National Park, Gabon. *Sociobiology, 51*(1), 207–218.

Mikissa, J. B., Jeffery, K., Fresneau, D., & Mercier, J. L., (2013). Impact of an invasive alien ant, *Wasmannia auropunctata* roger on a specialized plant–Ant mutualism, *Barteria fistulosa* Mast. and *Tetraponera aethiops,* Smith, F., in a Gabon forest. *Ecol. Ent., 38,* 580–584.

Morat, P., & Lowry, II P. P., (1997). Floristic richness in the Africa-Madagascar region: A brief history and prospective. *Adansonia, ser. 3, 19*(1), 101–115.

Moumbogou, C., Meunier, Q., Ogoula Ikinda, L. B., & Doucet, J. L., (2011). *Les 5 Arbres Protégés du Gabon,* d'après le décret 692/PR/MEFEPEPN du 24 août 2004.

Ndoutoume-Ndong, A., & Mikissa, B., (2007). Influence of the presence of the ant *Wasmannia auropunctata* (Roger 1863) (Hymenoptera: Formicidae) on other species of ants in the lope reserve (central Gabon). *Ann. Soc. Ento. de Fr., 43,* 155–158.

Pauwells, O. S. G., & Vande weghe, J. P., (2008). *Reptiles du Gabon.* Smithsonian Institute.

Robertson, H. G., (1999). Ants (Hymenoptera: Formicidae) of Mkomazi. In: Coe, M., McWilliam, N., Stone, G., & Packer, M. J., (eds.). *Mkomazi: The Ecology, Biodiversity and Conservation of A Tanzanian Savanna.* Royal Geographical Society, London, pp. 321–336.

Sosef, M. S. M., Wieringa, J. J., Jongkind, C. C. H., Achoundong, G., Azizet Issembé, Y., Bedigian, D., et al., (2006). Checklist of Gabonese vascular plants. *Scripta Botanica Belgica, 35,* 1–438.

Tutin, C. E. G., & Fernandez, M., (1984). Nationwide census of gorilla (*Gorilla g. gorilla*) and chimpanzee (*Pan t. troglodytes*) populations in Gabon. *American J. Primatology, 6,* 313–336.

Vande Weghe, J. P., (2006). *Les Parcs Nationaux du Gabon: Ivindo et Mwagna – Eaux noires, forets vierges et baïs.* Wildlife Conservation Society.

Vande Weghe, J. P., (2007). *Les Parcs Nationaux du Gabon: Loango, Mayumba et le Bas-Ogooué*. Wildlife Conservation Society.

Vande Weghe, J. P., (2011a). *Les Parcs Nationaux du Gabon: Akanda et Pongara – Plages et Mangroves*. Wildlife Conservation Society.

Vande Weghe, J. P., (2011b). *Les Parcs Nationaux du Gabon: Lopé, Waka et Monts Birougou – Le moyen Ogooué et le massif du Chaillu*. Wildlife Conservation Society et Parcs Gabon.

Walker, K. L., (2006). *Impact of the Little Fire Ant, W. auropunctata auropunctata*, on Native Forest Ants in Gabon. *Biotropica, 38,* 666–673.

Walsh, P. D., Abernethy, K. A., Bermejo, M., Beyers, R., De Wachter, P., Akou, M. E., et al., (2003). Catastrophic ape decline in western equatorial Africa. *Nature, 422,* 611–614.

Walsh, P. D., Henschel, P., & Abernethy, K. A., (2004). Logging speeds little red fire ant invasion of Africa. *Biotropica, 36,* 637–641.

White, L. J. T., (1994). The effects of commercial mechanized logging on forest structure and composition on a transect in the Lopé Reserve, Gabon. *J. Trop. Ecol., 10,* 309–318.

CHAPTER 4

Biodiversity in Ghana

MICHAEL KWABENA OSEI,[1] **LAWRENCE MISA ABOAGYE,**[2]
PATRICK OFORI,[3] **and BENJAMIN ANNOR**[1]

[1]CSIR-Crops Research Institute, P.O. Box 3785, Kumasi, Ghana, E-mail: oranigh@hotmail.com
[2]CSIR-Plant Genetic Resource Research Institute, Bunso, Ghana
[3]CSIR-Soil Research Institute, Academy Post Office, Kwadaso, Kumasi, Ghana

4.1 Introduction

Ghana abounds in rich biological diversity as a result of its location within three major bio-geographical zones. These are the Guineo-Congolian zone, which covers the southwestern part of the country; the Guineo-Congolian/Sudanian transition zone, covering the middle belt and the Sudanian zone covering the northern tip of Ghana. Globally, increasing human populations and poverty have accelerated the depletion of natural resources leading to environmental degradation and loss of biodiversity. All over the world, there has been increasing interest in environmental issues, especially the current deteriorating state of the global and local environment, largely as a result of undesirable human activities. The current increased rate of biodiversity loss is of special and immediate concern, because of its negative implications for human survival on earth. The loss of each species comes with the loss of potential economic benefits as well as a loss of ecosystem balance. Biodiversity conservation in tropical countries is of great importance due to high levels of endemism. Biodiversity loss and conservation have been a global focus for at least two decades.

Over the past century, Ghana has reserved over 300 ecologically important areas for biodiversity conservation and a national strategy for this purpose has been developed under the framework of the convention on biological diversity. Nevertheless, Ghana's biodiversity continues to dwindle because of different human activities, especially in agriculture, have degraded biological resources   . In the past years, more than the country's forest was near pristine condition in the forest zone. Today they are dominated by human settlement with only about 15% of forest patches covering the land. This can be attributed to the conversion of forestlands into agricultural landforms, a phenomenon which may lead to continuous loss of biodiversity if not managed properly. Moreover, there is inadequate capacity of resource managers with the requisite and up to date understanding of conservation and resource management to direct the conservation of our resources. Although information on Ghana's biodiversity tends to be generally incomplete, diffuse and inaccurate, much detailed and complete information have been documented on the genetic diversity of life forms in Ghana's terrestrial ecosystem than the marine and other aquatic ecosystems (MES, 2002). There is very little information available on microbial diversity in the terrestrial, marine and aquatic ecosystems of Ghana.

In contemporary times, it has become clear that this apathetic attitude towards resource management and use must change. Even though the change must come there are inadequate human resources coupled with insuf-

relevant information to meet the challenges posed by the deteriorating resource base. This chapter describes the geography and climatic conditions as well as biodiversity systems in Ghana. It further assesses the status of Ghana's biodiversity and genetic diversity of some cultivated plants in Ghana. Finally, the chapter deliberates on biodiversity loss and conservation legislation on biodiversity and international conventions related to biodiversity conservation in Ghana.

4.2 Geography and Climatic Conditions in Ghana

4.2.1 Location

Ghana is situated in the center of the countries along the Gulf of Guinea in West Africa. The country has an area of 238,530 square kilometers and lies between latitudes 4°44′ and 11°11′N and longitudes 01°12′ and 03°11′W (Figure 4.1). Administratively, Ghana is divided into ten regions. The population of Ghana is about 28 million (2017) with an average growth rate of 3.3% per annum.

4.2.2 Vegetation, Rainfall, and Crops

There are six main agro-ecological zones defined on the basis of climate, reflected by the natural vegetation and influenced by the soils (Figure 4.1). These are Rain Forest, Deciduous Forest, Transitional zone, Coastal Savanna, Guinea Savanna and Sudan Savanna zones. The Rain Forest is found in the southwestern part of the country. The annual rainfall of this zone is between 1,700 mm and 2,100 mm and there are two rainy seasons (March-July and September- November). The soils are heavily leached. The major crops grown are oil palm, rubber, coconut, rice, bananas,

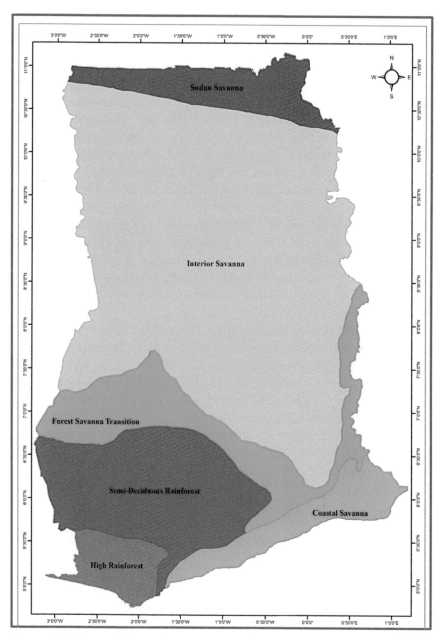

Figure 4.1 Map of Ghana showing agro-ecological zones.

plantains and cocoyam. The Deciduous Forest constitutes 21% of the land area of the country. It has two rainy seasons: March to July and September to November and a rainfall amount of 1,200–1,600 mm. Soils are more fertile than in the Rain Forest and are suitable for cocoa, coffee, oil palm, maize, plantain, cocoyam, cassava, rice and vegetables including eggplant, beans, pepper, and okro. The Forest-Savanna Transition zone has two rainy seasons with an annual rainfall of 1,300–1,800 mm. The major season is from April to June and minor from September to November. Soil fertility is high but the soil is liable to erosion. Major crops grown include maize, plantain, cassava, yam, cocoyam, cotton, tobacco, groundnut, tomato, pepper, eggplant, and cowpea. The Coastal Savanna forms 7% of the land area of Ghana. It has two rainy seasons. The major season is from March/April to June while the minor is from September to October. The average rainfall is 600–1,000 mm. Relief is gentle and soils are either heavy clay or light textured and underlain by clay. Among the crops grown are cassava and maize. Vegetables are grown on lighter soils while rice, cotton, and sugarcane are planted on the heavier soils. Coconut is found on the coastal fringe.

The Sudan and Guinea Savanna zones cover about 57% of the land area of Ghana. These zones are sometimes referred to as Interior. Unlike the rest of the country, the Interior Savanna has only one rainy season that is April/May to October. Annual rainfall ranges from 800 mm to 1000 mm. It declines from south to north. The principal food crops in Interior Savannah Zone are maize, cowpea, bambara groundnut, yam, vegetables, rice, sorghum and millet. Maize accounts for between 50%-60% of the total cereal production in Ghana. Vegetables include pepper, eggplant, tomato, okro, Roselle (*Hibiscus sabdariffa*), cocoyam leaves and beans.

4.2.3 *Topography and Soil*

The topography is predominantly undulating, with slopes less than 1%. Even though the slopes are gentle, about 70% of the country is subject to moderate to severe sheet and gully erosion. The soils have predominantly light textured surface horizons in which sandy loams and loams are common. Lower soil horizons have slightly heavier textures varying from coarse sandy loams to clays. Heavier textured soils occur in many valley bottoms and in parts of the Accra Plains. Many soils contain abundant coarse material either gravel and stone, or concretionary materials which affect their physical properties, particularly their water holding capacity.

4.3   Biodiversity Systems in Ghana

Ghana is located in Africa along the Atlantic coast about 400 miles north of the equator. The country has a mostly tropical climate with alternate wet and dry seasons. The northern part of the country has a rainy season from April to October and is dry and dusty from November through March. The southern half of Ghana has rain from April to July and from September through November. Ecosystems in northern Ghana are Guinea savanna and Sudan savanna. Southern Ghana, however, contains Deciduous Forest, Moist Evergreen Forest, Wet Evergreen Forest and Coastal Savanna. Ghana is endowed with enormous terrestrial and aquatic biodiversity at the ecosystem levels. The country has a large land mass made up of varied ecologies, including the Evergreen Rain Forest. The terrestrial ecosystems of Ghana include Forest, Savannah woodland, and Coastal savanna scrubs whereas the aquatic ecosystem includes rivers, floodplains, reservoirs, lakes, small irrigation impounds, other wetlands and aquaculture ponds.

4.3.1   *The Flora of Terrestrial Systems*

The plants in Ghana play an essential role in the development of the country. They offer medicines, food, recreation, and aid as a ritual object for the innumerable ethnic groups in the country. Together, aboriginal and introduced species have been deliberated in the assessments of the country's floral diversity. An overall of some 3,600 species of the major regional centers of endemism (White, 1965) epitomize the three major taxonomic groups. Floral diversity is more prominent among the angiosperms exemplified with well over 2,974 indigenous and 253 introduced species. The list of plant species in Ghana's terrestrial ecosystem has undergone tremendous changes with time. A report by the International Union for the Conservation of Nature (IUCN) indicates that the number of plant species from various collections in Ghana increased from below 600 in 1914 to 3,600 in 1996 (IUCN, 1994). A recent study by Hackman (2014), however, puts Ghana's total plant species richness at 5,429 out of which 5,217 are Angiosperms (1,257 monocots and 3,950 dicots) which exhibit the greatest floral diversity with over 2,974 indigenous and 253 introduced species. Forty-six (46) species of Bryophytes (35 mosses and 11 liverworts), 12 species of Lycophytes (club and spike mosses) and 12 species of Gymnosperms have also been described. According to IUCN (2013), a total of 121 plant species have been classified as threatened with 3 species, i.e., *Talbotiella gentii, Salacia fimbrisepala,* and *Anbregrinia taiensis* considered as critically endangered. Nineteen (19) species have also been classified as endangered with four near threatened and 95 vulnerable species.

Among the various vegetation types of the tropical rainforest, it is the wet evergreen forest type in the southwestern part of the country that reveals the highest level of endemism and species richness. Information on species diversity and endemism in the savanna biomes is very sparse. The forest areas of Ghana are known to show more endemism and diversity in plant species than the savannas. The tropical rainforest zone is dominated by plant species such as *Cynometra ananta, Tieghmella heckelii, Triplochiton scleroxylon,* and *Millicia excelsa.* The two major savanna biomes (Guinea and Sudan savannas) have some plant species common to both agro ecologies such as *Lophita lanceolata, Afzelia africana, P. clappertoniana, Adansonia digitata,* and *Ceiba pentandra.* There are also species belonging to the genra *Andropogon, Hyparrhenia, Pennisetum, Aristida,* and *Brachiara.* There is a decrease in plant species diversity and the level of endemism from the wet evergreen rainforest through the moist semi-deciduous forest to the southeast outliers (Hall and Swaine, 1981). Biological diversity of species in the savanna woodlands and gallery forests of the savannas may show greater species richness than the dry savannas. Within Ghana there are areas of high biological diversity, referred to as "biological hotspots." The most notable of such areas is the Ankasa and Nini-Suhien Conservation Area in the southwestern portion of Ghana (CI, 2002). The apparent climatic diversity is greater here. In West Africa, the Upper Guinea Forest Ecosystems is also recognized as one of the 34 global biodiversity hotspots. This is attributed to the fact that even though there is a very high concentration of biological diversity, the entire area has lost about 80% of the original forest cover and the remnants continued to be threatened with destruction. There is only one known gymnosperm, *Encephalartos barteri*, which is indigenous to Ghana. The few others growing in various ecological zones in the country are introduced species for purposes including aesthetics and economic. The third taxonomic group, pteridophytes, is well represented with 124 known species. Most studies on plant diversity in the past have been concentrated in the forest zones with very little information being documented on   A☐   -
man (2014), only three studies (Asase and Oteng-Yeboah, 2007; Tom-Dery et al., 2012, 2013) on plant diversity in the savanna zones have been conducted over the past decade.

A report by Ayensu et al. (1996) indicates that Ghana's coastline has lagoons and estuaries, which support the growth of mangrove vegetation with characteristic species including *Avicennia nitida, Rhizophora rac-*

Mangifera   *Adonidia memillii*   *Opuntia cochenillifera* *Murraya paniculata*

Figure 4.2   Some flora of Ghana.

emosa, and *Laguncularia* spp. Other notable plant species found in the coastal vegetation include *Sesuvium portulacastrum, Sporobolus virginicus, Phoenix reclinata,* and *Paspalum viginatum.* The strand vegetation on the sandy beaches along the coast of Ghana has woody species such as *Cocos nucifera* and *Sophora occidentalis* as well as herbaceous species including *Canavalia rosea, Ipomoea pes-caprae,* and *Cyperus maritimus* (Figure 4.2).

4.3.2   The Fauna of Terrestrial System

Ghana has a vast array of fauna and they are of great significance in the socioeconomic development of the country. They provide food, medicines, livelihoods, and serve as a sacred symbol for some tribes in the country. The fauna of the Ghanaian terrestrial ecosystem, comprise a diverse array of species plus several of conservation concern. Existing records show that there may perhaps be as many as 221 species of amphibians and reptiles, 724 species of birds, 225 mammalian species (with 93 recorded to inhabit the savanna ecological zone). Threatened species recorded in the country include four species of marine turtles and three species of crocodiles. Bird species of conservation concern include seven threatened species, including four species endemic to the Upper Guinea forest block and seven near-threatened species. Keystone species such as hornbills, parrots, and birds of prey are well represented in the country. Of the 728 birds species confirmed to be occurring, 408 are nonpasserines and 320 passerines, of which 498 are known or thought to be resident and 176 are regular seasonal migrants, including 100 from the Palearctic. Of the total number of species occurring, 180 restricted to the Guinea-Congo Forests Biome and 37 restricted to the Sudan-Guinea Savanna biome have been recorded (Ntiamoa-Baidu et al., 2000a,b, 2001). Furthermore, 11 of the 15 endemic bird species within the Upper Guinea Forest occur in Ghana. Six of the total species are considered threatened and 12 near-threatened (BirdLife International, 2000). The country is also important for water-birds being on the boundary of the east Atlantic Flyway and the Mediterranean Flyway (Smit and Peirsma, 1989; Ntiamoa-Baidu et al., 2001). Endemism among terrestrial fauna has been observed in three species of frogs, *Hyperolius baumanni*, *H. fusciventris* and *H. syslvaticus* and the lizard, *Agama sylvanus* found in the Bia Forest Reserve and the Atwema Range Forest Reserve. There is a high degree of butterfly endemism in Ghana where about 23 species are classified as endemic or near-endemic. As with floral diversity, "hot spots" for faunal diversity may be located in the high forest areas (accounting for 83% of the total number of species recorded), where canopy stratification and micro-climatic differentiation has provided habitats and niches for specific faunal organisms. Ghana is home to 84 known amphibian species: 78 frogs, 5 toads and caecilians. Ghana is an important country for dozens of vulnerable, threatened, endangered, critically endangered or near-extinct mammalian species including primates such as the *Pan troglodytes* and *Procolobus kirkii*, big cats such as the *Panthera leo* and *Panthera pardus*, elephants such as the *Loxodonta africana* and water-birds (Table 4.1), being located on the boundary of the east Atlantic Ocean Flyway and Mediterranean Flyway (Figures 4.3 and 4.4).

Table 4.1 Summary of Mammalian Species in Ghana and Their Conservation Status

Order	Species Richness	Endangered	Near Threatened	Vulnerable
Chiroptera	124	6	–	–
Rodentia	86	–	–	1
Carnivora	30	2	3	1
Artiodactyla	29	1	2	–
Primates	26	1	3	2
Soricomorpha	18	2	–	–
Pholidota	5	2	–	–
Hyracoidea	4	–	–	–
Lagomorpha	2	–	–	–
Erinaceomorpha	1	–	–	–
Proboscidea	1	–	1	–
Tubulidentata	1	–	–	–
Total	327	14	9	4

Source: Hackman (2014).

Figure 4.3 Fauna of Ghana.

Conraua derooi   *Crocodylus niloticus* Marine turtle (Chelonioidea)

Figure 4.4 Some threatened species of the Herpetofauna in Ghana.

4.3.3 Fresh Water (Aquatic) Ecosystems

Ghana's fresh water covers major river systems and lakes in the country. The country's fresh water fish fauna includes 28 families, 73 genera, and 157 species. About 121 species have been recorded from the Volta system within Ghana, which drains more than a third of the entire country. About nine species viz. *Barbus subinensis* (cyprinidae), *Irvinea voltae* (Schibeidae), *Chrysichthys walkeri* (Clarioteidae), *Synodontis arnoulti, S. macrophthalmus, S. velifer* (Mochokidae), *Limbochromis robertsi, Steatocranus irvinea* (Cichilidae) and *Aethiomastac embeluspraensis* (Mastacembelidae) are endemic to freshwater system of Ghana (Dankwa et al., 1999). Economically, 81 species are of food importance. Species of cultural importance include *Heterotis niloticus* (Osteoglossidae),

Clarias gariepinus, Heterobranchus longifilis (Claridae), *Chrysichthys nigrodigitatus* (Clariotiedae), *Oreochromis niloticus* (Chichlidae) and *Lates niloticus* (Centropomidae). Some species need to be protected because of their restricted distribution or their habitat degradation or destruction. Other freshwater ecosystems include the major rivers such as the White Volta, Black Volta, Lower Volta and Oti. Others are Pra, Tano, Ankobra, Bia, and Todzie-Aka. Also included are other impoundments serving as drinking water sources and/or for irrigation. It is estimated conservatively that about

4.3.4 Marine (Aquatic) Ecosystem

The marine ecosystem serves as the main source of protein to the people and the livelihoods of some of the citizens depend on it. Hence the life of the people of Ghana depends a lot on the marine ecosystem that stretches on the shores along the southern boundaries of the country. Indications of extremely high biodiversity of the benthos of the shallow waters of the continental shelf have been reported in recent studies by the Department of Oceanography and Fisheries of the University of Ghana. About 60% of the soft bottom benthic macrofauna encountered are believed to be new and unrecorded. There is virtually no information on meiofauna (dominated by worms, oligocheates, and crustaceans) and microfauna (such as ciliates, amoebas, and foraminiferans) organisms in benthic waters. About 392 marine species of organisms comprising 347 fish species belonging to 82 families have been recorded. There is also evidence available that the coastal waters of the country are being invaded by marine algae, a typical example is *Enteromorpha flexuosa*, which is believed to have drifted eastward from areas west of Ghana.

Although the entire Ghanaian coast environment is devoid of any living coral reefs it has been established that the entire continental shelf is traversed by the belt of dead madreporaria coral at a depth of 75 cm. The coastline of Ghana is lined with about 90 lagoons, several estuaries and rocky shore habitats that exhibit distinct array of biological diversities. Information on faunal, microbial and   [A
sites namely the lagoons of Keta, Songor, Sakum, Densu Delta, and Muni-Pomadze, where an appreciable amount of knowledge is available. The sixth Ramsar site, the Owabi Wildlife Sanctuary, is the only aquatic protected ecosystem. The Site protects the source of drinking water for Kumasi and its environs.

4.3.5 Lakes (Aquatic) Ecosystems in Ghana

Even though few lakes exist in Ghana, the biological resources they contain contribute to the socioeconomic development of the country. Lake ecosystems are scarce in Ghana. The only natural lake system is the Lake Bosomtwi, which covers an area approximately 50 square kilometers and has 11 fish species belonging to 9 genera, and 5 families. The Volta Lake created in 1964 and inundating some 4,840 square kilometers of pristine natural forest and the two dams on the Volta River at Akosombo and Kpong have indisputably altered the biodiversity and ecology of the river and adjacent areas. The original Volta River was found to consist of at least 100 fish species (Petr, 1967). Studies in

Yeji sector of the lake encountered 66 species representing 39 genera belonging to 19 families. Fish species including *Brycinus nurse, B. macrolepidotus, Eleotris senegalensis,* the bivalve, the Volta clam (*Egeria radiata*), the shrimp, *Macrobrachium* spp., the West African manatee, *Trichechus senegalensis, etc.* are under severe threat of extinction (Ofori-Danson and Agbogah, 1995).

4.3.6 *Microbial Diversity*

Variability in Ghana's microscopic organisms is supported by the diverse nature of its habitat. Cyanobacteria (blue-green algae) found in both terrestrial and aquatic habitats are classified under 3 orders, i.e., Chamaesiphonales with 1 genus and 2 species, Chrococcales (5 genera and 7 species), and the Nostocales represented by 11 genera and 26 species. These organisms form mutualistic associations with some plants and have the ability to fix atmospheric nitrogen into forms usable by plants.

A total of 1072 fungal species belonging to 324 genera have been recorded in Ghana. Some of these species are *Aspergillus flavus, Gibberella intricans, Macrophomina phaseolina,* and *Fusarium oxysporum* (Ayensu et al., 1996). Most fungal species in Ghana are members of the higher fungi, which consist of the subdivisions Ascomycota, Basidiomycota and Deuteromycota. Parasitic fungi are also found in Ghana and this includes members of the order Erisyphales (powdery and sooty mildews), Uredinales (rusts) and Ustilaginalis (smuts).

4.4 State of Crop Diversity in Ghana

Cocoa is the most important commercial crop in Ghana followed by oil palm, pineapple, cashew, and cotton. Other important commercial crops are ginger, coffee, rubber, kola, and shea nut. In addition, several important varieties of fruits, vegetables, cereals, root, and tubers as well as Legumes are cultivated in Ghana. The most commonly grown vegetables in Ghana are tomato, onion, hot pepper, okra, eggplant and shallots. Moreover, fruits such as bananas, papaya, citrus, and mangoes are generally grown in the country. Other important crops that are in cultivation in Ghana are cereals like barley, wheat, rice and maize. Root and tuber crops are also very important to the Ghanaian economy and are cultivated by several farmers across the country. The most important commercially grown root and tuber crops in Ghana include yam, cassava, cocoyam, potato and sweet potato with a total of 503 accession comprising: 39 accessions of cocoyams, 20 accessions of sweet potato, 121 accessions of Yam, 201 accessions of cassava, 55 accessions of Frafra potato and 30 accessions of taro were conserved under field conditions (Table 4.2) (Annual Report of PGRRI, 2016).

- *Sweet potato:* Sweet potatoes a crop with great potential for Ghana as a food and nutrition security, income generation, as well as raw material for industry. It has the versatility to be used in various food preparations. Despite this great potential to alleviate food and nutrition insecurity and poverty in Ghana, its level of utilization in Ghana is very low as compared to the other root and tuber crops. Twenty varieties of sweet potato landraces have been collected and conserved. A total of 12 improved varieties have been bred and released to farmers.
- *Cocoyam:* Thirty-nine accessions have been collected and conserved. Two main types are found in Ghana, the *white* and the *red*. Three improved varieties have been released to farmers.
- *Cassava:* Cassava is a very versatile crop with numerous uses and by-products. It is the most widely cultivated crop in Ghana, with over 90% of all rural households involved in its production. Ghana is the third largest producer of cassava in Africa. About 201 local accessions have been collected, characterized and conserved. A total of 18 improved varieties have been released to farmers.
- *Yam:* Yams (*Dioscorea* spp.) are annual or perennial vines and climbers with annual or perennial underground tubers. They belong to

Table 4.2 Genetic Resources of Root and Tuber Crops in Ghana

Crops	Number
Sweet potatoes	20
Frafra potatoes	55
Cocoyam	39
Cassava	201
Colocasia	30
Dioscorea rotundata	22
D. alata	50
D. bulbifera	8
D. cayennensis	22
D. dumetorium	7
D. esculenta	12
Total	466

Source: CSIR-Plant Genetic Resource and Research Institute.

the Dioscoreaceae family. The most important edible yams belong to only a few species, such as *D. rotundata* Poir. (widely known as white Guinean yam), *D. alata* L. (known as water yam, winged yam or greater yam), *D. cayennensis* Lam. (yellow yam or yellow Guinea yam; may be composed of a complex set of different species), *D. esculenta* (Lour.) Burkill (lesser yam, potato yam or Chinese yam), v *D. dumetorum* (Kunth) Pax (bitter yam or trifoliate yam) and *D. bulbifera* L. (aerial potato yam) (Eastwood and Steele, 1985; Aboagye, 2012). About 121 accessions of yams (*D. alata* – 50; *D. bulbifera* – 8; *D. cayennensis* – 22; *D esculenta* – 12; and *D- rotundata* – 22 accessions) have been collected and conserved and three improved varieties have been released to farmers (Figure 4.5).

Furthermore, legumes like soybeans, cowpea and groundnuts have long been cultivated across the country. Table 4.3 shows legume genetic resources under conservation in Ghana. About 1,513 accessions of legumes are under conservation in Ghana up of 11 species (Aboagye, 2012). The greatest number under conservation is *Vigna unguiculata* followed by *Arachis hypogaea* and

Figure 4.5 Genetic resources of some *Dioscorea* (yam) species in Ghana (Aboagye, 2011).

Table 4.3 Genetic Resources of Legumes in Ghana

Crop	Number under conservation
Vigna subterranea	190
Vigna unguiculata	796
Sphenostylis sternocarpa	52
Phaseolus vulgaris	35
Phaseolus lunatus	38
Mucuna pruriens	24
Kesrtingiella geocarpa	1
Glycine max	2
Canavalia ensiformis	18
Cajanus cajan	9
Arachis hypogaea	348
Total	1513

least in *Kesrtingiella geocarpa*. Among the legumes the following improved types have been released to farmers: Cowpea – 11, groundnut – 20 and soybean – 2. Figure 4.6 also shows other crops genetic resources including legume.

4.5   Losses of Biodiversity in Ghana

Factors leading to biodiversity loss include habitat loss, invasive species, and overuse of biological resources being attributed to increasing world population (Conservation International, 2002). There is an overall consensus among biodiversity stakeholders in Ghana that the major factors causing biodiversity loss in the country are: rapid population growth, effects of climatic change, human resources, scientific and technical, environment, international and local political factors (Fisher, 1971; Attuquaye and Fobil, 2005).

The major factors contributing to genetic loss(in order of importance) in the country include (i) the replacement of local varieties (Figure 4.7) (ii) land clearing (iii) pests/weeds/diseases (iv)population pressure, and (v) changing agricultural systems (Bennett-Lartey and Oteng-Yeboah, 2008).

4.6   Conservation of Biodiversity in Ghana

The current deteriorating state of the environment, principally as a result of undesirable human activities has increased policymakers' interest in environmental issues throughout the world. According to Attuquayefio and Fobil

Figure 4.6   Images of the different cultivars of vegetables, cereals, legumes and roots and Tuber crops in Ghana.

(2005), the loss of each species comes with the loss of potential economic benefits including natural products that the world food supply and the medicines depend on as well as a loss of ecosystems balance. Biodiversity has both direct (source of food, medicine, construction materials, raw materials for industries,

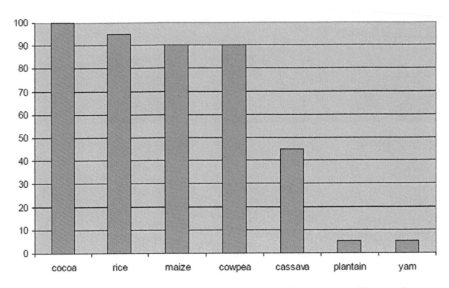

Figure 4.7 Estimated percentage of area sown to modern varieties (*Source*: Bennett-Lartey and Oteng-Yeboah, 2008).

recreation, etc.) and indirect (ecological research, education, etc.) therefore, it is imperative to conserve the available biodiversity by ensuring that biological resources are used in ways that do not diminish the variety of genes and species or destroy important habitats and ecosystems. In Ghana, biodiversity conservation is carried out both in situ (maintenance, protection, and management of variety of life in their original habitats), or ex situ (collection and maintenance of whole or parts of individuals of some species or their population and communities in facilities away from their original habitats) (Oteng-Yeboah, 1997; Attuquayefio and Fobil, 2005) which are achieved through the establishment of protected areas, gene banks, and zoological/botanical gardens.

4.6.1 Chronology of Biodiversity Conservation Initiatives in Ghana

The arrangements put together in order to sustain biodiversity in Ghana include the establishment of forest and wildlife reserves.

4.6.1.1 Permanent Protected Forests (Forest Reserves)

An estimated 70% of land in Ghana was made up of a forest cover (IIED, 1992; Attuquayefio and Fobil, 2005). However, 59% of the forest cover was destroyed before the twentieth century through cultivation, logging and bush fire caused by man. The remaining 11% of the total land surface area of Ghana which is currently made up of forest is protected through the establishment of over 280 forest reserves which are spread across the country with all identified vegetation types of the country represented (NRMP, 2001). The legislation empowered the colonial government to establish forest reserves. In 1948, a forest policy was adopted in Ghana to regulate the forest activities through: (i) the creation and management of permanent forest estates, (ii) research into all branches of scientific forestry, (iii) maximum utilization of areas not dedicated to permanent forestry, and (iv) provision of technical advice and cooperation in schemes for the prevention of soil erosion and in land use plans. Forestry Commission Act (Act 571), the act then established a Forestry Commission (FC) to regulate the utilization of forest and timber resources, manage the nation's forest reserves and protected areas, assist the implementation of forestry and wildlife policies and undertake the development of forest plantations as well as the restoration of degraded forest areas, expansion of the nation's forest cover and increase in production of industrial timber.

4.6.1.2 Wildlife Reserves

The department of game and wildlife was established from the game unit of the Forestry Department, with the mandate to manage the country's wildlife resources both within and outside forest reserves. The wildlife conservation areas include Mole National Park, Shai Hills Resource Reserve, Bui and Digya National Parks, Owabi Wildlife Sanctuary, Kogyae Strict Nature, Bomfobiri Wildlife Sanctuary, Kalakpa and Gbele Game Production Reserves, Nini-Suhien National Park, Ankasa Game Production Reserve, Bia Game Production Reserve (Ntiamoa-Baidu et al., 2001; Ampadu-Adjei, 2002).

4.6.2 *Approaches to Biodiversity Conservation in Ghana*

Approaches to biodiversity conservation in Ghana have currently focused on the modern scientific strategies (research and environmental education) and cultural or traditional methods. These methods cover the broadest range of society and are mainly useful in less-developed countries like Ghana where the majority of the population lack formal education.

4.6.2.1 Traditional Biodiversity Conservation

The African traditional concept of land ownership charged the living to manage and conserve the environment for the future generation, while accounting for such stewardship to their ancestors (Abayie, 1997). According to Attuquaye-fio and Fobil (2005), before the conventions on biodiversity conservations in Ghana came into force, traditional African societies maintained complex religious and cultural beliefs systems that guided the conservation of biodiversity using traditional norms such as taboos and myths to ensure the preservation of certain critical and finite resources. In some rural and even urban areas of the country, taboo days for farming, fishing and hunting widely observed presently, with many water bodies being worshiped as deities. Such traditional practices enabled the protection of biological resources from human disturbances and overexploitation. Moreover, several sacred groves (small patches or traditional-protected tracts of land) are being managed across the country (Kingdon, 1989). These small sacred areas have ensured that forests surroundings and protecting watercourses and other environmentally sensitive areas were demarcated and protected as shrines. They range in size from hundreds of hectares of forest to single trees or a few stones (Gordon, 1992), and because of their perceived links to some deities or ancestral spirits are referred to variously as "nananom mpow" (ancestral groves) (Adarkwa-Dadzie, 1997), "abosompow/asoyeso" (shrine), "mpanyinpow" (ancestral forest), and "nsamanpow" (burial grounds) by the Akans (Ntiamoa-Baidu, 1995). Sacred groves serve important ecological and sociocultural functions by preserving virgin forest, being important refuges for rare and useful local biodiversity, and being sources of herbs for medicinal, social and religious purposes (Dorm-Adzobu et al., 1991; Decher, 1997). There are about 2,000–3,200 sacred groves in Ghana of which 80% are located in the southern part of the country (Gordon, 1992). Disobedience to the traditional laws (taboos) on the sacred forest usually attracts severe punishments such as sacrifices and performing of certain rites to avert any mishaps, ill-health and death.

4.6.2.2 Modern Approaches to Biodiversity Conservation

4.6.2.2.1 In Situ Conservation

Sites for *in situ* conservation are both in wild and farmed areas and they contain most of Ghana's forest and wildlife resources. Ghanaian farmers play an essential role in conserving on-farm, and its very important for the country's food security (Abayie, 1997). Crop wild relatives are wild species that are related to crop plants (Ntiamoa-Baidu, 1995). Crop wild relatives biodiversity is very important reservoirs of traits and characteristics useful for improving crop varieties. Crop wild relatives are best conserved *in situ*, in their natural habitats to allow evolution and adaptive changes. In situ conservation approaches uses all the legal protected areas such as forest reserves, wildlife conservation areas and Ramsar sites in Ghana. Currently, there are 280 large-scale forest reserves and 21 legally constituted wildlife conservation areas (Decher, 1997; MES, 2002; Attuquayefio and Fobil, 2005) which are playing a major role in protecting the many plants and animal species from extinction. Other methods of in situ conservation recently adopted in the country are the Globally Significant Biodiversity Areas (GSBAs), Important Bird Areas (IBAs), and the Community Resource Management Areas (CREMA). The Forestry Division of the Forestry Commission of Ghana has redesignated about 29 of the existing forest reserves, which covers a total area of about 117,322 ha, as GSBAs. This was attributed to the fact that those areas harbor a high concentration of biodiversity of global conservation importance (NRMP, 2001; MES, 2002). The IBAs concept uses birds as indicators of habitat quality. Birdlife International in collaboration with the Ghana Wildlife Society have identified 36 IBA's of global significance which are made up of protected areas and forest reserves covering an area of about 11,494 km² (4.8% of the country's land surface area). Moreover, the Wildlife Division of the Forestry Commission under the Protected Areas Development Program (PADP) initiated the community participation in wildlife management concept known as CREMA. The CREMA allows those communities close to the protected areas, to manage and sustainably use the wildlife resources within a defined area through community participatory approach. This concept allows the local communities to actively participate in the conservation of wildlife outside the forests and protected areas systems (Wildlife Division, 1998).

4.6.2.2.2 Ex- Situ Conservation

The ex-situ conservation method uses the zoos, botanical gardens, and genetic resource centers (gene banks) to preserve some important species which could later be introduced or restock diminishing natural populations. Currently, there are seven major ex-situ conservation facilities in Ghana. These facilities are located at (i) University of Cape Coast (Botany Department Herbarium), (ii) University of Ghana (Zoology Department Entomology Museum, Botany Department Herbarium and Botanical Gardens and Noguchi Memorial Institute of Medical Research), (iii) Kwame Nkrumah University of Science and Technology and the Technology Botanical Gardens and Forestry Herbarium, (iv) Kumasi Zoo, (vi) Aburi Botanical Gardens, (vi) CSIR-Plant

Genetic Resource Research Institute, and (vii) Akropong Centre for Scientific Research in to Plant Medicine (MES, 2002; Attuquayefio and Fobil, 2005).

The CSIR-PGRRI, however, holds the national genebank. It was established in 1964 and has a total of 4412 accessions conserved at Bunso currently. At CSIR-PGRRI there are two types of *ex situ* collections: orthodox seeds under cold storage (Figure 4.8) holding 3240 accessions and 876 under condition (Figure 4.9). These seeds are from cereals, vegetables, cucurbits and legumes and are kept under short, medium and long-term storage conditions. Other institutions such as the CSIR – Crops Research Institute (CRI), the Ghana Grains and Legumes Development Board in Kumasi, and CSIR – Savanna Agricultural Research Institute (SARI) all have short-term storage facilities for the conservation of grains, mainly of cereals and legumes.

4.6.2.2.3 In-Vitro Conservation

In vitro facilities also exist in some institutions for the conservation of plant genetic resources. At CSIR-PGRRI, 259 accessions of root and tuber crops

Figure 4.8 Seed conservation.

Figure 4.9 Field conservation.

are under in-vitro conservation. The Biotechnology and Nuclear Agricultural Research Institute (BNARI) at Kwabenya has *in-vitro* facilities for germplasm conservation, conserving pineapple, plantain, cassava and yams. The Department of Botany and Environmental Biology (DBEB) of the University of Ghana, has in-vitro materials such as cocoyam, cassava, frafra potato (*Solenostemon rotundifolius*), yams, sweet potato and pineapple. The DBEB also has cryoconservation facilities. Other institutions like the Crop Science Department of the Kwame Nkrumah University of Science and Technology (KNUST), Kumasi, the School of Agriculture of the University of Cape Coast (UCC), the Cocoa Research Institute of Ghana (CRIG), Forestry Research Institute of Ghana (FORIG) and the CRI all have in vitro facilities for research and conservation of germplasm. There are several institutions in the country that conserve tree crops and other living collections in field gene banks. CSIR-PGRRI is conserving citrus, mango, avocado, *Musa* species and various palms and spices with duplication at the University of Ghana – Agricultural Research Centre (ARC) at Kade. Other institutions also conserve various tree crop species *ex situ*. CRIG conserves its mandate crops namely cocoa, cola, coffee, cashew and shea tree (*Vitellaria paradoxa*) in field genebanks. OPRI conserves collections of oil palm (*Elaeis guineensis*) and coconut (*Cocos nucifera*) as living plants, seeds and pollen. CSIR-CRI also maintains some field germplasm, which include *Musa* spp., citrus, mangoes, yams and cassava. CSIR-SARI conserves root and tuber crops in the field. These include yams, cassava, sweet potato and Frafra potato.

There are also institutions in the country that have arboreta and clonal banks where various tree species, medicinal plants and timber species are conserved. These include: CSIR-FORIG, CSIR-PGRRI the Centre for Plant Medicine Research (CPMR). The CSIR-Animal Research Institute (ARI) maintains museum of forage plants. The Aburi Botanical Garden was established in the nineteenth century for the conservation of local and exotic plant species, including ornamental plants for educational and esthetical purposes. The DBEB houses the Herbarium of Ghana a collection of preserved plant parts.

4.6.2.3   Biodiversity Conservation of Forest Species

A portion of Ghana's high forest zone forms part of one of the world's biodiversity "hotspots": the Upper Guinean forest zone. This is also the region where the forest is disappearing rapidly (Hawthorne, 1990) and contributing to the alarming rate at which the world is losing biological resources. In the past, attempts to conserve Ghana's biological resources have included the establishment of Protected Areas, Forest Reserves, Zoos and Botanical Gardens for both conservation and educational purposes, and the conduct of research on species diversity in some of the vegetation zones of Ghana.

Attempts have been made to conserve Ghana's biological resources with a recommendation by MES (1995), which resulted in the   A categorization of vulnerable areas, provenance protection areas,   A location of the country's "hot spots" and prioritization of each species for conservation. Ghana has 280 forest reserves, covering a total area of about 23,729 km^2, or 11% of the total land area of Ghana, of which about 75% of these reserves have been designated production reserves and the remaining

Table 4.4 displays wildlife conservation areas in Ghana. These include: national parks, Wildlife sanctuary, Ramser sites, and strict nature reserves.

> ***National Parks****:* There are seven national parks of which the largest is the Mole National Park in the Northern region, covering 4840 km^2 and the least is the Bia National Park in the Western region (78 km^2).

Table 4.4 Wildlife Conservation Areas in Ghana

Type	Name	Area (km²)	Location/Region
National Park	Mole	4,840	Northern
	Digya	3,478	Volta
	Bui	1,821	Northern Brong-Ahafo
	Kyabobo	360	Volta
	Kakum	207	Central
	Nini-Suhien	160.2	Western
	Bia	78	Western
Wildlife Sanctuary	Bomfobiri	*53*	Ashanti
	Owabi	13	Ashanti
	Buabeng-Fiema	4.4	Brong-Ahafo
	Agumatsa	3	Volta (Proposed)
Resource Reserve	Gbele	565	Upper West
	Ankasa	343	Western
	Kalakpa	320	Volta
	Bia	228	Western
	Assin-Attandaso	140	Central
	Shai Hills	49	Greater-Accra
Strict Nature Reserve	Kogyae	386	Ashanti
Ramsar Site	Keta Lagoon Complex	1,200	Volta
	Songor	330	Greater-Accra
	Muni-Pomadze	90	Central
	Densu Delta	70	Greater-Accra
	Sakumo	*35*	Greater-Accra
		13	Ashanti
Total Area		14,173	

Source: FDMP (2016).

Wildlife Sanctuaries: Currently there are three wildlife sanctuaries at Bomfobiri (53 km²); Owabi (13 km²); Buabeng-Fiema (4.4 km²). A fourth one is being proposed at Agumatsa (3 km²) in the Volta region. **Resources reserves:** There are six resource reserves with the biggest at Gbele (565 km²) in the Upper West region and the smallest in the Shai hills (49 km²) of the Greater Accra Region.

Ramsar Sites: There are six Ramser sites, three in the Greater Accra region and one each in the Volta, Central and Ashanti regions, with the Keta Lagoon Complex being the biggest (1200 km^2).

Strict Nature Reserve: There is one strict nature reserve at Kogyae in the Ashanti region with a total area of 386 km^2.

4.6.3   Challenges and Prospects of Biodiversity Conservation in Ghana

Ghana like many other countries across the globe is faced with the present global biodiversity loss (Myers et al., 2000; Brooks et al., 2006). However, the quest to confront the crisis through biodiversity conservation in the country is hindered by the following crisis.

4.6.3.1   Rapid Population Growth

The population of Ghana has over the years been increasing rapidly and this has become a major contributory factor to the threats of biodiversity conservation. Between 1960 and 2017, the population of Ghana is more than tripled with an average annual population growth of 3.3%. Presently, the country's population has increased from 6.6 million in 1960 to 28.7 million in 2017 (Worldometers, 2017). This increase in population has led to the destruction of many reserved areas for settlements, agriculture and other developmental projects.

4.6.3.2   Poverty

The highest percentage of people living in abject poverty is found in Africa (Mekasha, 2001). According to the Ministry of Lands and Forestry (2000), 31% of Ghana's population are living in poverty. This has forced several people into illegal practices such as illegal mining (galamsey), logging, and winning and other activities that cause serious threats to the ecosystem.

4.6.3.3   Over-Reliance on Agriculture

Agriculture is the largest and most important sector of Ghana (Benin and Barbier, 2001), employing over 50% of the population and contributing about 20.1% of the country's GDP (Index Mundi, 2017). Its expansion brings remarkable benefits to the nation. However, the impact of agricultural expansion on the country's biodiversity may be quite disastrous as indigenous species are threatened, displaced and replaced with most often fast-growing and genetically inferior introduced varieties. According to Benin and Barbier (2001) and Hackman (2014), it has been the major agent of deforestation and land-use intensification since the beginning of the nineteenth century, posing a major threat to biodiversity conservation. Moreover, intense grazing of livestock has led to habitat degradation and subsequent loss of local fauna and flora. This is a major contributory factor to habitat degradation and biodiversity loss.

4.6.3.4 Financial

Despite the importance of the of the biodiversity conservation in Ghana, most of the policies and methods especially the in situ conservation methods, has over the years suffered some administrative problems as a result of low budgetary allocation, infrastructure and field equipment (Attuquayefio and Fobil, 2005).

4.6.4 *Prospects*

Over the past few decades, the government of Ghana has been trying hard to transform the country to middle-income status where people live in harmony with the natural environment and derive optimum benefits through the sustainable use of the country's rich biodiversity. This is in recognition of the important role that biological resources play in national efforts towards the realization of socioeconomic and cultural growth and development (MES, 2002). The prospects of Ghana's biodiversity conservation initiatives are therefore, quite promising, especially with several environmental NGOs and other stakeholders assisting in government's efforts including local communities, traditional authorities, NGOs and government institutions are in the right direction.

Poverty alleviation has also been receiving serious government's attention through several initiatives in the country. Initiatives such as the Ghana Poverty Reduction Strategy (GPRS), the emergent social relief program, District assemblies common fund, village infrastructure project, planting for food and jobs as well as the one district one factory are all aimed at providing jobs and improving the living conditions of Ghanaians, thereby eliminating/reducing the reliance on environmentally unfriendly activities such as illegal mining, logging etc.

Without funding and willingness of policymakers to implement the various biodiversity conservation initiatives, no success could be achieved.

However, presently there is potential for obtaining funding for biodiversity conservation programs both locally and internationally.

4.7 National Legislation on Biodiversity and International Conventions Related to Biodiversity Conservation in Ghana

Ghana's growth and socioeconomic development in times past has been achieved with much recourse to its abundant natural resources. Ghana like many other countries in sub-Saharan Africa is faced with the challenge of biodiversity loss (Brooks et al., 2006) as a result of environmental degradation arising from the unsustainable exploitation of its natural resources in response to the pressures of rapid population growth. Good policies that are easy to implement, enforceable legislative arrangements and strengthened institutions are some of the key factors to be considered in ensuring a decline in the rate of biodiversity loss and strengthening of its potential of being managed and conserved for the benefit of future generations.

4.7.1 National Legislation on Biodiversity

The socioeconomic policies of the government such as the Economic Recovery Program, Structural Adjustment Program and Trade Liberalization Program have contributed significantly to the accelerated degradation and loss of the country's natural resources including biodiversity. The Forest Policy for instance, which was passed in 1948 promoted the conversion of off-reserve forests to nonforest lands and resulted in massive degradation and loss of biodiversity. A new Forest and Wildlife Policy was therefore promulgated in the year 2012 to correct the anomalies in the previous one. This new policy is aimed at "conservation and sustainable development of forest and wildlife resources for the maintenance of environmental stability and continuous flow of optimum benefits from the sociocultural and economic goods and services that the forest environment provides to the present and future generations, whiles fulfilling Ghana's commitments under international conventions and agreements" (MLNR, 2012).

Other Legislation and Policies on Biodiversity Conservation in Ghana include: Forestry and Wildlife Policy of Ghana; International Biodiversity Convention; Marine, Coastal and Environmental Policy of Ghana; Mining and Mineral Laws of Ghana; Traditional Knowledge and Customs; Forest Protection Amendment Act; Control and Prevention of   □A
(PNDCL 229); Farmlands (Protection) Act (1962); Administration of Stool

Lands Act, 1994 (481); Forest Improvement Fund (Amendment) Act 1962; Forest Fees Amendment Regulations 1993 (LI 1576); Fisheries Law, 1991 (P.N.D.C. L 256); Fisheries Commission Act 1993, (Act 457); Land Planning and Soil Conservation Amendment Act 1957; Lands Commission Act 1994(Act 483); Water Resources Commission Act 1996, (Act 522);Wild Animals Preservation Amendment Law 1983 (P.N.D.C.L55); Wildlife Conservation (Amendment) Regulation 1983 (LI, 1983); Economic Plants Protection Decree 1979; Timber Resources Management Act 1997 Act 547; The Act Establishing the Environmental Protection Agency of Ghana.

□A

empowered to implement these policies and their related laws. For example, the Forestry Commission of Ghana is the agency responsible for the management of Ghana's forest resources. This was established under Act 405 – Ghana Forestry Commission Act, 1980 purposely for the coordination of the activities of the forestry sector institutions, i.e., the Forestry Department, Department of Game and Wildlife, Forest Products Research Institute and Ghana Timber Marketing Board. Under Section 6 of the Act, the Commission is mandated to regulate and manage the utilization of all forestry and wildlife resources of Ghana and also coordinate the policies in relation to forest resources (FC, 1994).

Although a number of legislation exist to regulate the use of Ghana's natural resources, the management and conservation of these resources still leaves much to be desired. Most of the legislations are for the regulation of terrestrial resources while a few are in place for the marine environment. This has therefore resulted in the misuse and abuse of the marine environment where it serves as a point of discharge of both domestic and industrial waste materials.

4.7.2 International Conventions Relating to Biodiversity Conservation in Ghana

Ghana is a signatory to numerous international conventions and agreements on environment and has ratified most of these agreements. Ghana is a signatory to the following conventions and agreements on environment and biodiversity: International Convention for the Prevention of Pollution of the Sea by Oil; Convention on the African Migrating Locust; International Convention for the Conservation of Atlantic Tunas; Africa Convention for the Conservation of Nature and Natural Resources; Convention on Wetlands of International Importance, Especially as Waterfowl Habitat; Convention on International Trade in Endangered Species of Wild Flora and Fauna; Convention for the Conservation of Migratory Species of Wild Animals; United

Nations Convention on the Law of the Sea; International Tropical Timber Agreement; Convention to Combat Drought and Desertification; Framework Convention and Climate Change and the Convention on Biological Diversity.

Ghana was the 157th country to sign onto the CBD agreement in 1992 and has subsequently   A
vention is a legally binding agreement on the use and conservation of biological diversity. Pursuant to Article 6 of the convention, which entreats signatories to develop a national strategy for the conservation and sustainable use of their biodiversity, a national biodiversity study and action plan document has been produced (MES, 2002). Similarly, a biodiversity country study of Ghana has been conducted as advocated by Article 7 of the convention, which enjoins parties to undertake biodiversity inventories to obtain information on the abundance and distribution of biodiversity.

4.8   Conclusions

Ghana abounds in enormous biodiversity of flora and fauna, due to its strategic location on the African continent. However unsustainable exploitation had had a great toll on its biodiversity. Undoubtedly, greater efforts have been made in conserving the biodiversity of plant genetic resources for food and agriculture, but little on fauna of the terrestrial, marine, fresh water ecosystems and microbes. To a large extent, the biodiversity of food crops and forest species are well documented and conserved. The variability of the biodiversity has added impetus to our economic fortunes by providing food, medicine and income generation for the citizenry and in some cases foreign exchange earned through ecotourism activities. Threatened species of flora and fauna that have been identified need to be given utmost attention. It is recommended that the overexploitation of biodiversity will require a multi-stakeholder and multi sectorial approach for its sustainable management.

In Ghana, institutional arrangements are in place for the conservation of biodiversity. However, there are constraints, which need attention for sustainable conservation, including the necessary human capital for the management of the biodiversity, in this era of the ever-changing environmental conditions, in which the very biodiversity exists, not forgetting the needed

Notwithstanding the conservation efforts made by stakeholders, there is a continuous loss of biodiversity through the activities of man, which will require the commitment of government to halt this impunity and restore sanity in the biodiversity and human interface and to restore degraded environments, on which humankind will rely and continue to rely, for their various needs, now and in the future. Ghana should be mindful of the numerous conventions and protocols it has assented to internationally and to implement policies and programs locally that will ensure the sustainable conservation of biodiversity.

Keywords

• conservation

• flora and fauna

References

Abayie, B. A., (1997). *Traditional Conservation Practices: Ghana's Example*. Paper presented at a UNESCO MAB seminar on biosphere reserves for biodiversity conservation and sustainable development in Anglophone Africa (BRAAF), Cape Coast: Ghana.

Aboagye, L. M., (2011). Plant genetic resources of some yam species in Ghana. *Genetic Resources Research Institute Handbook, CSIR-PGRRI/HB/11/02*, p. 13.

Aboagye, L. M., (2012). Genetic resources of some legumes in Ghana: Origin, characteristics and conservation. *CSIR-Plant Genetic Resources Research Institute Handbook, CSIR-PGRRI/HB/11/01*.

Adarkwa-Dadzie, A., (1997). *The Contribution of Ghanaian Beliefs to Biodiversity Conservation*. Paper presentation, UNESCO-MAB seminar on biosphere reserves for biodiversity conservation and sustainable development in Anglophone Africa (BRAAF), Cape Coast: Ghana.

Akromah, R. (1993). Economic, sociological and cultural importance of yams in Ghana. *Paper presented at the 1st Symposium of the Yam Network*, Cotonou, Benin 26–28th October, 1993.

Ampadu-Adjei, O., (2002). *The Role of Totems in the Bushmeat Extinction Prevention Campaign in Ghana*. Conservation International: Ghana.

Asase, A., & Oteng-Yeboah, A. A., (2007). Assessment of plant biodiversity in Wechiau Community Hippopotamus Sanctuary in Ghana. *J. Bot. Res. Inst. F. Texas*, 1, 549–556.

Attuquayefio, D. K., & Fobil, J., (2005). An overview of biodiversity conservation in Ghana: Challenges and prospects. *West African J. Applied Ecol.*, 7, 1–18.

Ayensu, E. S., Adu, A., & Barnes, E., (1996). *Ghana: Biodiversity and Tropical Forestry Assessment*. A report submitted to USAID mission in Ghana, Accra, Ghana.

Benin, J. K. A., & Barbier, E. B., (2001). The effects of structural adjustments program on deforestation in Ghana. *Agric. Res. Econ. Rev.*, 30(1), 66–80.

Bennett-Lartey, S. O., & Oteng-Yeboah, A. A., (2008). *Ghana Country Report on the State of Plant Genetic Resources for Food and Agriculture*, FAO, Rome .

BirdLife International, (2000). *Threatened Birds of the World*. Barcelona and Cambridge, UK, Lynx Edicions and BirdLife International.

Brooks, T. M., Mittermeier, R. A., Fonseca, G. A. B., Gerlach, J., Hoffmann, M., Lamoreux, J. F., et al., (2006). Global biodiversity conservation priorities. *Science, 313*, 58–61.

CBD (2009), *CBD – Fourth National Report* – Ghana Secretariat of the Convention on Biological Diversity, Montreal, Canada.

Conservation Interantional (CI) (2002). *Zero Biodiversity Loss: A Handbook for the Designing and Managing conservation Strategies.* Conservation International Regional Strategic Planning Department, Washington DC.

Convention on Biological Diversity (CBD) (1992), *A Report on the United Nations Conference on Environment and Development*, Rio de Janerio.

Dankwa, H. R., Abban, E. K., & Teugels, G. G., (1999). *Freshwater fishes of Ghana: Identification, Distribution, Ecological and Economic Importance.*

Decher, J., (1997). Conservation, small mammals, and the future of sacred groves in West Africa. *Biodiv. Conserv., 6,* 1007–1026.

Dorm-Adjobu, C., Ampadu-Adjei, O., & Veit, P. G., (1991). *Religious Beliefs and Environmental Protection: The Malshegu Sacred Grove in Northern Ghana.* World Resources Institute, Washington DC, African Centre for Technology Studies (ACTS) Press, Nairobi, Kenya.

Eastwood, R. B., & Steele, W. M., (1978). Conservation of yam in West Africa. *Plant Foods for Man, 2,* 153–158.

FC, (1994), *Forest and Wildlife Policy.* Forestry Commission of Ghana, Accra, Ghana.

Fisher, J., (1971). Wild life in danger: In: Detwyler, T. R., (ed.). *Man's Impact on the Environment.* McGraw-Hill, New York, pp. 625–653.

Gordon, C., (1992). Sacred groves and conservation in Ghana. *Newsletter of the IUCN SSC African Reptile and Amphibian Specialist Group, 1,* 3–4.

Hackman, K. O., (2014). The state of biodiversity in Ghana: Knowledge gaps and prioritization. *Intern. J. Biodivers. Conserv., 6*(9), 681–701.

Hall, J. B., & Swaine, M. D., (1981). *Distribution and Ecology of Vascular Plants in a Tropical Rain Forest: Forest Vegetation in Ghana.* Junk, W., (ed.), The Hague.

Hawthorne, W. D., (1990). *Field Guide to Forest Trees of Ghana.* Natural Resources Institute, Chatham.

Index Mundi, (2017). *Ghana Economy Profile 2017.* Retrieved from http://www.indexmundi.com/ghana/economy_profile.html.

International Institute for Environmental and Development (IIED) (1992), *Environmental Synopsis of Ghana.* Overseas Development Administration (ODA): London.

IUCN (1994), *Red Data Book, IUCN/WCMC,* Gland, Switzerland.

IUCN (2013), *IUCN Red List of Threatened Species.* Version 2013. 2.

Kingdon, J., (1989). *Island Africa: The Evolution of Africa's Rare Animal and Plants.* Princeton University Press, New Jersey.

Martin, C., (1990). *The Rainforest of West Africa: Ecology, Threats and Conservation.* Birkhauser Verlag, Basel.

Mekasha, A., (2001). Microfinance institutions and their impact on wealth creation. *ADB Bull., 4*(5), 5–7.

MES, (1995). *The National Biodiversity Strategy for Ghana.* Ministry of Environment and Science National Biodiversity Country Study Report.

Ministry of Environment and Science (MES), (2002). *National Diversity Strategy for Ghana,* Accra.

Ministry of Lands and Forestry (2000). *Forest and Wildlife Policy*. Government of Ghana Accra.

MLNR (2012). *Forest and Wildlife Policy of Ghana*. Ministry of Lands and Natural Resources, Accra, Ghana.

Myers, N., Mittermeier, R. A., Mittermeier, C. G., Fonseca, G. A. B., & Kent, J., (2000). Biodiversity hotspots for conservation priorities. *Nature, 403*, 853–858.

NRMP (2001). *Implementation Manual of the Natural Resources Management Programme (NRMP)*, Ministry of Lands and Forestry, Ghana.

Ntiamoa-Baidu, Y., (1995). *Indigenous Versus Introduced Biodiversity Conservation Strategies: The Case of Protected Area Systems in Ghana*. Biodiversity Support Programme (Issues in African Biodiversity No. 1). Washington DC.

Ntiamoa-Baidu, Y., Asamoah, S. A., Owusu, E. H., & Owusu-Boateng, K., (2000a). Avifauna of two upland evergreen forest reserves, the Atewa Range and Tano Offin, in Ghana. *Ostrich, 71*, 277–281.

Ntiamoa-Baidu, Y., Owusu, E. H., Asamoah, S., & Owusu-Boateng, K., (2000b). Distribution and abundance of forest birds in Ghana. *Ostrich, 71*, 262–268.

Ntiamoa-Baldu, Y., Owusu, E. H., Daramani, D. T., & Nuoh, A. A., (2001). Ghana. In: Fishpool, L. D. C., & Evans, M. I., (eds.). *Important BirdAres in Africa and Associated Islands*. 'Priority sites for conservation. Newbury and Cambridge, UK, Pisces Publications and BirdLife International (BirdLife Conservation Series No 11). pp. 367–402.

Ofori-Danso, P. K., & Agbogah, K., (1995). *Survey of Aquatic Mammals in Ghana*. Institute of Aquatic Biology Technical Report, 143. A Report to UN Environment Programme, Oceans and Coastal Areas.

Oteng-Yeboah, A. A., (1997). *Modern Concepts of Biodiversity: An Overview of Agenda 21*. Paper presented at a UNESCO MAB Seminar on biosphere reserves for biodiversity conservation and sustainable development in Anglophone Africa (BRAAF), Cape Coast: Ghana.

Petr, T., (1967). Food preference of the commercial fishes of the Volta Lake. *VBRP Tech. Rep. 22*, University of Ghana, Legon.

PGRRI Annual Report (2016), CSIR Plant Genetic Resources Research Institute, Bunso, Ghana.

Smit, C. J., & Piersma, T., (1989). Numbers, mid-winter distribution and migration of wader populations using the East Atlantic Flyway. In: Boyd, H., & Pirot, J. Y., (eds.). *Flyway and Reserve Networks for Water Birds*. IWRB Special Publication No. 9. IWRB. Slimbridge. UK, pp. 24–64.

Tom-Dery, D., Dagben, Z., & Cobbina, S., (2012). Effect of illegal small-scale mining operations on vegetation cover of arid Northern Ghana. *Res. J. Environ. Earth Sci., 4*, 674–679.

Tom-Dery, D., Hinneh, P., & Asante, W. J., (2013). Biodiversity in Kenikeni forest reserve of Northern Ghana. *African J. Agri. Res., 8*(46), 5896–5904.

White, F., (1965). The savanna woodlands of the Zambian and Sudanese domain. *Webbia, 19*, 651–681.

Wildlife Division, (1998). *Community Participation in Wildlife Management*. Wildlife Division. Plan, Ghana. Wildlife Dept. Ghana.

Worldometers, (2017). *Ghana Population*. Retrieved from http://www.worldometers.info/world-population/ghana-population/ (accessed on 21 June 2018).

CHAPTER 5

Biodiversity in Libya

MOHAMMED H. MAHKLOUF[1] and KHALED S. ETAYEB[2]

[1]Department of Botany, Faculty of Sciences, Tripoli University, Libya,
E-mail: mahklouf64@yahoo.com

[2]Department of Zoology, Faculty of Sciences, Tripoli University, Libya,
E-mail: khaledetayeb@yahoo.com

5.1   Introduction

Libya occupies a part of northern Africa from 20 to 34° N and 10 to 25° E. It is bounded in the east by Egypt (1,150 km), in the west by Tunisia (459 km), and Algeria (982 km), Mediterranean Sea in the north, and by Sudan (383 km), Chad (1,055 km), and 'Niger (354 km), in the south (CIA, 2004). It has an important physical asset by its strategic location at the midpoint of Africa's northern rim. The total area of Libya is about 1.76 Million km²; it ranks fourth in the area among all countries of Africa and 15th among all countries on earth (McMorris, 1979). More than 95% of Libya is desert, which is a part of Sahara, that is the most extensive area of severe aridity. The aridity of the central and eastern Sahara is due to its domination by continental tropical air all-year, which is continually descending from the upper levels of the atmosphere where, in these latitudes, anticyclone conditions are permanent.

The cultivable areas are estimated slightly over 2% of the total area (Ben Mahmoud, et al., 200). The fertile lands of Jifara Plain in the northwest, Jebal Al-Akhdar in the northeast and the coastal plain east of Sirt receive
A
Libyan population resides there. Between the productive lowland agriculture zones lies the Gulf of Sirt that stretches 500 km along the coast, from where deserts extend northward to the sea. Libya's total population was at 5.9 million in 2004 including more than 500,000 non-nationals; Almost 95% of the population lives in the coastal region in the north, and the rest in widely scattered oases in mid- and southern Libya.

According to the population distribution in Libya based on 2001 estimation, people concentrate on two centers, the   A
Plain) where about 60% of all Libyans live, including Tripoli city – the capital of Libya – where more than 1 million people live, and the second center in northeastern Libya (Ben-Ghazi Plain).

5.2   Floral Diversity of Libya

Libya is characterized by a beautiful plant cover, especially during spring. As the traveler moves along the coastal highway, he observes numerous plant species with attractive flower colors. Also, the hills, mountains, the desert with its oasis, and the valleys dissecting these huge lands are as much wonderful as the coastal area. Each area has its own attraction and flowers.

Libya consists of three main local botanical habitats: the coastal, the mountainous, and the desert habitats with their crossing valleys from south to north and from west to east. More than 1800 plant species are   -ing in these habitats. These plant species form a vegetation type with variable features (Boulos, 1972). The original vegetation of the coastal area is dominated by wormwood (*Artemisia campestris*) and white broom (*Retama raetam*) with some early spring   plants such as *Senecio gallicus, Hussonia pinnata, Eruca sativa, Chrysanthemum segetum, Malva sylvestris,* and *Erodium laciniatum,* in addition to the perennial herb *Echium angustifolium.* Aljabal Alakhdar area comprises the most wonderful plant diversity in Libya. This area is characterized by the red alluvial soil (terra rosa), relatively good rainfall (up to 600 mm/annually in certain locations), and closeness to the sea. The dominant plant species are *Arbutus pavarii, Juniperus phoenicea, Olea europaea* var. *oleaster, Pistacia lentiscus, Phlomis floccosa,* and *Cupressus sempervirens.* The western mountain is dominated by the true esparto (halfa) *Stipa tenacissima* with *Asphodelus microcarpus.* Remnants of *Pistacia atlantica* exist in certain locations. The desert, well known by its dryness, constitutes the most area of Libya with low plant cover of poor diversity especially the sand dune covered areas lacking crossing valleys. The sand sea south of Jaghboub along the Egyptian border and sand dunes of Rebyana north east of Kufra, the Murzuq basin are examples of huge areas covered with sand. Here, the rainfall is few millimeters or nil for several successive years.

The plant species in the desert valleys are totally different from those in the coastal or mountainous areas. The life signs are clear in the oasis and valleys of the desert represented by date palm trees (*Phoenix dactylifera*), *Tamarix* shrubs, white broom (*Retama raetam*), lotos (seder) *Ziziphus lotus,* European boxthorn (Ausej) *Lycium europaeum, Acacia* trees (*A. tortilis*), and the perennial grasses such as the panic turgid grass (*Panicum turgidum*), Sbut (*Stipagrostis pungens*), *S. plumosus,* and members of the Compositae as *Francoueria crispa* (crisp   A*Artemisia judaica, Hyoscyamus muticus* (the Egyptian henbane) of the Solanaceae, and *Zilla spinosa* (shobrum) of the Cruciferae. The coastal areas are well known by its   -

ing colored spring. During the spring season, many annual plants appear dominating large spots in a short time after the winter rainfall. These annuals
          □A
Examples of some early spring   plants are *Senecio gallicus, Hussonia pinnata, Eruca sativa, Chrysanthemum segetum, Malva sylvestris,* and *Erodium laciniatum.* These annuals shed fruits and dry just before summer. This phenomenon is applicable to ephemerals of the desert habitats where the rain usually falls in late summer or early autumn.

Salt marshes occur along the coastal strip and dominated by halophytes. These are mostly succulent plant species belong mainly to the family Chenopodiaceae, few other plants belong to the families Zygophyllaceae and Plumbaginaceae. Examples of halophytes are *Suaeda* sp., *Atriplex halimus* (qataf)*, Limoniastrum monopetalum* (zeta)*, Limonium pruinosum* (sea lavender or thrift)*, Nitraria retusa* (ghardaq or sea plum)*,* and *Zygophyllum album* (belbal)*.* The winter rain lowers the salinity rate and compensates the evaporated water during summer allowing some salt tolerable annuals such as branching century *Centaurium pulchellum* to appear.

5.2.1 *Floristic Composition*

The floristic composition of plants in Libya is still comparatively unknown as far as in-depth ecological and botanical studies go (Pergent and Djellouli, 2002).

The analysis of   according to Jafri and Ali (1981) and Klopper et al. (2007) (Feng et al., 2013) resulted in occurrence of about 2,118 species belonging to 864 genera and 161 families in Libya, of them 2,088 species, 844 genera and 145 families, are Angiosperms, 15 species of 8 genera and 6 families are Gymnosperms and 15 species of 12 genera and 10 families are Pteridophyta.

From Table 5.1 it can be seen that Libyan plants are comparatively rich in number. The great majority of the families are widely spread (Aqciteex, 1985; Hammer et al., 1988; Keith, 1965). The dominant families in Libya are Asteraceae (237 species), Gramineae (228 species), Leguminosae (200 species), Brassicaceae (100 species), Rubiaceae (90 species), Labiatae (63 species), Caryophyllaceae (62 species), Boraginaceae (53 species) and Chenopodiaceae (49 species). The dominant families encompass 51.8% of the species found. Libya's dominant genera are *Euphoria* (27 species), *Astragalus* (25 species), *Silene* (23 species), *Trifolium* (22 species), *Allium* (18 species), *Medicago* (18 species), *Erodium* (15 species), *Lotus* (15 species), *Ranunculus* (14 species), and *Helianthemum* (14 species) (Table 5.2).

Table 5.1   Statistics on Generic and Species Number in Families of Plants in Libya
(Genera: Species)

>50 species (8 families)
Asteraceae (97:237), Gramineae (93:228), Leguminosae (42:200), Brassicaceae (59:100), Rubiaceae (50:90), Labiatae (22:63), Caryophyllaceae (18:62), Boraginaceae (23:53)
50–31 species (4 families)
Chenopodiaceae (23:49), Liliaceae (15:42), Scrophulariaceae (10:34), Euphorbiaceae (5:32)
30–21 species (9 families)
Ranunculaceae (8:29), Geraniaceae (4: 28), Cyperaceae (7:26), Rosaceae (19:25), Zygophyllaceae (8:25), Solanaceae (10:24), Malvaceae (10:22), Polygonaceae (5:22), Cistaceae (4:22)
20–11 species (20 families)
Anacardiaceae (11:20), Crassulaceae (3:18), Convolvulaceae (3:18), Alliaceae (1:18), Orchidaceae (5:17), Plumbaginaceae (3:16), Plantaginaceae (1:16), Illecebraceae (5:715), Agavaceae (6:14), Rutaceae (5:14), Cucurbitaceae (9:14), Myrtaceae (6:13), Iridaceae (5:13), Tamaricaceae (2:13), Orobanchaceae (2:13), Resedaceae (5:13), Oleaceae (6:12), Urticaceae (4:11), Amaranthaceae (4:11), Capparaceae (4:11)
2–10 species (61 families)
Valerianaceae (3:10), Fumariaceae (1:10), Papaveraceae (4:9), Rhamnaceae (5:9), Aizoaceae (5:9), Lythraceae (4:8), Verbenaceae (6:8), Dipsacaceae (2:8), Asclepiadaceae (8:8), Juncaceae (1:8), Potamogetonaceae (1:17), Caesalpiniaceae (5:7), Apocynaceae (6:6), Casuarinaceae (1:6), Mimosaceae (2:6), Campanulaceae (3:6), Hypecoaceae (1:6), Primulaceae (5:6), Commelinaceae (4:5), Salicaceae (2:5), Moraceae (2:5), Frankeniaceae (1:5), Lauraceae (4:5), Fagaceae (4:5), Molluginaceae (4:5), Gentianaceae (1:5), Clusiaceae (1:5), Bignoniaceae (4:4), Nyctaginaceae (3:4), Cuscutaceae (1:4), Amaryllidaceae (2:4), Arecaceae (3:4), Buddlejaceae (2:3), Bombacaceae (2:2), Ericaceae (2:3), Araceae (3:3), Caprifoliaceae (2:3), Tiliaceae (2:3), Thymelaceae (2:3), Leonticaceae (2:2), Cactaceae (1:3), Oxalidaceae (1:3), Nymphaeaceae (1:2), Najadaceae (1:2), Lentibulariaceae (1:2), Lemnaceae (1:2), Callitrichaceae (1:2), Acanthaceae (2:2), Alismataceae (2:2), Vitaceae (1:2),
Portulacaceae (1:2), Cannaceae (1:2), Onagraceae (2:2), Linaceae (2:2), Typhaceae (1:2), Polygalaceae (1:2), Santalaceae (1:2), Juncaginaceae (1:2), Saxifragaceae (1:2), Globulariaceae (1:2), Strelitziaceae (1:2)
1 species (42 families)
Coridaceae (1:1), Theligonaceae (1:1), Violaceae (1:1), Neuradaceae (1:1), Cynomoriaceae (1:1), Vahliaceae (1:1), Salvadoraceae (1:1), Dioscoreaceae (1:1), □   □   □   □ Pedaliaceae (1:1), Elatinaceae (1:1), Ceratophyliaceae (1:1), Sterculiaceae (1:1), Sapindaceae (1:1), Meliaceae (1:1), Myoporaceae (1:1), Tropaeolaceae (1:1),

Table 5.1 (Continued)

Juglandaceae (1:1), Pittosporaceae (1:1), Tetragoniaceae (1:1), Menispermaceae (1:1), Ruppiaceae (1:1), Zannichelliaceae (1:1), Araliaceae (1:1), Balsaminaceae (1:1), Simarubaceae (1:1), Celastraceae (1:1), Sapotaceae (1:1), Aristolochiaceae (1:1), ☐ ☐ ganiaceae (1:1), Aceraceae (1:1), Aquifoliaceae (1:1), Polemoniaceae (1:1), Musaceae (1:1), Begoniaceae (1:1), Phytolaccaceae (1:1), Hydrocharitaceae (1:1)

Table 5.2 Dominant Families in Libya

Global		Local		
Species	Genera	Species	Genera	Family
1100	25000	237	97	Asteraceae
700	11000	228	93	Poaceae
500	2000	200	42	Fabaceae
350	3000	100	59	Brassicaeae

The dominant genera include only 9.15% on the species level, but these all belong to large and widely spread genera in arid zones (Szafer, 1964) (Table 5.3).

5.2.2 Life Forms

Life forms are given in Table 5.4. The life form distribution among Libyan plants was characterized by a high proportion of herbs (annual to perennial). The low number of woody (tree and shrub) species in our dataset reflects the defensive capabilities of the vegetation in bad conditions (such as drought), i.e., the lack of moisture in Libya. It seems that the herb life form is the preferable strategy in the temperate deserts of the studied area (Table 5.4).

This is not only a ☐ A ☐ of highly adapted, drought-resistant species. These xerophytes are widely distributed in the subhumid and semiarid tropics and play major economic and ecological role. Therefore, these are very successful species, capable of stabilizing mobile sands by their rapid growth and long roots (Higgins et al., 1997). We should pay more attention to the matching ability between the protective effect of different life forms and the time of occurrence of strong winds and sandstorms during ecological "building" of vegetation so that the ☐ ☐ ☐ ☐ ☐

Table 5.3 Dominant Genera in Libya

Genus	Global No	Libyan No
Euphorbia	2,000	26
Astragalus	1,500	25
Silene	500	23
Trifolium	300	22
Medicago	60	18
Lotus	100	15
Erodium	90	15
Convolvulus	250	14
Stipagrostis	50	13
Echium	40	13

Table 5.4 Statistics and Comparison of Life Forms of Plants in Libya

Life form	Tree	Shrub	Liana	Parasitical plant	Annual herb	Perennial herb	Total
No. of species	133	234	44	14	858	805	2088
Percentage of species	6.4	11.2	2.1	0.7	41.1	38.6	100

5.2.3 *Geographical Elements*

5.2.3.1 Geographical Elements of Family Level

Floristic elements have been considered as useful tools in phytogeographical analysis (Preston and Hill, 1997). This is an important method in floristic research to divide the distribution into different area types. According to the Chinese botanist Wu's documentation (Wu et al., 2003), the distribution type in Libya at the family level was counted and is presented in Table 5.5 as the percentage, relative contribution to the family.

5.2.3.2 Geographical Elements of Generic Level

Szafer (1964) and Wu (1991) have proposed that geographical elements of genera are the greatest contributors to the analysis of flora. Meanwhile, some genera contain species that have the same origin and similar evolutionary

Table 5.5 Geographical Elements of Family Level of Plants in Libya

No.	Distribution types	No. of family	Percentage in family (%)
1	Widespread	52	35.9
2	Pantropic	47	32.4
3	Tropical and Sub Tropical	8	5.5
4	Old World Tropic	1	0.69
5	Tropical Asia to Tropical Australasia and Oceania	1	0.69
6	Tropical Asia to Tropical Africa	6	4.14
7	Tropical Asia	1	0.69
8	North Temperate	17	11.7
9	East Asia and North America disjunction	1	0.69
10	Old World Temperate	4	2.76
11	Temperate Asia	5	3.45
12	Mediterranean, West Asia to Central Asia	1	0.69
13	Extra T A disjunction	1	0.69

trend. Thus, from the viewpoint of phytogeography, genera accurately reflect plant systematics, evolution and regional characteristics. In order to demonstrate the floristic characteristics of the flora of Libya, the genera have been studied and classified according to the Wu system (1991). Statistics show that the geographical elements are multiple since there are 16 area-types in this region.

Among them, the type of tropical distribution (2–7 types) compose 282 genera, or 33.4% of the total genera; such genera include *Aristida, Impatiens, Euphorbia, Paspalum, Phaseolus, Heliotropium, Acacia, Celosia,* etc., and shows the highest percentage among plant types. The Mediterranean type includes genera such as *Alhagi, Anabasis, Bassia, Calligonum, Cardaria, Cistanche, Nitraria, Triticum, Haloxylon,* etc. (20.9%), and shows the second highest percentage group. Next is the type of widespread distribution, such as the genera *Xanthium, Senecio, Salsola, Phragmites, Ranunculus, Carex, Astragalus, Lepidium,* etc. (15.2%), immediately followed by the North Temperate types, *Acer, Allium, Avena, Capsella, Alopecurus, Cirsium, Iris, Malva, Potentilla, Populus,* etc., which contribute 13.9%. Only four are endemic to Libya, such as the genera *Oudneya* and *Pseuderucaria,* all in Brassicaceae. Undoubtedly, the tropical and Mediterranean elements were the main part of local   A

ⒶⒶ

and tropical nature. It is fully proved that the methods of quantitative clas-

Ⓐ

☐ ☐   ☐   ☐ Table 5.6).

5.3 Survey of Main Ecological Systems in Libya

5.3.1 Coastal Ecosystems

Coastal ecosystems are from 25–100 km wide in the northern regions of Libya. In this area, the annual rainfall is about 200–250 mm. Over 75% of vascular plants are distributed in the coastal areas, such as *Acacia* spp.,

Table 5.6 Geographical Types of Genera of Plants in Libya

No.	Distribution types	No. of genera	Percentage in genera (%)
1	Widespread	128	15.2
2	Pantropic	161	19.1
3	Tropical Asia and Tropical America disjunction	33	3.9
4	Old World Tropics	31	3.7
5	Tropical Asia and Tropical Australasia	12	1.4
6	Tropical Asia to Tropical Africa	38	4.5
7	Tropical Asia	7	0.8
8	North Temperate	117	13.9
9	East Asia and North America disjunction	12	1.4
10	Old World Temperate	96	11.4
11	Temperate Asia	5	0.59
12	Mediterranean, West Asia to Central Asia	177	20.9
13	Central Asia	11	1.3
14	East Asia	12	1.4
15	Endemic to Libya	4	0.47

Borassus, Phoenix, etc. The ecotype consisted of Mediterranean groups of xerophytes that protect shorelines from erosion and storms.

5.3.2 Mountain Ecosystem

The mountain ecosystem is located in the western mountains of Libya, Nafosa Mountain, and Green Mountain. It ranges from dry mountain forests at low elevations to mountaintop vegetation. Only 0.1% of the land (about 217,000 m²) is woodland, with an annual rainfall of about 200–300 mm. However, these forests should be more accurately called Mediterranean coast shrub. The native forest is the richest; the biodiversity index is the highest. The dominant species were *Cupressus sempervirens, Eucalyptus camaldulensis, Melia azedarach,* and *Olea europaea.* Semi-desert ecosystem regions are located in the transitional zone between the mountain and desert zones, with an annual rainfall of about 50–150 mm.

5.3.3 Desert Ecosystem

Since 90% of Libya is desert, this ecosystem is the most characteristic. The desert landscape ecosystem is made up of three landscape types: rocky desert, sandy desert, and congenital desert. Here, the climate is hot and dry, the desert ecosystem has sparse vegetation and small biomass, and the ecoenvironment is fragile. Due to human activity, the ecoenvironment is severely degraded. Pioneer species formed a unique form of adverse environment, with strong resistance to drought and barren stress. Such forms are *Haloxylon schweinfurthii, Acacia flava, Aristida acutiflora, Euphorbia abyssinica, Calligonum comosum, Acacia senegali, Cordia africana, Tamarix mannifera,* and *Salsola tetrandra* (Feng et al., 2013).

5.3.4 Important Plant Areas in Libya (IPAS)

Botanically and in terms of threatened habitats, there are five general important plant areas in Libya, representing the coastal belt, mountainous, and desert habitat types. Those are Jabal Alakhdar, Jabal Nafusah, Tawuorgha wetland on the coast, the Messak area at the southwestern part, and the Alaweinat at the southeastern corner on the borders of Egypt and Sudan, with a further six areas that require study to confirm their status as internationally significant sites for plants (Alheesha, Farwa Island, Mamarica, Jabal Al Harouj and Benghazi Coast, Msellata Natural Resesrve). The oil exploration, hunting, overgrazing, and tourism are the main threatening factors for the biodiversity of those habitats.

Tawuorgha is characterized by a hot spring making a small lake. Its water runs in open canals providing wet habitats for many aquatic plant species. The *Phyla nodiflora* of the Verbenaceae family was discovered in 2002 to grow there after reporting it as an extinct species from Libya. For Jabal Nafusah (the western mountain) which extends for 500 km from the Tunisian borders to Niggaza area on the seashore. Sha afeen protected area (or national bark) is a recently established protected area characterized by the newly recorded species *Bupleurum jibraltaricum* Lam. (Umbelliferae) which represents the western limit of its range of distribution (starting from south Spain). *Ebenus pinnata* Ait. (Fabaceae) is endemic to North Africa existing in the reserve and other locations on the mountain.

There are some efforts towards making the Akakoos desert mountainous area, close to the Messak, as a national reserve by cooperation with Catania university of Italy. During October 2001, the French Total oil company conducted an excellent environmental baseline study for the Messak resulted in lists of biodiversity components of the area. Jabal Alharuj al Aswad needs more investigation; even it is subjected to a lot of misuses, before identifying it as an important plant area. Oil companies exploring the area are required to do environmental impact assessment studies before they get permission to start seismic work and drilling. However, strict following up is much needed.

The Environmental General Authority (EGA) is planning to conduct an investigation for the Alaweinat at the southeastern corner of the country. Of course, there are previous studies for the area conducted by Boulos (1972). Libyan IPAs face a number of threats including the development of tourism infrastructure, overgrazing of livestock, forest cutting for wood and charcoal and the spread of invasive alien species. Unregulated development at the coast is a particular threat. Planning processes are erratic and environmental impact assessments (although required by law) are seldom completed or adhered to.

Here, we will focus on Al-Jabal Al-Akhdar IPA (The Green Mountain), which is a priority IPA.

5.3.4.1 Jabal Akhdar

Jabal Akhdar is located in the northeastern part of the country bordered by the sea from the north and west, the Marmarica Plateau at the east, and the desert at the south (320–330 N and 200–230 E). Al Jabal Al Akhdar IPA (The

Green Mountain) in the Cyrenaica region of northeast Libya is the largest and most significant IPA in Libya, it is characterized by distinct ecological features since it is the only evergreen forest area of unique type along the coastal belt from Atlas Mountains in the west to Palestine-Syria-Lebanon in the east. Jabal Akhdar is similar in its biodiversity to the areas south and east of the Mediterranean. The ancient civilizations established on Jabal Akhdar area were correlated with its unique natural vegetation. Although the area of Jabal Akhdar is not more than 1% of the total area of the country, it is distinguished by its high percentage of the country's biodiversity. It contains about 1,300 plant species, which encompasses 70–80% of the Libyan flora. The general topography of the area consists of three levels of escarpments differs little bit climatically. The first level adjacent to the seashore up to altitude 200 m represents a plain with Mediterranean climate, the second level is from 200 m to 600 m, and the third level is from 600 to 800 (the highest is 882 m (a.s.l.) is characterized by cool winter and hot summer. Temperature means ranges from about 7°C during the winter months to about 27.6°C during the summer period in Shahat area on top of mountain range. The mountain slopes gradually towards the desert (south) and northern slopes facing the sea are very steep.

Also, the difference in the geographical locations of the subareas of the Jabal Akhdar led to notable variation in average annual rainfall. The maximum annual rainfall reaches 600 mm at Shahat (Cyrene) and this annual average drops as we move to the east, west, or south to reach the lowest 200 mm south of the limits of the forest range. These climatic and topographical differences are   on the types and characters of the plant cover in these areas and types of soils (mostly red alluvial *terra rosa*) spreading within which produced ecological variation allowed agricultural activities along the year in different locations.

The vegetation communities are (from sea level): the coastal plain, coastal escarpment, central plateau and upper escarpment, and upper plateau. The coastal plain consists of the sandy beaches, salt marshes, and rocky coasts. As in the majority of Mediterranean dune communities *Elytrigia juncea* (L.) Nevski subsp. *juncea* is common, its association on Jabal Akhdar with *Centaurea pumilio* L. and *Silene succulenta* is unusual. Endemics of the dunes include *Helianthemum cyrenaicum, Thapsia garganica* var. *sylphium, Anthemis taubertii, Teucrium zanonii,* and *Plantago libyca.* The salt marshes are analogous to others in the Mediterranean with endemic species such as *Frankenia syrtica* and *Limonium teuchirae.*

The coastal escarpments are dominated by *Juniperus phoenicea* scrub/ forest. Endemic species include *Cyclamen rohlfsianum, Micromeria conferta* and *Stachys rosea*. The wadis are poorly known; the vegetation comprises *Juniperus phoenicea* scrub/forest on the slopes, with dense semi-deciduous mixed woodland in the channels dominated by *Quercus coccifera, Pistacia lentiscus, Arbutus pavarii, Ceratonia siliqua, Olea europaea* and *Cupressus sempervirens*. These wadis are rich in endemic plant taxa, e.g., *Arum cyrenaicum, Erica sicula* subsp. *cyrenaica, Onosma cyrenaica* and *Nepeta cyrenaica*.

The central plateau of Al Jabal Al Akdhar is used heavily for agriculture. The vegetation of this area is a mixture of *maquis* and a shrubby *batha community* in areas of grazing and/or shallow soils. Patches of dense woodland also occur on the upper escarpment above the central plateau. The tree layer here is dominated by *Cupressus sempervirens, Juniperus phoenicea, Olea europaea, Quercus coccifera, Ceratonia siliqua,* and *Pinus halepensis*.

The upper plateau is also heavily used for agriculture, with only small patches of *Juniperus* remaining. *Batha* forms a major plant community in this area, again often dominated by *Sarcopoterium spinosum, Phlomis floccosa, Pallenis spinosa* and a rich diversity of grasses and ruderal species. The upper plateau site of Sidi Al Hamri is one of two known locations for *Pachyctenium mirabile*. Many areas of Al Jabal Al Akhdar lack botanical data, which greatly hinders conservation planning. Four areas within Jabal Akhdar were investigated during the compilation of this report and are described in more detail within the national report (see references): Ain Estowa, Dabbusia spring, Morcus Valley and Spring and El Kouf Valley. The principal threats to the conservation of this IPA are: heavy grazing and inappropriate development and agricultural activities. There is poor environmental planning and management, and the coastal zone is being developed without detailed environmental impact studies. Deforestation is occurring for domestic fuel and charcoal and there is die-back of *Juniperus* forest.

5.3.5 *Endemism in Libya*

The total endemic plant species in Libya about 80–81 species, distributed in four centers of endemism in Libya: (i) Jabal Akhdar with 44 species, about 50% of endemic plants in the country, (ii) coastal belt including the Jabal Nafusah and Marmarica plateau with 26 endemic species, the center of the Sahara with 8 to 9, and the plateau of Ghat, Tebesti, and Aweinat with 2 species. Generally, the flora of Libya possesses a low percentage of endemic plants, not more than 7% because of the similar topography and harsh environment. However, the endemic plants constitute a unique genetic diversity limited to the flora of the country. There are three endemic genera represented by one species each: *Pachyctenium mirabile* (Umbelliferae), *Oudneya africana* (Cruciferae), and *Libyella cyrenaica* (Gramineae). Examples of endemic plants in Libya are *Cyclamen rholfsianum* (Primulaceae), its picture on the series of the Flora of Libya's cover, *Arum cyrenaicum* (Araceae), *Teucrium cyrenaicum* (Labiatae), *Linaria tarhunensis* (Scrophulariaceae), and *Tourneuxia varrifolia* (Compositae). *Sedum cyrenaicum* (Crusulaceae), *Thapsia garganica* var. *sylphium* (Apiaceae), *Cupressus sempervirens* var. *horizontale* (Cupressaceae).

5.4 Faunal Diversity in Libya

Libya is mostly characterized by arid climatic conditions, except the coastal strip and the northern hills toward the east and the west, while the rest of the country is located under the conditions of desert and semidesert because of its geographical location in terms of latitude (Essghaier et al., 2015). This resulted in the presence of environments with distinct characteristics in terms of temperature, humidity, and rainfall that reflected on the biological components of the plants and the animals that are able to coexist in various ways with those difficult environmental conditions (Hufnagel, 1972). In Libya, there are a lot of ecosystems that range from the coastal environment with all its scattered salt marshes along the coastline, to green plains in the northeastern region and northwest highlands (which include Nafusa Mountains), to desert and semidesert ecosystem showing its content of oases and valleys (Toschi, 1969). The desert is ecologically sensitive and very important in terms of wildlife (flora and fauna), which coexist in this habitat in spite of the harsh living conditions as much heat, especially during the summer months in addition to water scarcity and drought. However, these systems include a few diversity and abundance of species particularly those that have the capacity to live under these circumstances and some of them are endemic.

5.4.1 Components of Biological Diversity

Depending on climatic and topographic biological data, the main four basic bio-geographic ecoregions can be distinguished (EGA, 2010) as given in the following subsections.

5.4.1.1 Coastline

Coastline varies in width; it is generally between 5 and 25 km, and about 100 km in some areas such as Al-Jfara plain. The Libyan coast extends over a distance of about 1970 km, to occupy the majority of the southern coast of the Mediterranean Sea. Moreover, this coastline contains many ecosystems; gulfs, rocky and sand beaches, coastal lagoons, islands and Sabkhas (Salt marshes).

5.4.1.2 Mountains

Mountains two main regions are located south of the coastal strip, one at the eastern part called Al-Jabel Al-Akhder (Green Mountain), with the highest peak of 880 m, and the other at the western part of the country called Jabel Nafusa (Al-Jabel Al-Gharbi). These mountains contain many valleys. The average rainfall in Al-Jabel Al-Akhder mountains is between 250–600 mm, and less than that in Jabel Nafusa (200–300 mm).

5.4.1.3 Arid or Semi-Desert Region

Arid or semidesert region, which lies directly south of the mountainous regions, and in parallel to it. The average precipitation is between 50 to 150 mm. This area is often used as pastures and some agricultural activities that carried out by some of the Bedouin in some valleys.

5.4.1.4 Sahara

Sahara occupies about 90% of Libya, a barren land includes sand dunes and the vegetation is very scarce. Very low precipitation or almost nonexistent.

5.4.2 Basic Features of Animal Biodiversity in Libya

Globally, Libya is a poor country in term of living species space that occupies its area. From a total of 1.7 million described, Libya has about 2,135 plants and 4,590 animals (EGA, 2010).

Libya also has a large diversity of habitats and ecosystems such as sea and beaches, forests, mountains, steppes, grasslands, a variety of wetlands land and desert.

The diversity of wetlands in Libya:

1. Salt marshes (Sabkhas), such as Sultan, Abo Kemmash and Benghazi sabkha.

2. Coastal lagoons, such as Al-Ghazala and Azzayana.
3. Water springs, such as Tawergha and Ain Kaam.
4. Desert oases, such as Gaberoun, Bzimah Oasis.
5. Dams, such as Almjenin dam Wadi Attot.
6. Artificial reservoirs, such as Made River reservoirs.
7. Water treatment plants.

These ecosystems have a great economic importance, they are also shelters for many species. So, any disturbance, threats or destructing of these habitats will negatively affect the components of biodiversity of these areas. Furthermore, they have a great value as touristic and recreational zones. They also have a crucial role in the   A □
kidneys in nature).

5.4.3 Libyan Fauna

The number of animals in Libya according to preliminary estimates 4,590 species. The most important of these taxa in terms of number is the insects (81%), followed by birds (7%). However, animals diversity in Libya still needs further taxonomic studies to be well documented.

Estimates of marine animal and plant species are about 1,500 species, for instance, 560 species of marine algae, and three species of endangered seagrasses in the Mediterranean Sea, and about 100 species of   [A
species of marine reptiles (turtles).

- No available information on the Protista.
- A total of 139 species of Molluscs.
- The number of Arachnids is about 170 species.
- Insects are the majority of Libyan animal diversity. The approximate number is 3,763 species.
-       □   □   □   □   □
- The smallest number of Libyan animals is three of amphibian species.
- The Libyan reptiles are 25 (Frynta et al., 2000).
- The recent documentary publication has   that the total number of birds in Libya is 350 species (100 are currently breeding in Libya) (Isenmann et al., 2016).
- Libyan mammals are 76 species, including 4 endemics and 12 threatened (EGA, 2010).

Keywords

- animal biodiversity
- Libyan fauna

References

Aqciteex, A. O., (1985). Status of plant genetic resources in the Libyan Arab Jamahiriya. *Plant Genetic Resources, 62*, 25–27.

Boulos, L., (1972). Our Present Knowledge on the Flora and Vegetation of Libya. Bibliography. *Webbia, 26*, 365–400.

CIA, (2004). *Libya, World Fact Book*, www.cia.gov/cia/publications/factbook /geos/ly.html (accessed on 10.01.2005).

Environment General Authority (EGA), (2010). *The Fourth National Report on the Implementation of the Convention of Biological Diversity* (CBD).

Essghaier, M. F. A., Taboni, I. M., & Etayeb, K. S., (2015). The diversity of wild animals at Fezzan province. *Biodiversity Journal, 6*(1), 253–262.

Essghaier, M. F. A., Taboni, I. M., & Etayeb, K. S., (2015). The diversity of wild animals at Fezzan Province. *Biodiversity Journal, 6*(1), 253–262.

Feng, Y., et al., (2013). Composition and characteristics of Libyan flora. *Arch. Biol. Sci., Belgrade, 65*(2), 651–657,.

Frynta, D., Kratochvil, L., Moravec, J., Benda, P., Dandova, R., Kaftan, M., et al., (2000). Amphibians and reptiles recently recorded in Libya. *Acta. Soc. Zool. Bohem., 64*, 17–26.

Hammer, K., Lehmann, C. O., & Perrino, P., (1988). A check-list of the Libyan cultivated plants including an inventory of the germplasm collected in the years 1981, 1982 and 1983. *Kulturpflanze, 36*, 475–527.

Higgins, S., Rogers, K. H., & Kemper, J., (1997). A description of the functional vegetation pattern of a semi-arid floodplain, South Africa. *Plant Ecol, 129*, 95–101.

Hufnagel, E., (1972). *Libyan Mammals*. The Oleander Press.

Isenmann, P., Hering, J., Brehme, S., Essghaier, M., Etayeb, K., Bourass, E., & Azafzaf, H., (2016). *Oiseaux de Libye. Birds of Libya*. Paris, SEOF/MNHN, ISBN: 2-916802-04-5, p. 302.

Jafri, S. M. H., & Ali, S. I., (1981). *Flora of Libya*, Tomus (1–145). Published by Department Botany, Al-Faateh University in Tripoli.

Keith, H. G., (1965). *A Preliminary Check List of Libyan Flora*, vol. 2. Ministry of Agriculture and Agrarian Reform, London.

Klopper, R. R., Gautier, L., Chatelain, C., Smith, G. F., & Spichiger, R., (2007). Floristics of the angiosperm flora of sub- Sahara African: An analysis of the Africa Plant Checklist and Database. *Taxon, 56*, 201–208.

McMorris, D. S., (1979). Society and its environment. In: Nelson, H. D., (ed.). *Libya a Country Study 'Foreign Area Studies', 3rd edn*. The American University, Washington.

Pergent, G., & Djellouli, A., (2002). Characterization of the benthic vegetation in the Farwà Lagoon (Libya). *Journal of Coastal Conservation, 8*, 119–126.

Preston, C. D., & Hill, M. O., (1997). The geographical relationships of British and Irish vascular plants. *Botanical Journal of the Linnean Society, 124*, 1–120.

Szafer, W., (1964). *General Plant Geography*. Warszawa. PWN Polish Scientific Publishers.

Toschi, A., (1969). Introduzione alla ornitologia della Libia. *Supplemento alle ricerche di zoologia applicata alla caccia, 6,* 1–381.

Wu, Z. Y., Zhou, Z. K., Li, D. Z., & Peg, H., (2003). The areal-types of the world families of seed plant. *Acta Botanica Yunnanca, 25*(3), 245–257.

Wu, Z., (1991). The areal-types of Chinese genera of seed plants. *Acta Botanica Yunnanica, Suppl. IV,* 1–139.

Appendices

Some of the endemic species in Libya

A- *Cupressus sempervirens var. horizontale*   B- *Sedum cyrenaicum*

C- *Linaria tarhunensis*   D- *Thapsia garganica var. sylphium*

Some photos from ecosystems and flora in Libya.

Some photos from ecosystems and flora in Libya.

CHAPTER 6

Biological Diversity in Morocco

MOHAMED MENIOUI

Scientific Institute of Rabat, Mohammed V. University, Morocco,
E-mail: mohamed.menioui@gmail.com

6.1 Introduction

"Biodiversity" (or biological diversity) is considered in Morocco neither as a formally individualized development sector nor as a clearly defined environmental theme, but rather as a concept and a support/ecological and environmental constituted by all the biological, ecosystem resources, Genetic, landscape and abiotic conditions governing these resources. The International Convention on Biological Diversity (CBD, Rio de Janeiro, 1992) clearly defined its content as: "The variability of living organisms from all sources, including, inter alia, terrestrial, marine and other aquatic ecosystems and the ecological complexes of which they are a part. This includes diversity within and between species and ecosystems."

However, alongside the inventory of the various components of biodiversity, the emphasis is on the notion of "interactivity" between these three different levels of organization.

Biodiversity is important in many ways: ecological, economic, sociological,   A

rization of biological resources distinguishes between direct values (direct consumption value, productive value, recreational value) and indirect values (ecological value, option value, existence value).

Biodiversity in Morocco is very rich (Figure 6.1) and the importance of the Kingdom of Morocco in terms of Biodiversity lies in its geographical location (Figure 6.2), where it meets major geostrategic and ecological complexes: the Mediterranean Sea to the north, the Atlantic Ocean to the west and the desert front of the Sahara at the Southeast. This particular geographical position gives it a remarkable range of bioclimates ranging from humid and subhumid to Saharan and desert through the arid, semiarid and high mountain climate in the Rif, Middle and High Atlas, where the altitudes exceed 2,500, 3,000 and 4,000 meters, respectively. To this diversity of relief and climate there is a great diversity of bioecology, species, ecosystems and genetic resources, as well as a wide range of natural environments: woody forest formations, pre-Saharan and Saharan formations, steppes, matorrals,

Wetlands, natural or   A
coastal areas (lagoons, estuaries, saline.), etc. with   and faunistic pro-

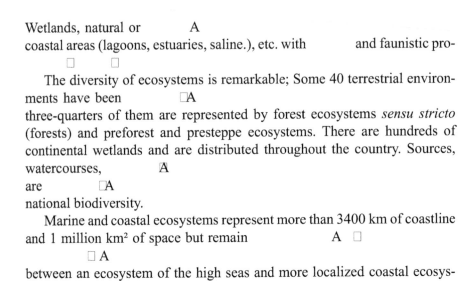

The diversity of ecosystems is remarkable; Some 40 terrestrial environments have been   A
three-quarters of them are represented by forest ecosystems *sensu stricto*
(forests) and preforest and presteppe ecosystems. There are hundreds of
continental wetlands and are distributed throughout the country. Sources,
watercourses,   A
are   A
national biodiversity.

Marine and coastal ecosystems represent more than 3400 km of coastline
and 1 million km² of space but remain   A
            A
between an ecosystem of the high seas and more localized coastal ecosystems such as lagoons and estuaries.

            A   -
nomic importance, with more than 31,000 animal species and 7,000 plant
species and an overall endemism rate of 11% for wildlife, and more than
20% for vascular plants, almost unparalleled in relation to the entire Mediterranean basin and which places Morocco in second place after Turkey. As part
of the development of a comprehensive national environmental vision, which
aims to integrate the environment into the country's various socioeconomic
sectoral activities and lead to the development of a National Environment
Strategy (SNE), the public authorities (UNEP/Department of Environment,
1997) have initiated a diagnostic analysis of the national environment in the
main sectors, including "biodiversity."

Regarding genetic diversity, Morocco has always been considered, as a
biological wealth, as a preferred destination for bioprospecting. The terroir
varieties of agrobiodiversity or the medicinal and aromatic plants that are
distributed in the steppe and mountain areas of the Atlas and Rif. In addition, the marine environment, approaching the 3,500 km of coast, conceals a
diversity of resources whose inventory is only at the beginning. As regards
wild and domestic animal resources, Morocco has an important genetic heri-

Genetic/biological resources are associated with a wide range of traditional knowledge and practices. For example, the local wealth in aromatic
and medicinal species is linked to a diversity characteristic of traditional
knowledge and pharmacopeias, which are able of feeding new ideas in the

cosmetics industry and modern pharmacy. In the Arganeraie local knowledge about the varied uses of all the products of this mythical tree results from the cumulative effect of the experiences transmitted from generation to generation by the local communities which have the right of usufruct.

To this end, the present work constitutes a contribution to a better knowledge of this component of the national environment that is "Biodiversity." It is organized around four main sections: the   part introduces the socioenvironmental context and the challenges of national biodiversity. The second part presents an analysis of the environmental   of Moroccan biodiversity using the DPSIR model (Driving Forces, Pressures, State, Impacts, Responses). The third part is based on an analysis of the responses to these "pressures – states – impacts" in relation to the different objectives of the national policy. It highlights the achievements, the results achieved and their environmental and socioeconomic impacts and sets out the technical, institutional and regulatory constraints that need to be removed. Finally, part 4 proposes strategic axis for the orientation of action within the framework of the SNE so as to be able to take the necessary measures to better capitalize and enhance this heritage in the socioeconomic development of the country (Figures 6.1 and 6.2).

6.2   General Context and Issues

6.2.1   General Context of Biodiversity in Morocco

6.2.1.1   Characteristics and Structure of Biological Diversity

The particular geographical position of Morocco gives it a remarkable range of bioclimatic peculiarities ranging from perhumid and humid stages to arid and desert floors. Geomorphological and bioclimatic diversity corresponds to a wide bio-ecological and, therefore, ecosystemic, specific and genetic diversity, which can be ordered in three main types of ecosystems/distinct environments: (a) terrestrial ecosystems, (b) marine and coastal ecosystems; and (c) ecosystems of inland waters.

It is very   to comment on the number of national ecosystems as the   □A
referred to as "ecosystem" by some is only "habitat" for some or "biocoenose" for others, and so we can speak of the "forest ecosystem" as we can speak about 40 ecosystems within the Moroccan forest. Moreover, there is not a national consensus around this notion of ecosystem, nor what must be contained in this space.

Figure 6.1   Fresco on Moroccan biodiversity (Haut-Commissariat aux Eaux et Forêts et à la Lutte contre la Désertification: HCEFLCD).

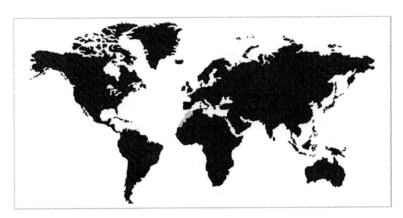

Figure 6.2   Location of Morocco (PNUE, DE, 1998).

Nevertheless, Morocco enjoys a great wealth of natural areas as homogeneous ecological units ranging from snowy high mountain areas to desert, coastal wetlands, Continental wetlands (springs, oueds, temporary lakes), the marine and coastal environment., and many others. In terms

of   □A
and successive adjustments show that the species richness of the terrestrial
domain remains the highest with a large predominance of Invertebrates and
more particularly insects (Figures 6.3 and 6.4). Moroccan terrestrial
is estimated to be close to 7000 taxa, largely dominated by terrestrial species (about 4,500 species according to the national biodiversity study 1998,
5,211 species and subspecies (Fennane and Ibn Tattou, 2012) that are the
best known and the best studied.

The terrestrial species richness is numerically followed by that of the
marine and coastal environments, with more than 7,800 species including
7,145 animal species. The aquatic fauna of the inland waters has more than
2,000 taxa.

Genetic diversity is also vitally important for ecological, socioeconomic
and human well being. An analysis by Birouk et al. (2012) under the aegis
of the Department of the Environment allows to brew the potentials of the
Kingdom in terms of genetically exploitable resources.

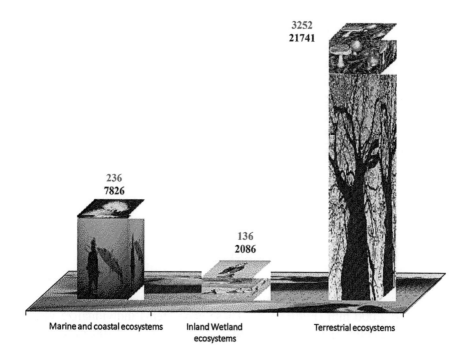

Figure 6.3   Number of species by type of environment: marine, continental wetlands
and terrestrial (Black numbers) and numbers of endemic species (Red
numbers).

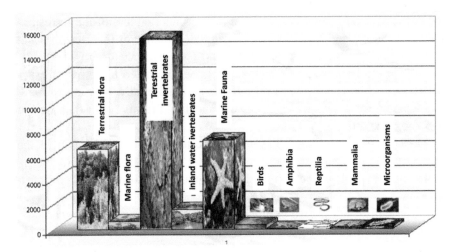

Figure 6.4 Structure of species diversity by taxonomic group and habitat.

The list of these genetic resources, which were   could be consulted in the above-mentioned study and it is   at the present time to draw up a synthetic table to visualize the relative importance of these resources, (i) forest trees and aromatic and medicinal plants; (ii) genetic resources for food and agriculture; (iii) wildlife resources; and (iv) marine genetic resources.

It is a reservoir of species, subspecies, races, varieties and various species to preserve for their supposed potential. Indeed, Morocco has a particular importance in the conservation of biodiversity within the broader Mediterranean area and is considered to be a major biodiversity hub.

In the following, a summary of the characteristics of national biodiversity as presented in the many national reports will be presented.

6.2.1.2   Terrestrial Ecosystems

6.2.1.2.1   Forest and Steppe Ecosystems

These ecosystems occupy in Morocco a very wide range of Mediterranean bioclimatic and their variants in a range of annual precipitations ranging from 50 to 2,000 mm (Figure 6.5). These ecosystems individualize vegetation communities that take turns from sea level up to 2,700 m altitude, following a very original succession. They are made up mainly of natural formations of hardwoods (green oak, cork oak, Tauzin oak, argon tree, etc.), softwood (pine, cedar, etc.) together occupying 5,762,900 Mha, Of the Alfa

Figure 6.5 Cedar forest with snow in Morocco.

steppes (3,318 Mha). If only forest plant communities are considered, there are more than 60 tree species (Mhirit et al., 1999).

6.2.1.2.2 Oasis ecosystem

The Moroccan oases cover almost 107,324 sq. km, or about 15% of the national area (Figure 6.6, Table 6.1). It is a space that houses 1.6 million inhbitants corresponding to nearly 5.3% of the national population.

For many centuries, this ecosystem has been built around the date palm that has provided its fruits, palms and wood to serve man and his livestock, in many uses born of his creativity. It has given birth to a model of tiered agriculture that groups together fruit trees and vegetable crops, depending on the water resources available to satisfy the population and its agricultural speculation. The importance of oases at the national level is given in Table 6.1.

6.2.1.2.3 Agricultural Ecosystems or Agrosystems

Agricultural biodiversity occupies the Useful Agricultural Surface (UAE) of Morocco, which extends over 8.7 million ha divided into different

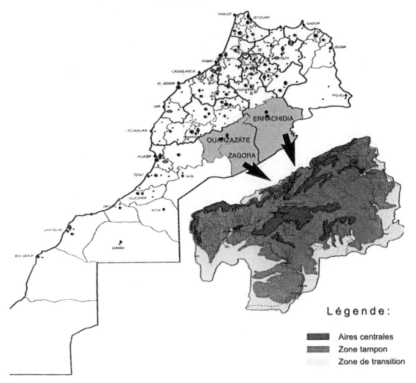

Figure 6.6   Map of situation and zoning of the biosphere reserve of oasis in southern Morocco (Sedra, 1994), Central area, Amortization zone, Transition zone.

Table 6.1   Importance of Oasis in Morocco (Sadra, 1994)

Province	Number of feet	Percentage (%)
Ouarzazate	1,873,000	40.7
Errachidia	1,250,000	28.3
Figuig	125,000	2.8
Tiznit	139,600	3.1
Guelmim	138,000	3.1
Tata	800,000	19.7
Marrakech	100,000	2.3
Total	**4,425,640**	**100**

agro-ecological zones: (i) the rain-fed agriculture zone or favorable zone (> 400 mm of rain); (ii) the intermediate chamber (300 to 400 mm); (iii) unfavorable Bour (200 to 300 mm); (iv) the mountain (400 to 1,000 mm), and (v) the oasis spaces (<200 mm).

This ecosystem comprises two very contrasting subsectors. The traditional, mobilizes the 9/10th of the rural population, characterized by the predominance of the small subsistence farms. The second, modern, is about 1,500,000 ha where agriculture is mechanized, intensive and oriented towards the market and agro-industry. It also includes exploitable grazing land covering an area of 21 million ha out of ten different large ecological complexes.

In these environments, the most important aspect of biodiversity remains local varieties and breeds (local products) as well as traditional know-how. In spite of its invasion by imported varieties and breeds, which are more lucrative in the short-term but less or not adapted, the agroecosystem always shows some resistance to this invasion and is distinguished by numerous plant and animal forms (phytogenetic and zoogenetic resources, particulary in remote areas of high mountains and desert areas).

6.2.1.2.4 Terrestrial Species and Genetic Diversity

On the taxonomic level, the terrestrial environment, with its various components, constitutes by far the richest environment in terms of species with more than 21,700 taxa identified there (PNUE/DE, 1998). Recent research, more or less numerous, in particular, the development on capacity building needs in taxonomy (DE, 2012), has led to the identification of only a limited number of species, About 30, which does not affect the general structure of the composition of the established Moroccan biodiversity.

Vascular   is physiologically the most represented in forest ecosystems where nearly two-thirds of the plant species live in the country; the remaining third being divided between the steppe formations and the wet biotopes, all forming part of the forest domain. Figure 6.3 presents the current state of biodiversity in forest ecosystems.

Fungi and lichens are also relatively well represented with, respectively, 820 and 700 species. The high mountains of the Rif and the Atlas are the most important areas of endemic species. The forest   comprises several major forest species, such as the Cedar of the Atlas (*Cedrus atlantica* (Manetti ex Endl.) Carrière), cork oak (*Quercus suber* L.), oak (*Quercus rotundifolia* Lam.), Fir (*Abies maroccana* Trabut), argan tree (*Argania spinosa* (L.) Skeels), thuja (*Tetraclinis articulata* (Vahl) Masters), *Junipe-*

rus thurifera (*Juniperus thurifera* L.) (*Juniperus oxygenus* L.), carob tree (*Ceratonia siliqua* L.), pistachio of the Atlas (*Pistacia atlantica* Desf.), Lithic pistachio (*Pistacia lentiscus* L.), red pine (*Juniperus phoenicea* L.) The Atlas cypress (*Cupressus atlantica* Gaussen), yew (*Taxus baccata,* L.), acacia (*Acacia raddiana* Savi), pines (*Pinus halepensis* Miller and *Pinus pinaster* Aiton).

These tree species provide several products: timber,   cork, etc. Forest trees also play a very important role as forage species, honey-producing species and Moroccan forest ecosystems according to biocli-matic, vegetation levels and soil (Mhirit et al., 1999, Mhirit and Et-tobi, 2010)

It is within these forest ecosystems that the main aromatic and medicinal plants (PAM = Plantes Aromatiques et Médicinales) are also concentrated, constituting a real reservoir and a potential for lucrative socioeconomic activities for the development of the populations living in these forest or peri-forest regions.

In these forest areas, there are many plants with aromatic or medici-nal virtues among the few 5211 vascular plants in Morocco (Fennane and Ibn Tattou, 2012). Some of the most abundant of these PAM, used for the extraction of essential oils, are Myrtle (*Myrtus communis* L.), Rose-mary (*Rosmarinus officinalis* L.), White Artemisia (*Artemisia herba-alba* Asso), penguin mint (*Mentha pulegium* L.), oregano (*Origanum campac-tum* Benth., and *Origanum elongatum* Emb. & Mayor), and bay leaf (*Lau-ris nobilis* L.), etc. The socioeconomic role of WFP is hardly dismantled; Export earnings generate more than 615 million DH and offer over 500,000 working days with an overall income of 25 million DH according to the HCEFLCD (2013).

The phylogenetic resources that can be exploited for agricultural pur-poses are numerous and according to the latest report of the Department of the Environment (DE, 2012) to more than 405 taxa. These include pastoral resources, ornamental plants, industrial plants and wild relatives of culti-vated forms.

6.2.1.3 Marine and Coastal Ecosystems

6.2.1.3.1 Marine Ecosystems

Marine ecosystems would be more diverse and richer than the entire Medi-terranean, including the Black Sea, with their biotic components (virtually all zoological groups, different types of algae, marine phanerogams, etc.)

and virtually the majority of ecosystems (Sandy, muddy, rocky, with or without metaphytes, coralligens, etc.).

The Moroccan marine domain has physicochemical, hydrological and biogeographical features that make it one of the most   countries on a planetary scale. Figure 6.7 shows the main upwelling areas in the world where Morocco occupies an important place.

6.2.1.3.2 Coastal Ecosystems (Lagoons and Estuaries, Bays, Saline, etc.)

Coastal ecosystems are generally brackish environments (Figure 6.8), more or less closed or softened by continental waters. A large number of species develop, breed, feed, or simply rest.

The main estuaries of Morocco are those of the Oued Moulouya, on the Mediterranean coast and the Wadis Sebou and Oum-Er-Rbiâ, on the Atlantic coast. The largest lagoon systems in the Mediterranean are the Nador lagoon (115 km²) extended by the Arekmane salt pans and the Restinga-Smir lagoon, while on the Atlantic coast, the most important ones are the lagoon of Moulay Bousselham, Merida de Sidi Boughaba, the lagoon complex of Oualidia-Sidi Moussa, the   A]
The majority of these ecosystems are highly affected by anthropogenic activities.

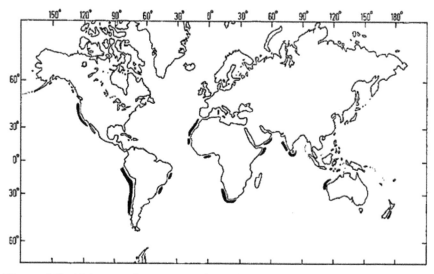

Figure 6.7 Major upwelling areas in the world (www.fao.org).

Figure 6.8   Khnifiss lagoon, example of coastal wetland in southern Morocco

6.2.1.4   Marine Species and Genetic Diversity

The species diversity in the marine environment is one of the richest in the Mediterranean region. More than 7820 species have been identified, enriched until 2013 by a dozen new species, which in no way changes the structure initially described of this specific marine biodiversity. It is a biodiversity that is largely dominated by arthropods, especially crustaceans, and then vertebrates with fish and then mollusks (Figure 6.9).

The Department of the Environment's recent study on capacity-building needs in the area of genetic resources shows that in the marine environment, and in contrast to the terrestrial ecosystem, Genetically, chemically or biochemically valuable species that are exploited elsewhere and that it is possible to capitalize on local or national development actions.

6.2.1.5   Inland Water Ecosystems

According to the National Wetlands Report (RNZH = Rapport National sur les Zones Humides), inland water ecosystems which occupies an area of about 200,000 ha, including lagoons and estuaries (HCEFLCD, 2015). The main continental aquatic environments are discussed in the following subsections.

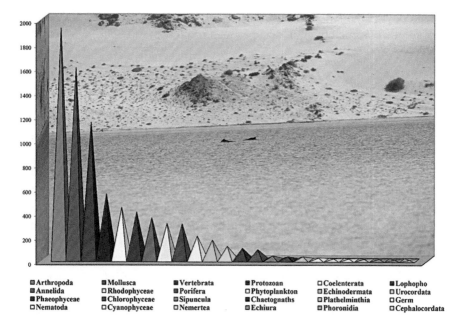

▣ Arthropoda	▪ Mollusca	▪ Vertebrata	▪ Protozoan	▢ Coelenterata	▪ Lophopho
▪ Annelida	▢ Rhodophyceae	▪ Porifera	▢ Phytoplankton	▣ Echinodermata	▢ Urocordata
▪ Phaeophyceae	▪ Chlorophyceae	▣ Sipuncula	▪ Chaetognaths	▢ Plathelminthia	▢ Germ
▢ Nematoda	▢ Cyanophyceae	▢ Nemertea	▣ Echiura	▢ Phoronidia	▢ Cephalocordata

Figure 6.9 Taxonomic structure of Moroccan marine biodiversity.

6.2.1.5.1 Lakes/Dams

These are natural stagnant (lakes) or artificial (dams) water ecosystems. The lakes are mainly concentrated in the Middle Atlas, the largest of which is Lake Aguelmame Sidi Ali, reaching 300 ha in area and 40 m in depth (92 m for Lake Isly). Natural lakes can combine artificial habitats (dam reservoirs with over 100 large dams) as well as many very specific environments in the south and southeast of the country, which are relatively desert oases.

6.2.1.5.2 Rivers, Springs and Caves

The most important streams, springs, and caves are located in the mountain ranges. The Middle Atlas gives birth to the three main rivers of the country (Oueds Moulouya, Oum-er-Rbiâ and Sebou) and the High Atlas with oueds Dades, Ghériss, Guir, Souss, Tensift and Ziz, and partly Draâ. The High Atlas also feeds the great tributaries of the Oum-er-Rbiâ (Oueds Abid, Lakhdar and 9 explains the processions of endemic species, each of them dependent on them.

The Caves constitute a particular wet ecotype. Several dozen caves exist in Morocco, many of which are of prehistoric interest, in addition to their bioecological interest. These environments, characterized by the stability of their abiotic parameters, include a particular fauna, mainly based on inverte-

☐   ☐   ☐   ☐ ☐   ☐   ☐

6.2.1.5.3   Wetland species and genetic diversity

Much work has been done on inland waters; However, given the size of the country, the dispersal of these wetlands throughout the Kingdom, and the difficulty, for example, in estuaries, of differentiating between "saltwater" and "salt water," especially in brackish environments (PNUE/DE, 1998, Dakki et al., 2015). What is "marine water," it is impossible to decide on a definitive list of the freshwater species listed so far. It is believed, however, that at least 3,290 species including 1,477 plants (algae and phanerogams) and more than 1,812 animal species have been recorded in the inland waters of the country (Figure 6.10).

Very little data are available on the genetic values of inland water biodiversity; perhaps, the ichthyological characteristics of this heritage. It is a group largely dominated, qualitatively and quantitatively, by the barbs, which seems to testify to the predominance of warm eutrophic waters. The diversity of phenotypes does not seem to correspond, at least to the current

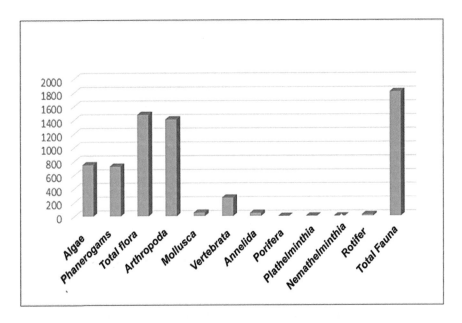

Figure 6.10   Species richness of continental and brackish waters (PNUE/DE, 1998).

state of knowledge, to genotypic diversity, despite the studies carried out in this ⬜A

for example, are different from those of Lake Isly, which were erected as different from those known in the Atlas. This diversity ⬜A

⬜ ⬜ ⬜

6.2.2 Morocco's Commitment to Biodiversity

In Morocco, biodiversity is a vital sector for its socioeconomic development. It has never spared its efforts to engage in bilateral or international activities or agreements to contribute to the conservation of his natural, but also global, natural resources. For example, today, as in a few countries, it is planning to implement one of the most recently signed protocols by the international community, the Nagoya Protocol on Genetic Resources.

It signed in 1992 and A ⬜

Protocol (CBD) setting up all the structures for its implementation (creation of a State Department, a national committee and a biodiversity cell).

Morocco has honored all its procedural commitments concerning the CBD, from the organization to the reports, to the establishment of a national strategy and action plan for the conservation and sustainable use of biodiversity, Sustainable use of its biological resources.

Apart from the CBD, Morocco is engaged in many other activities and agreements for the conservation of its natural heritage, including CITES, Ramsar, ACCOBAMS, Barcelona, IUCN, as well as a host of Reforestation Plans and Programs, Aquaculture, soil conservation, phylogenetic research, etc.

6.2.3 Major Issues

The scientific community is unanimous in recognizing that the services rendered by biodiversity called "natural services," ecosystem services, or "environmental goods and services" are numerous in the neighborhood of 40. Indeed, biodiversity offers many fundamental benefits to humans, which go beyond the simple supply of raw materials for housing, clothing, and so on (Figure 6.7).

Biodiversity supports many processes and services of natural ecosystems, such as air quality, climate regulation, water   pest and disease control, pollination and erosion prevention. The well-being – and survival – of humans is hardly conceivable without a   biodiversity. Food systems are highly dependent on biodiversity and a considerable proportion of drugs are directly or indirectly biologically derived.

Whole the country's economies depend on biodiversity. Therefore, loss of biodiversity has adverse effects on several aspects of human well-being, such as food security, vulnerability to natural disasters, energy security and access to clean water and materials   A □
tions and freedom of choice.

National biodiversity is of particular ecological importance, with more than 24,000 animal species and 7,000 plant species and an overall endemism rate of 11% for wildlife and more than 20% for vascular plants, equal to the entire Mediterranean basin. The diversity of ecosystems is also remarkable; In addition to coastal and marine, Mediterranean and Atlantic ecosystems, some 40 continental areas have been   □A
versity, nearly three-quarters of which are represented by strict forest eco-systems and preforest and steppe ecosystems.

Morocco's biological wealth is also of vital socioeconomic interest to the country. Exploited biological resources contribute to a   proportion of national wealth in different sectors of the economy, such as agriculture (8,456,000 ha of arable land, over one third of the active population, 20% of the country's total exports, 25% of GDP), livestock (one third of the PIBA, 40% of rural employment), forestry (9 MHa 969,600 ha, 114,000 permanent jobs, 1.5 MUF/year, Wood, cork, hunting) and   (total production of around 750,000 tons per year, exports worth nearly 5 billion DH, almost 80,000 jobs, according to ENB (PNUE/DE, 1998).

Terrestrial and aquatic ecosystems contain many other occult potentials whose consumption or direct use value and option values are not always eas-ily estimated: Vegetable potentialities (medicinal plants, lichens, aromatic plants, carob, acorns, Edible mushrooms, ectomycorrhizal mushrooms, truf-
      wild onions, etc.), animal products (apiculture, game, snails, turtles, birds, continental   □A
scape) and genetic diversity (endemic species, local varieties and breeds, spontaneous species related to crops). These components play a
role, providing income that is sometimes   to estimate in economic terms, but whose social role is very important.

Morocco, as any other country, draws most of the elements necessary for its development in its natural resources, the main ones, apart from phos-phates, being agriculture, forests, livestock trails,   resources, and diversity of its landscapes (tourism). All these biological resources are based on the notion of race, variety, species, and ecological environment.

However, serious threats, derived mainly from the multiple activities of man, weigh on biodiversity in Morocco, as in many other countries of the planet. Ecosystems are more or less affected by direct or indirect human activities linked to economic development and population growth (intensive agriculture, overgrazing, deforestation, excessive ⬚A lution, urbanization, etc.).

In extreme cases, the negative impact of these activities leads to the irremediable disappearance of animal or plant species and irreversible degradation of certain ecosystems, as in the Central Rif, where the cork oak ecosystem has been virtually annihilated. This impact is particularly important in mountain regions, where the largest areas of biodiversity are concentrated, most of which are ⬚A and Maghnouj, 1997).

The third "REEM" (MDCE, 2015) report highlighted the different categories of threats to the components of national biodiversity. Threatened species are 1617 for plants and 610 animal species including 86 marine species, 98 bird species and 18 mammals. The forecasts point to the risk of extinction of nearly 22% of national biodiversity by 2050.

Morocco has a long tradition in the ⬚ of nature protection, which may not always have been applied in the same terms and according to the current perception of biodiversity. The existence of age-old community practices, the antiquity of certain legislative texts, dating back to the year 1917 on the conservation and exploitation of forests, and the establishment of national parks since the 1940s.

Similarly, the signature of the CBD and its ⬚A ⬚ August 1995 were followed by important initiatives. In this sense, awareness and political will have resulted in the construction of sustainable development processes, in particular:

1. Strategic political-economic choices in the areas of economic liberalization, regionalization and decentralization, education and training, and poverty alleviation;
2. National strategies and action programs:

   • In the socioeconomic ⬚A
     Drinking Water Supply Program (PAGER); Comprehensive Rural (PERG); Social Priorities Program for Basic Education (BAJI); National Strategy for the Protection of the Environment and Sustainable Development (DE, 2006); Rural Development Strategy 2020 (DAT, 1999);
   • In the area of natural resource management, biodiversity management and ⬚ control: National Irrigation Program

(NIP); National Watershed Management Plan (PNABV, 1995, 2002); Reforestation Master Plan (RDP, 1997); Master Plan for Protected Areas (PDAP, 1996); Management Plan for Land Conservation; Rangeland Development Strategy (RPS); National Forestry Program (NFP, 1999), National Program of Action to Combat De-
(NAPCP, 2001, 2011), National Biomass Energy Plan (GNPP 1998), and Master Plan for Prevention and Control Forest Fires (PDPCI, 2001), Urban and Peri-Urban Forest Strategy.

3. Establishment of framework, institutional and regulatory instruments: National Charter for Education and Training: Water Code, Labor Code, Family Code, Commercial Code, Agricultural Investment Code, Charter National Land Use Planning, etc., National Charter of the Environment (SEDD, 2011), which are all levers and factors of success for the policy of sustainable development.

As a follow-up to these political and economic choices and the corresponding national strategies and action programs, Morocco has implemented an ambitious plan, in coherence with international processes (UNCED, ADM, WCO, WSSD, etc.). National Action Plan for the Protection of the Environment (PANE, 2002), which comprises seven federative programs, and promulgated three new environmental protection laws in 2003.

In the same vein, in 2005, it developed a multidisciplinary prospective study on the 50th anniversary of the country's independence, "50 Years of Human Development in Morocco and Perspectives for 2025" (IRES, 2006; www.rdh50.ma.) To support the country's strategic framework and planning tools for development, including an important component: "natural setting, environment and territory," focusing on the management and development of natural resources and areas.

Concerning biodiversity, a national strategy and three thematic strategies (terrestrial biodiversity, marine and coastal biodiversity, and wetland biodiversity) have been developed with the aim of conserving and using sustainably the different components of the national natural heritage. These strategies are translated as part of a National Biodiversity Plan and supported thematic action plans for their implementation through a series of indicators on: forest biodiversity; Agricultural biodiversity; Arid zone biodiversity; Marine biodiversity; Biodiversity of wetlands. Other   A☐ tors are developed for threatened and vulnerable species.

Despite the importance given by the public authorities to this problem, as demonstrated by the initiatives described above, the strategies and corresponding action plans such as the NEAP have not been followed (PDAP,

PDR, PNABV, NFP, PDT, etc.) and a few   A☐
of forestry). The inclusion of certain sites of ecological interest in international lists, albeit in very different settings from those suggested by the CBD.

Generally speaking, while efforts to protect natural resources have been undertaken since the beginning of the twentieth century, they have not completely stopped the growing pressure on biological resources. It goes without saying that the main issues of biodiversity remain in relation to:

1. preserving biodiversity;
2. the valorization and sustainable use of its constituent elements;
3. the fair and equitable sharing of benefits arising from the operation of Genetic resources;
4. the Improved knowledge on the functioning of terrestrial and marine ecosystems and their diversity.

6.3 Relationship Between Biological Diversity and Human Well-Being in the Country, Between Other, Through Ecosystem Services

6.3.1 Well-Being Through Ecosystem Services

According to Messouli (2008), the services offered by the National Biodiversity are listed in the following subsections.

6.3.1.1 Procurement

In rural areas, procurement services are of paramount importance in supporting the livelihoods of the poor, and therefore restrictions on the provision of such services could lead to increased vulnerability and worsening poverty. The opposite situation was also observed, that is, poor households use supply services for income generation to cope with poverty. There are case studies attesting that the poorest sectors of society make the greatest use of supply services, both for domestic consumption and for income generation (Table 6.2).

The degradation of the local environment undermines the essential security assurance function by leaving poor households extremely vulnerable to natural disasters. The dependence of poor rural populations on ecosystem services is rarely measured and therefore generally neglected in statistics and poverty assessment, resulting in inappropriate strategies that do not take into account the role of the environment in poverty reduction.

6.3.1.2 Regulation

Given that regulatory services are not directly marketable consumables to generate income, their role in supporting livelihoods and buffering against poverty is less easily demonstrated than for supply services. Their roles in local livelihoods are less direct, but the case studies and evidence show that they are just as important. They are rarely taken into account in biodiversity conservation policies and action plans.

Because poor people often live in marginalized areas, both in rural and urban areas, they are most vulnerable in situations where regulatory services are undermined, such as   drought, poor Areas where the incidence of disease is higher, and degraded or depleted soils.

The trends are given as follows:

- advancement of sand dunes;
- change in precipitation;
- ☐
- changes in wetlands;
- increase in temperature;
- increased coastal erosion; and
- increased spread of diseases and parasites among wildlife.

6.3.1.3 Cultural

It is particularly difficult to measure the impact of the loss of cultural services, but this impact is of significant importance to many people. Human culture, knowledge systems, religions, and social interactions have been strongly influenced by ecosystems. A number of global assessments conducted in the MS at intermediate scales have shown that the spiritual and cultural values of ecosystems are as important as other services for many local communities, both in developing countries (e.g., the importance of sacred forest groves in India) and industrialized countries (e.g., the importance of parks in urban areas).

There is excellent qualitative information that documents local community rituals and respect for the environment, and their concerns and fears of consequences when these services are diminished or lost. The moussem of the betrothal of Imilchil, the festival of the roses in KelaatM'gouna, the moussem of cherries Sefrou, the Hamat, etc., are examples that develop cultural services that stimulate the tourism potential of the regions, attracting an increasing number of national and international tourists. Tourism is often a contributor to national GDP.

The trends are given as follows:

- changes in the availability and quality of traditional or country foods can affect cultural traditions (Imilchil, pink moussem, cherry, dates, etc.);
- increased frequency of algal blooms in many lakes, affecting recreational activities;
- increased number of stewardship initiatives on private lands;
- increased number of terrestrial protected areas;
- little progress in marine protected areas;
- sublimation (mountain skiing); and
- the decline of the bird population.

6.3.2 Well-Being Through Ecosystems, Species, and Terrestrial Genetic Resources

The earth's natural heritage is nearly 9.2 million hectares (ha) of arable land, of which only 13% is irrigated, 5.8 million ha of forests, 3 million ha of alfalfa and 21 million ha of exploitable land. The agricultural sector mobilizes 9/10th of the rural population (Figure 6.11).

Despite the   of the Moroccan economy, agriculture is still a vital sector, contributing between 12% and 24%, according to agricultural campaigns, to national GDP, providing 80% of rural employment and more than 40% of employment at the national level. In 2006, it accounted for nearly 13% of national GDP. The agricultural sector also plays an important role in national foreign trade. During the 1990s, agricultural imports accounted for an average of 19% of the value of aggregate imports, while the share of agricultural exports in total exports was 18%. This vital role of terrestrial, agricultural and forest ecosystems is sometimes deeply affected by the many above-mentioned threats, which have more or less direct consequences on the quality of life and the well-being of populations. Among these consequences are:

- *Reduced availability of natural resources.* It seems clear that the reduction of forest areas, pastoral areas, and soil fertility can only have negative consequences on the availability of natural resources and ser-

Table 6.2　Ecosystem Services Provided by National Biodiversity (Messouli, Unpublished)

Services	Evolution
Provisioning Services	
Culture	+
Breeding	+
Fruits	−
Fish farming / aquaculture	+
Wild foods	−
Construction wood	+/−
Cotton, hemp, silk	+/−
Firewood	−
Genetic resources	−
Biochemicals, medicines natural and pharmaceutical products	+
Water	−
Regulatory Services	
Regulation of air quality	−
Global climate regulation	−
Regional and local climate regulation	−
Regulation of the water cycle	+/−
Regulation of erosion	−
Purifcation of water and treatment of waste	−
Regulation of diseases	+/−
Regulation of parasites	−
Pollination	−
Regulation of natural risks	−
Cultural Services	
Spiritual and religious values	−
Aesthetic values	−
Recreation and ecotourism	+/−

vices provided by these ecosystems (wood, forest by-products, Cereals, vegetables, fodder units, livestock, etc.). It is also evident that a shortage of these products can only be translated into the　　　by less income and revenue for the population, fewer days and jobs, more unemployment and so on.

These are consequences that can only worsen with the adverse climatic conditions that the country has been experiencing for decades.

- *Extension of poverty.* Poverty is a cause of degradation of natural resources; but poverty is also a consequence of all these threats, both "natural" and anthropogenic. This is more important because the Moroccan population is predominantly rural, and it is precisely in this rural environment that the forestry and agricultural resources are concentrated, and therefore the natural resources that support the needs of these populations.

Indeed, poverty remains a predominantly rural phenomenon, as 72% of the poor live in rural areas, where this phenomenon has in-
in recent decades. Thus, apparent national productivity per agricultural worker per year has steadily declined, falling from DH 10,652 per worker per year in the period 1993–1997 to DH 6740 per worker per year in the following period, However, while the overall poverty rate declined from 25.5% in 1985 to 17.7% in 2001 at the national level, rural areas still account for nearly three-quarters of the poor in Morocco. According to the RDH50 report "Poverty and social exclusion factors," all studies show that poverty in Morocco is a predominantly rural phenomenon, for various reasons. Since independence, rural areas have   little from public investment. However, the beginning of an urbanization of the phenomenon has begun to manifest itself for at least a decade. It will undoubtedly have social implications to be taken into account when drawing up any strategy to combat poverty.

If the rural world is generally poorer, poverty is not evenly distributed, making the averages displayed at times unrepresentative. Some regions have rates up to three times higher than others. The Northwest and Central regions have the lowest rates. Moreover, poverty does not have the same meaning or effect in urban and rural areas. The same analysis, conducted according to two dimensions (residence and economic regions), shows, more critically, the disparity of the phenomenon on the national territory. Indeed, for some surveys, the poverty rate rises from less than 3% in urban areas in certain regions (e.g., Chaouia-Ouardigha) to almost 40% in rural areas of other regions such as Meknes-T

- *A rural decline in marginal areas.* Knowing that the State, for want of resources, could not maintain   A
rural people in   A

lems, it is possible to think that serious doubts weigh on the future of the ecosystems Terrestrial forests and also agricultural whose cereal production alone offers nearly 80 million t/year and is practiced in nearly two thirds of the UAA. These threats and the reduced possibilities of converting the areas sown to cereals will encourage exodus if alternative activities are not developed. They would become urgent from 2010 onwards and be enforceable by 2020. At this time, it is conceivable that the agricultural sector is deployed in less than half of the current number of farms. The whole problem becomes to mobilize employment niches capable of absorbing the   □A of rural people that the transformation of the agricultural landscape will release. In the end, everything seems to indicate that agricultural detriment and its corollary, rural-urban migration, is inescapable and is part of a societal dynamic. This decline is determined by the low profitability of agricultural labor, which in turn refers to the narrowness of production support and the weakness of agricultural production in marginal areas with low ecological potential. These changes will have

Figure 6.11 Women's cooperative working on the fruit of the Argan tree for the extraction of Argan oil.

the foreseeable consequence of placing nearly 1 million rural households outside the market economy.
- **Increased food dependence.** It is a consequence essentially linked to
  the national agrosystem and, through the pastoral system, it also affects the forest, Alfatier and Saharan ecosystems. Morocco has made
  food   □A

  after independence. This objective has only partially been achieved
  due to the continuing increase in domestic demand and the limited
  progress in productivity growth. Thus, for animal products, the trends
  were positive with coverage ranging from 87% for milk to 100% for
  red and white meats, respectively. On the other hand, for oil, sugar and
  cereals, coverage rates have deteriorated considerably.

6.3.3   Well-Being Through Ecosystems, Species and Marine Genetic Resources

It is certain that marine biodiversity does not have the socioeconomic weight
of the agricultural system, for example, especially since the Moroccan citizen
is not a "major consumer" of seafood, except perhaps for sardines Sought
after by many destitute families. However, the number of jobs generated by
the maritime sector may very well be affected by a worsening of threats and
negative trends in the evolution of this heritage.

The national marine ecosystem provides much of the animal protein; it
ensures direct employment and more or less stable incomes for a large percentage of the national labor force (seamen, collectors, civil servants, inves-
tors, etc.). The sea also supplies a large share of raw material for certain
industries (fertilizers,   □And   A□
livestock feed, etc.); unfortunately, it also serves as an outlet for more than 1
billion cubic meters of untreated sewage).

The national maritime space, which is larger than the land part, also
plays a strategic role in economic and social terms. Its Atlantic façade plays
the role of a structuring pole of the national economy, given its demographic and economic weight and its function in the organization of the
national space (61% of the urban population of large cities, 80% of the permanent workforce of industries, 78% of the total industrial investment of
the country, 67% of value added, 53% of tourist capacity, 92% of maritime
□

It concentrates the main urban centers in the country (Casablanca, Rabat,
Kénitra, Agadir,   Tangier, Tan Tan, Laâyoune, Dakhla, etc.), the highest
urban and rural demographic densities, infrastructure and communication

networks more dense, as well as the main economic activities. However, the strong coastal situation in Morocco over the last few decades has led to a major dysfunction and deep degradation of the marine environment. In this large national maritime area, the exploitable biological potential has been estimated at 500,000 tons for demersal species and 1,500,000 tons for pelagic species.

From these few socioeconomic data, it seems important to infer that the degradation of marine resources would certainly have an impact on the well-being of the people who live there. It is very   to give precise on the extent of this impact but some signs are very revealing and concern for example:

- sardine processing plants, which are no longer "in full swing";
- redundancies in these factories;
- the constantly rising price of   ⌐A
  consumed sardine in the country;
- ☐   ☐

6.3.4 *Well-Being Through Ecosystems, Species and Genetic Resources of Continental Waters*

From the analysis of the graph (Figure 6.12), it is clear that the volume of water per inhabitant and per year will decrease over time, which is not conceivable without negative impacts of course on populations and their various activities (Agoumi and Debbagh, 2005).

Water is   and foremost a vital part of life, including   ☐ - life, human populations and their livestock. Water is also a structuring element of socioeconomic development: industry, agriculture, urbanization, tourism, etc. A decrease in precipitation and, therefore, water resources will always be felt to be   A ☐ shortage requires large-scale programs that have mobilized over the years increasingly   budgetary envelopes. For example, the proposed drought response program for the 2001 crop year cost as much as MAD 6.5 billion, of which two-thirds was devoted to rural employment creation (Figure 6.12).

It is clear that the threats to continental wetlands are likely to affect the availability of water resources, which by 2020 would be reduced by an average of around 15%. Meeting Morocco's water needs at this time, estimated at 16.2 billion cubic meters, would require heavy investments to mobilize available water resources.

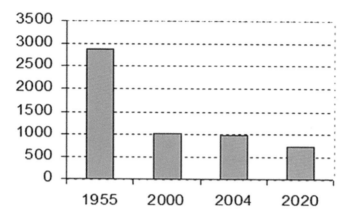

Figure 6.12 Evolution of water capital in Morocco (Agoumi A et Debbagh A., RDH50, 2006).

6.4 Biodiversity and Genetic Resources Valorization Strategy

A strategy is based on several key concepts, the most important of which are the Vision, the objective, which can be defined as priority or specific objectives, strategic axis or orientations and activities that are detailed in a strategic direction (SEDD, 2017).

6.4.1 Strategy Vision

The national vision for the enhancement of genetic resources seeks to identify all the ingredients necessary for the sustainable utilization and valorization of genetic resources and related traditional knowledge.

From the evaluation of its genetic resources and its AOCs, to the satisfaction of national needs in natural molecules, and to conservation programs (in situ and ex situ), programs of innovative and targeted   A oriented towards Biotechnology and genomics, Morocco aims to capitalize all of these ingredients to make this natural heritage contribute to the socio-economic development of Morocco and the well-being of its population.

6.4.2 Objective

This objective aims to ensure that Morocco retains its valuable raw materials, transforms them, valorizes them before exporting them, if need be, but with a surplus value of at least the third or fourth degree. It wants to make Morocco not a producer of raw materials but a supplier of valued products and natural molecules and derivatives.

6.4.3 *Strategic Axis for the Valuation*

Among the interests and potentialities of the NP in the GRs and CTAs, which are at the same time potential strategic orientations of valorization, and which are hidden between the lines of the objective of the Protocol and its scope. It should be mentioned:

- The interest and the economic potential, which is evident even in the creation of wealth (Strategic axis 1: Economic valuation of the GRs).
- ⌐interest and potential of CTA RGs in the discovery of other knowledge and other peculiarities useful for mankind and its future
    ☐   ☐   ☐
- Environmental interest and its potential to help in the   against environmental degradation and biodiversity loss (Strategic axis 3: environmental valuation of GR).
- Social and societal interest and its potential in job creation, the against poverty and the promotion of the gender dimension (strategic axis 4: social valuation of GR).
- The interest of this capital and this national intangible heritage which is the Associated Traditional Knowledge and their hidden potential that should be   ⌐A
    ation of the Traditional Knowledge associated with GR).

6.4.3.1 Strategic Axis 1: Scientific Valuation of Genetic Resources

Plant and animal genetic resources as well as the traditional knowledge associated with these resources constitute a natural heritage, whether or not shaped by man, and which remains the basic material for the development of new pharmaceutics, medical, cosmetic, As well as many environmental varieties of interest in the fight against desertification, drought, diseases and pests. Better value for biodiversity implies a better understanding of the nature, extent, structuring and evolutionary dynamics of the diversity of genetic resources.

This knowledge is brought about by the conjugation between research and apprehension of traditional knowledge. The contributions of

taxonomy, in situ and ex situ collections, genetic improvement, genomics, etc. are essential for a   characterization of genes and alleles of agronomic, medicinal, cosmetic and other interest which is a prerequisite for a more effective selection of favorable combinations and potential uses. They are also used for the   A ☐   -
tal and adaptive forces, whether natural or domestic, which affect the genetic diversity of cultivated and high species,   A ☐
regions or even genes controlling the variability of traits of medicinal, cosmetic, agricultural interest. But   A ☐   -
nation to farmers, foresters and various users and decision-makers require in-depth knowledge of the diversity of genetic resources in situ and ex situ,
☐   ☐ ☐   ditional uses.☐

It is in these terms that the Nagoya Protocol emphasizes, in its Article 23, is the need to "foster the development and strengthening of a sound and viable technological and   A
of the Convention and this Protocol" (Art. 23), including "bioprospecting, associated research, and taxonomic studies" (Art. 23).

It is therefore clear that, according to the Nagoya Protocol (and according to the logic itself, it is essential for a developing country like Morocco rich in biodiversity to extract as much as possible all the   A ☐
tion contained in the resources (Cryopreservation, sequencing, ex situ and in situ conservation, etc.) so that these resources can be used in the framework of a sustainable development of the Moroccan population.

Many species, varieties or races, apparently "not very interesting," are thus studied and scrutinized in their smallest details, characterized, hierarchized and their potentials highlighted so that potential users/investors tempted by the exploitation of these potentials can be used in different development sectors.

One example among many others on the   A
resources is that of "laboratories on chips" corresponding to miniaturized laboratories, which are biosensors or sensors based on algae or bacteria   and which respond to toxins present in seawater by emitting light. It is evident in this example that the primary and immediate interest is neither economic, social nor environmental but its application could draw   A
of various types of pollutants by these same biosensors.

Traditional knowledge associated with these resources is full of valuable, ancestral and sometimes unique information that should be "passed on to the laboratory" to verify its truthfulness and timeliness for any use   to the populations holding this knowledge, but also to the whole country.

6.4.3.2   Strategic Axis 2: Economic Valuation of Genetic Resources

Between the advent of the Mother Convention on Biological Diversity and that of the Nagoya Protocol, the economic stakes linked to advances in biotechnology have gained ground and are now concentrating interest not in the first two objectives of CBD (conservation and sustainable use), on large units such as landscapes, species and relationships between living things and their environment, but on the infinitesimal smallness of genetic diversity and its exploitation. The latter is indeed a source of food, cosmetics, medicines, information for genetic engineering, generating colossal sums through biotechnologies.

We have thus passed from a beautiful emotional and ethical image that inspires us the landscapes, the diversity of species and their ecological services, to a restrictive commercial and commercial vision limited to the invisible (DNA, proteins, derivatives.) almost virtual which becomes an ingredient in the manufacturing processes of various products.

It was a process that began in the early 1990s and was, among other things, at the origin of the CBD, with the progress of molecular engineering, when the living became an economic and commercial stake. Genetic resources, which are more abundant in the countries of the South, are then perceived by northern biotech countries as deposits of biomolecules that are normally subject to the laws of supply and demand and which generate exorbitant   for Private companies working in the   [A maceuticals and cosmetics (Table 6.3, Laird Sa et al.). This is followed by a
   [A
bilateral or multilateral free trade agreements, and so on.

Thousands of samples crossed the borders of the South, including Morocco, to the North, sometimes with derisory prices, sometimes in the framework of   cooperation," rather purely and simply a biopiracy aggravated by patent   on genetic resources from other countries and the traditional knowledge of local populations without being consulted or remunerated. Then came the Nagoya Protocol to put more visibility into these transactions and allow suppliers and holders of traditional knowledge to better value their own wealth, a term which, paradoxically, does not appear in the corpus of its text, nor even in the CBD or the Bonn guidelines.

Many vicariates of Moroccan species are exploited on a global scale and contribute to the creation of wealth and employment in the countries where they are exploited. Other species, varieties or breeds have shown, through evaluation studies, that they have a potential for very   economic valorization, sometimes unique.

Table 6.3 shows that the market involving genetic resources is immense and lucrative. Of course, care must be taken to ensure the sustainability of resources in order to conserve the national heritage, which can also guarantee the supply of raw materials to different industries.

6.4.3.3   Strategic Axis 3: Environmental Valuation of Genetic Resources

All species, without exception, have a primordial role in the ecological balance. Each species, whether microbial, plant or animal, has its own chemical, biochemical, genetic and other peculiarities. To communicate with the other components of the same ecosystem and to become one of the other species in the same space. It is in its interrelations with other species in its environment that each taxon contributes in its own way to this ecological balance, which is besides of existential importance for man.

The understanding and control of these characteristics by the latter are thus   ☐A
and in contributing to these species as genetic resources and as indispensable elements in their respective environments, To restore that balance, vital to man, but very intensely affected today. It is, therefore, a question of using the particularities of living species which are potential genetic resources for environmental actions or, again, to exploit these genetic resources in the protection of the environment and biodiversity.

Table 6.3   Examples of Activities and Industries Based in Part on Genetic Resources and the Benefits Generated

Industry	Global Markets (US$)
Pharmaceutical	$955.5 billion (2011)
Cosmetics	$426 billion (2012) – natural component $26.3 billion
Food and Beverage	$11.6 trillion (2009) – functional beverages $23.4 billion
Seed	$45 billion (2011)
Crop protection	$40 billion (2010)
Industrial Biotech	$65-78 billion (Including biofuels, 2010) - industrial enzymes $3.3 billion
Botanicals	$84 billion (2010)

Source: Laird et al. (2013), series of SCBD fact sheets and policy briefs.

In recent decades, environmental degradation and the erosion of biodiversity and ecosystem services are increasingly emerging as major global concerns that have a major impact on human well-being on this planet. All means would be appropriate to reduce the adverse effects of different human activities that are not environmentally friendly on the latter and ecosystem services and on the services provided by biodiversity and genetic resources for humanity.

There are many examples of the roles that certain species and genetic resources can play in maintaining or even restoring the natural ecological balance. Here are some examples of how some species, after research, are used to combat environmental degradation and ecosystem services provided to human populations.

In agriculture, for example, despite the economic   A □
and the development of highly effective synthetic molecules, the history of pesticide use would have demonstrated that strict management based on the use of chemical insecticides will remain always fragile as resistance to pests, environmental pollution and the presence of residues of these molecules on food and groundwater remain, not to mention the very ephemeral commercial lifetime of these insecticides. The destruction of these obsolete products also requires sophisticated and costly approaches.

Forest entomology is certainly quite different from agricultural entomology, but in terms of the environment, there is unanimous agreement that the extension of chemical control methods in forest environments is just as harmful. Especially, in intensive monoculturally intensive silviculture as in intensive agriculture, by genetic poverty and by the impoverishment of the environment in biodiversity.

These chemicals also have a negative effect on plants and microorganisms or other regulatory factors, naturally capable of controlling possible pests such as in primitive forests. In these fragile forests, many successful attempts and experiments have resorted to biological control of forest pests. The *Hylobius abietis* of the Atlas cedar or the pine processionary wreak havoc in the Moroccan forests, whereas these pests are effectively fought in other countries thanks to an integrated biological control based on natural enemies of the ant wood and machine.

                    based on the use of macrophytes or
on microorganisms are nowadays well known in various regions of the world to combat the pollution of inland waters. These valuations are all the more interesting in countries or regions suffering from rainfall damage and therefore drinking water, considered as a vital asset and one of the main

vectors of sustainable development. *Scirpus, Eleocharis*, Joncs, *Phragmites,* and *Typha* are some of the plants found in Moroccan wetlands and have

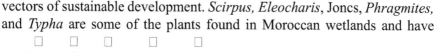

Among the microorganisms used in environmental protection, scrubbers, *Acinetobacter* sp., *Desulfovibrio* sp., *Nitrobacter* sp. *Nitrosomonas* sp. or *Methanococcus* sp. which are some of the most cited in the treatment of various types of pollution and whose use depends on the nature of the pollution, the polluting load and the quality of the treatment desired.

In the marine environment, one of the many examples of RG valorization is the successful application of polymeric substances produced by the marine bacterium *Pseudoalteromonas* sp. to   dirt. Hydrocarbons are also treated in oil tanker accidents by bacteria, fungi and marine yeasts, which, as evidenced by their contribution, contribute   to the conversion of hydrocarbons into oxidized products with an oxidizing power up to 1 gram of hydrocarbon per square meter per day, which is enormous.

In 2017, a team of researchers (PlastiCure) was able to combine the potential of two bacteria harvested in the marine environment, *Escherichia coli* and *Pseudomonas putida*, for the degradation of plastic (PET) in seawater and, in addition, the released energy by this degradation is used for the production of electricity.

6.4.3.4   Strategic Axis 4: Social Valorization of Genetic Resources

In its introduction, the Nagoya Protocol, it is clearly intended to contribute to "enhancing the contribution of biological diversity to sustainable development and human well-being" and, to "the extent that this protocol is resource specific genetic resources and has only genetic resources in its field of application, this contribution can only be made through genetic resources."

The diversity of landscapes due to the diversity of ecosystems and their services, which are themselves dependent on the diversity of species and, above all, on the genetic diversity within these species, directly conditions the major development activities Socio-economic and employment in sectors as diverse as agriculture,   A
metics, horticulture, agri-food, biotechnology, and tourism.

The exploitation of ABS genetic resources is at the heart of sustainable development issues. The protocol's   A
cited in the scope of this protocol is as follows: "Research and development activities on the genetic and/or biochemical composition of resources Genetic resources, in particular through the application of biotechnology."

And when the protocol refers to sustainable development, it is essentially in the context determined by the Johannesburg Summit, in other words the report of the UN World Commission on Environment and Development (Brundtland Report, 1987), which defines it as "development that meets the needs of the present without compromising the ability of future generations to meet their own needs" (Figure 6.13). Its main pillars are:

- economic   □A
  ment, without prejudice to the environment and to humans;
- social equity means meeting the basic needs of humanity in terms of housing, food, health and education, reducing inequalities between individuals, respecting their cultures and;
- environmental quality means preserving natural resources in the long-term, maintaining the main ecological balances and limiting environmental impacts.

Social development is also about everything related to human well-being, including health, food security, and the   against unemployment, precari-

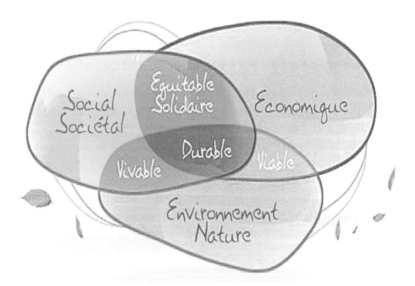

Figure 6.13 Diagram of the main elements of sustainable development (https://www. google.com/search?rlz=1C1UXZO_enMA643MA643&biw=915&bih=635& tbm=isch&sa=1&q=d%C3%A9veloppement+durable%2C+image&oq=d%C 3%A9veloppement+durable%2C+image&gs_l=psy-ab.3.0j0i24k1.3293938 .3303486.0.3303770.28.25.0.0.0.0.998.4916.0j11j4j1j1j1j1.19.0.0.1.1.64. psy-ab.9.19.4879.0.ArPvLy3-BTY#imgrc=UMKocYM94y2ZoM)..

ousness, human rights and education. To remain in conformity with the text of the Nagoya protocol, this strategy focused solely on the terms used in this protocol, namely health, food security, poverty alleviation, gender and education and training. Actions are thus proposed for the valorization of the genetic heritage in this sense.

6.4.3.5   Strategic Axis 5: Valuation of Traditional Knowledge

Traditional knowledge associated (TKA) with genetic resources is an intangible and ancestral national heritage transmitted from one generation to another through time immemorial. These TKA encompass the knowledge, innovations, know-how and practices of local communities. The result of an experience built up over the centuries and adapted to the environment and local culture, TKA are transmitted orally from one generation to another and by practice (e.g., Figure 6.14). It is a collective inheritance that takes various forms: agricultural practices, including the development of plant and animal species, fishing practices, breeding, selection, extraction, improvement, medication, etc. Traditional culture is essentially practical, especially in areas such as agriculture, fisheries, health, medicine, horticulture, and forestry.

Today, there is a growing appreciation of the value of traditional knowledge, which is important not only to the communities that own and use it for generations, but also capitalizes these valuable informations to save on time and resources. Many pharmaceutical, medical, cosmetic or food products based on plants or microorganisms are the result of traditional knowledge that contributes to Sustainable Development.

The Kingdom of Morocco, a millennial country, having built its civilization on a great geographical and cultural diversity, is full of knowledge, practices and traditional knowledge relating to medication, pharmacopeia, culinary recipes, irrigation, conservation of Fauna and   A☐ Throughout the millennia, these societies have cultivated and raised, used, designed preparations and managed biodiversity in a rational and sustainable manner. They have been able to preserve not only the different elements of national biodiversity, but also empirical techniques which are a source of valuable information about the human community in general and constitute a useful model for the design of innovative biodiversity policies.

TKA with genetic resources is the second main area of application of the Nagoya Protocol. They must be valued not only because it is a recommendation of the NP, but because it is a very valuable national heritage, they are

Figure 6.14   Traditional use of Henné.

identical and can bring a great deal of cultural,   A ☐

6.5   The Moroccan Strategy for Conservation and Sustainable Use of Biodiversity

The development of a national strategy is a requirement of Article 6 (a) of the CBD, which requires all Parties to the Convention to develop a National Biodiversity Strategy and Action Plan (NBSAP) or its equivalent. This strategy is advocated as a roadmap to guide decision-makers towards major initiatives to conserve the country's living natural resources. The Moroccan NBSAP (MDCE, 2016) has been developed are listed in the following subsections.

6.5.1   Vision and Strategic Directions

The vision and strategic objectives agreed upon and set out as the basis of this version of the NBSAP are aimed at making biodiversity a pillar of sustainable development and the well-being of Moroccan society:

"By 2030, biodiversity is conserved, restored, valued and rationally used, ensuring the maintenance of ecosystem services for the   of all, while contributing to the sustainable development and well-being of Moroccan society."

6.5.2   Strategic Axis of the NBSAP

National strategic axis numbers of operational objectives are defined as:

1. Strengthening the conservation of species, ecosystems and services they provide, is the primary objective of the CBD. Maintaining biological diversity and ecological services for the welfare of society is therefore at the heart of this first strategic axis of the NBSAP 2011–2020 (with 5 goals);
2. Ensure the sustainable use of biodiversity and biological resources. The NBSAP 2016–2020 also aims to reduce direct pressures on biodiversity and encourage its sustainable use (with 5 goals);
3. Contribute to the improvement of the living conditions of the populations through the effective implementation of the NBSAP and which aims to make biodiversity, its values and services national and local development tools to fight against poverty and improve the living conditions of the populations (with 3 objectives);
4. Consolidating the governance of biological diversity is one of the main priorities of Morocco to achieve its national objectives. It focuses on strengthening and consolidating the governance of national biological diversity, considering that the notion of biodiversity affects several environmental themes, if not all, and concerns more than one government department (with 7 objectives);
5. Enhance and share knowledge on national biodiversity, which is essential for implementing national biodiversity guidelines. Making this knowledge available is also a prerequisite for informed decision making (with 4 objectives);
6. Promote among citizens the desire to change attitudes towards the national biodiversity heritage and which is a priority approach, which aims to go beyond awareness and which must be translated into the

daily behavior of individuals, communities, and businesses (with 2 goals).

Keywords

• general context and issues   • valorization strategy

References

Agoumi, A., & Debbagh, A., (2006). Resources en eau et bassins versants du Maroc: 50 ans de development. *RDH50*. pp. 1–50.

Birouk, A., Sghir, T. M., Saidi, S., Sehar, E., Menioui, M., & Sbai, L., (2012). Evaluation des besoins de renforcement des capacités nationales en matière d'APA. Dép*artement de l'Environnement Maroc.*, pp. 1–382.

Dakki, M., Menioui, M. & Amhaouch Z., (2015). et Amhaouch Zouhair., Stratégie Nationale des zones humides. Stratégie et Plan d'Action, *HCEFLCD*, pp. 1–56.

DAT, (2011). Stratégie du Développement Rural. *Direction de l'Aménagement du territoire*, pp. 1–46.

DE, (2002). Plan d'Action National sur l'Environnement. Secrétariat d'Etat chargé de l'Environnement. *Département d'Etat*. pp. 1–22.

DE, (2006). Stratégie pour la protection de l'environnement et du développement durable. *Département de l'Environnement*.

DE, (2012). Plan d'Action National sur l'Environnement. Secrétariat d'Etat chargé de l'Environnement. *Département de l'Environnement*. pp. 1–112.

Fennane, M., & Ibn Tattou, M., (2012). Statistiques et commentaires sur l'inventaire actuel de la flore vasculaire du Maroc. *Bulletin de l'Institut Scientifique, Rabat, section Sciences de la Vie, 34*(1), pp. 1–9.

HCEFLCD, (2013). *Le Programme d'Action National de Lutte Contre la Désertification: Actualisation et adaptation aux spécificités zonales.*

HCEFLCD, (2015). Rapport national pour la COP12 de Ramsar.

IRES, (2006). 50 ans de Développement Humain – Perspectives 2025. Ins*titut Royal des Etudes Stratégiques*.

Laird, S. A., & Wynberg, R. P., (2013). *Bioscience at a Crossroads: Access and Benefit Sharing in a Time of Scientific, Technological and Industry Change*. Secretariat of the Convention on Biological Diversity. https://www. cbd. int/abs/policy-brief/default. shtml/.

MDCE, (2015). Rapport d'Etat sur l'Environnement au Maroc. *Ministère Délégué Chargé de l'Environnement.*, p. 187.

MDCE, (2016). Stratégie et Plan d'Action National sur la Biodiversité. *Ministère Délégué Chargé de l'Environnement*.

Messouli, M., (2008). *Assessment of Oases Ecosystem Health: Vulnerability and Adaptation to Climate Change in Southern Morocco*. International Forum EcoHealth, Merida, Mexico.

Mhirit, O., & Et-tobi, M., (2010). *Les écosystèmes forestiers face au changement climatique: Situation et perspectives d'adaptation au Maroc*. IRES Publications, pp. 1–260.

Mhirit, O., & Maghnouj, M., (1997). Stratégie de conservation des ressources génétiques forestières au Maroc. Ressources Phytogénétiques et Développement durable. *Actes Editions (IAV), Rabat. Maroc.,* pp. 313–318.

Mhirit, O., Blerot, P., & Benzyane, M., (1999). *Le grand livre de la forêt marocaine.* Sprimont: Mardaga, pp. 1–282.

Millennium Ecosystem Assessment, (2005). *Ecosystems and Human Well-being: Synthesis.* Island Press, Washington, DC.

PNUE/DE, (1998). Etude Nationale sur la biodiversité – Rapport de synthèse. *PNUE/ Département of Environment*, pp. 1–156.

SEDD, (2011). *Charte Nationale de l'Environnement et du Développement Durable.* Secrétariat d'Etat Chargé de l'Environnement, pp. 1–10.

SEDD, (2017). *Stratégie Nationale pour la valorisation des ressources génétiques et des connaissances traditionnelles associées.* Secrétariat d'Etat Chargé du Développement Durable. (Inédit).

SEDRA, My, H., (1994). *Diversité et amélioration génétique du palmier dattier dans les oasis marocaines. Séminaire sur – L'agronomie saharienne, atouts et contraintes.* 2–7/12194. Errachidia. Maroc.

CHAPTER 7

Biodiversity in Senegal: Terrestrial Ecosystems Flora and Great Fauna

GUEYE MATHIEU,[1] B. A. TAÏBOU,[2] and DIOP SALL AMINATA[3]

[1]Laboratory of Botany, Department of Botany and Geology, UMI 3189, IFAN Ch. A. Diop BP 206 Dakar, Senegal, E-mail: gueye_guirane@yahoo.fr, mathieu.gueye@ucad.edu.sn
[2]Ecological Monitoring Center, Rue Leon Gontran Damascus, Fann Residence, BP 15532 Dakar, Senegal, E-mail: taibou@cse.sn
[3]National Parks Directorate, Senegal, E-mail: aminat71@yahoo.fr

7.1 Introduction

With an area of 196,720 km², of which 75% is located at an altitude of less than 50 meters, Senegal, which has 14 administrative regions, is bordered on the north and northeast by Mauritania, on the southeast by Mali, south by Guinea and Guinea Bissau and to the west by the Atlantic Ocean (Figure 7.1)

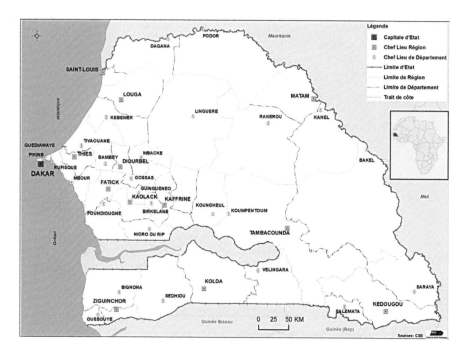

Figure 7.1 Administrative map off Sénégal (*Source:* CSE, 2014).

It is a Sahelian country with a semiarid tropical Sudano-Sahelian climate with a single rainy season extending from July to October. Rainfall is of great spatial and temporal variability. It has declined sharply over the last 40 years, ranging from less than 300 mm/year in the north to over 1000 mm/year in the south, a 100–200 mm shift of isohyets southward from normal 1931–1960 (Figure 7.2). Due to its geographical position, the country is exposed to a very particular wind regime modifying certain climatic parameters. Compared to what is observed in the interior of the continent at equal latitude, the decrease in rainfall is more rapid from south to north, but the saturation   and temperature maxima during the dry season are lower, especially near the coast.

These characteristics   forest stands distributed between the northern latitudes 12° 30' and 16° 30' in the Guinean, Sudanian, and Sahelian domains, characterized by a very marked   A□ (GUEYE, 2000).

Senegalese territory is rich in animal and plant diversity and water potential (CSE, 2002). For the conservation of ecosystems housing these riches,

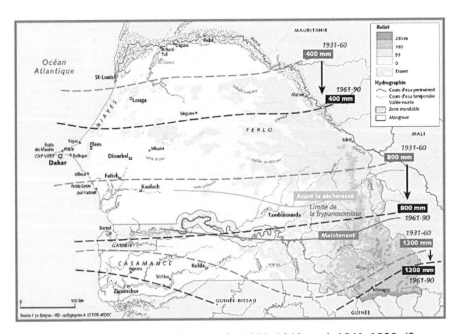

Figure 7.2 Isohyetes from the periods 1931–1960 and 1961–1990 (Source: Institut de Recherche et Développement (http://www.cartographie.ird.fr/SenegalFIG/secheresse.html).

the country has erected a system of Protected Areas (PA) including 6 national parks, 3 wildlife reserves, 20 silvopastoral reserves, and 213   A ⬜ ests. This PA network covers a total area of 11, 934,663 ha (MEPN, 1993), more than 40% of the national area, including four biosphere reserves, two UNESCO World Heritage sites and four wetlands of international importance (Ramsar).

However, the country is poorly endowed with quality soil resources. With regard to the   ⌈A for agriculture (47%), the percentage of poor soils is 36% (Gueye, 2003).

Agriculture remains one of the most important sectors of economic activity. According to the Poverty Reduction Strategy Paper (PRSP, 2002), agriculture, which accounts for more than half of the population, accounts for 10% of GDP and absorbs on average about 10% of the program public investments. The agricultural sector also plays a major role in the economy through its contribution to food security, by supplying raw materials to agribusiness (peanuts, cotton, etc.), by absorbing part of the economy. The production of the industrial and semi-industrial sector and crafts (fertilizers, pesticides, agricultural equipment).

According to the last census of the population (ANSD, 2014) Senegal has, in 2013, 12,873,601 inhabitants with 50.1% of women and 49.9% of men; the under 18s represent 50% of the population.

The majority of the poor is rural and depends on agriculture for their livelihoods. Land resources are their main factor of production when they own it, or use it.

For several decades, these resources have continued to deteriorate, due to   climatic conditions (repeated droughts) and the pressure of human activities (poor agricultural practices, deforestation, overgrazing, excessive use of pesticides, etc.), exacerbating the competition for this natural resource and land   ⌈A of latent   A ⬜ tices such as conservation management methods (sacred woods, prohibition of cutting species such as *Kadd – Faidherbia albida*, etc.).

7.2 Main Ecosystems in Senegal

Ecosystem and/or biodiversity: there is no need to oppose or distinguish these two concepts. Although they do not designate exactly the same realities, they are nonetheless related. The definitions of these two concepts show that biodiversity is fundamental to the existence and functioning of an ecosystem (without diversity of living organisms, there is no ecosystem). They also show that the ecosystem includes biodiversity, that biodiversity is one of the building blocks of an ecosystem. The ecosystem is not reduced to biodiversity. In addition to the diversity of living things, the ecosystem includes the inorganic environment, the habitat, a fundamental spatial dimension. As the ecosystem embraces biodiversity, the term ecosystem has been chosen because biodiversity will be addressed.

An ecosystem is a spatially   dynamic complex of plant, animal, and microorganism communities and their interacting inorganic environment as a functional entity (adapted from Mobius, 1877 and Tansley, 1935).

Several   have been given to the notion of ecosystem. Almost all the   take into account the biotic and abiotic factors as well as the interactions between the biotic components on the one hand, between the abiotic parameters on the other hand and   between the biotic and abiotic parameters. However, any ecosystem varies in space and time and is also contained in a larger system that can be considered an ecosystem. Thus, the   A ☐
it is necessary to   A⟧
to   A⟧
describe its structure or its functioning. It is therefore, necessary to adopt an approach that focuses both on the structure and functioning of the ecosystem. According to a   A ☐
ecosystems and aquatic ecosystems are distinguished. In Senegal, other biogeographical considerations have made it possible to distinguish ecogeographical zones (Figure 7.3).

In other West African countries, there are other names such as agro-ecological, bioclimatic, ecoclimatic, pedoclimatic zones, etc. Regardless of the denomination, ecological or ecoclimatic zones or ecoregions are generally   on the basis of biogeographic and/or phytogeographic characteristics.

By referring to the occupation/land use criterion, we would have obtained agricultural ecosystems, pastoral ecosystems, forest ecosystems (in relation to logging activity,   A⟧
on the fact that the severity of the disturbances caused by human activities would not be   A ☐
Convention on Biological Diversity, human activities have the property of eliminating or   reducing the diversity of living organisms and resulting in diminishing relations of Harmonious interdependence between living organisms (essential criterion for the existence of an ecosystem), by modifying the characteristics of the inorganic environment.

In Senegal, we thus distinguish globally:

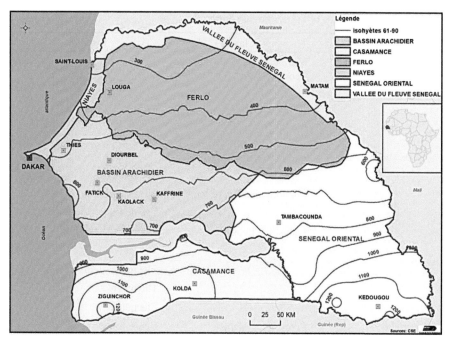

Figure 7.3 Ecogeographical areas in Senegal (*Sources*: DEFFCCS/CSE).

- Sahelian ecosystems;
- Sudan's ecosystems;
- sub-Guinean ecosystems;
-
- marine and coastal ecosystems; and
- mountain ecosystems.

These ecosystems are   ⌐A
restrial ecosystems and aquatic ecosystems (natural ecosystems), to which
we add the   ⌐A
ronments strongly   ☐ A
especially by human activities. Then, these three large sets are subdivided as
given in Table 7.1.

7.2.1 Terrestrial Ecosystems

The main terrestrial ecosystems of Senegal are steppes, savannahs, and forests.

Table 7.1 Main Types of Ecosystems in Senegal

Ecosystems	Classes	Biotope
Aquatic	Marine and Coastal Ecosystems	The sea
		The littoral/beach
	☐	☐
		☐
		☐
	River Ecosystems	
		The rivers
	Fluviomarine Ecosystems	The estuaries
		The deltas
	Lacustrine Ecosystems	Lakes
		The lagoons
		The backwaters
		The bowls
		The pound
Terrestrial	Forest Ecosystems	Dense forest
		Clear forest
		Gallery forest
	Savannah Ecosystems	Wooded savannah
		Tree savannah
		Shrub savannah
		Grassy savannah
	Steppe Ecosystems	Forested steppe
		Shrub steppe
		Herbaceous steppe
	The parks	Le parc à *Cordyla pinnata*
		Le parc à *Adansonia digitata*
		Le parc à *Faidherbia albida*
		Le parc à *Borassus aethiopium*
		Le parc à *Vitelleria paradoxa*
☐	Crops	Cropland
		Market gardening, irrigated culture
	Planting	Forest plantation

7.2.1.1   The Steppes

Steppes consist of scattered shrubs (shrub steppe) or scattered trees and shrubs (treed steppe) with a density of less than 10 feet/ha (Figure 7.4). They cover the northern third of Senegal and consist of a herbaceous carpet very often discontinuous with species such as *Borreria verticillata*, *Indigofera oblongifolia*, *Chloris prieurii*, *Schoenofeldia gracilis,* and other herbaceous species of the genus *Aristida* and *Cenchrus*. The steppes are often dotted

Figure 7.4   Shrub steppe (A) and forested steppe (B).

Figure 7.5   Shrub savannah (A) and shrub savannah with trees (B) (Photo Credit, CSE, 2016).

with thorny woody species such as *Acacia raddiana, Acacia senegal, Acacia seyal,* and *Balanites aegyptiaca.*

7.2.1.2   The Savannahs

They are formations with a continuous herbaceous carpet, composed of shrubs and trees whose spatial variation makes it possible to distinguish the shrub savannah, the shrub savannah with trees, and the wooded savannah[1].

In Senegal, they are the third center of the territory with wooded shrub savannas in the North (Figure 7.5) and wooded savannas in the South. The woody to shrub savannas is characterized by woody species such as *Cordyla pinnata, Ficus sycomorus, Diospyros mespiliformis, Dichrostachys cinerea, Acacia macrostachya, Combretum* spp., *Ziziphus mauritiana, Sclerocarya birrea, Neocarya macrophylla.* The wooded savannas are dominated by species such as *Sterculia setigera, Lannea acida, Sclerocarya birrea, Pterocarpus erinaceus, Parkia biglobosa, Terminalia macroptera,* and *Daniellia oliveri.* In these plant formations, the herbaceous layer is marked by species of the genera *Andropogon, Hyparrhenia,* and *Digitaria.*

---

[1] Shrub savannah: Formation dominated by undifferentiated trunks with a canopy cover of less than 20%. The herbaceous carpet is almost continuous. Average density 200 feet/ha.

Wooded Savannah: Formation mainly composed of trees greater than 5 m, the recovery of which does not exceed 25%. Average density 50 feet/ha. The herbaceous carpet is included almost continuous.

Woodland savannah: Formation mainly composed of trees greater than 5 m, covering between 25 and 50%. Average density 200 feet/ha.

7.2.1.3   The Forests

The extent of land covered by a woody vegetation whose tops are more or less joined; the trees are at least 8 meters long and can be made up of several layers. They can be dense with over 70% cover, 40–70% clear, and in Senegal they are generally found in the southern part of the country with clear forests, dense dry forests and gallery forests (Figure 7.6). The clear forests located in Upper and Middle Casamance are characterized by *Pterocarpus erinaceus, Khaya senegalensis, Daniellia oliveri, Ceiba pentandra,* and *Terminalia macroptera.* The dense dry forests mainly located in Lower Casamance in the form of relics, are dominated by *Erythrophleum guineense, Detarium senegalense, Malacantha alnifolia, Parinari excelsa, Pentaclethra macrophylla, Raphia sudanica,* and *Carapa procera.* The gallery forests occupy the valleys and are characterized by species such as *Elaeis guineensis, Erythrophleum guineense, Khaya senegalensis, Carapa procera,* and *Alchornea cordifolia.*

9.2.1.4   The Fluvio-Lacustrine Ecosystems

These are mainly the ecosystems encountered in the water surfaces of the country, including rivers, lakes and rivers with a rich and varied biocenose is most often distributed in the floodplains as evidenced by relict forests Gonakier north of the Senegal River (Figure 7.7).

7.2.1.5   The Artificial Ecosystems (Humanized)

They correspond to natural environments strongly modified by human activities mainly for agricultural or logging activities. There are essentially five

Figure 7.6   Clear forest (A) and gallery forest (B) (Photo credit, CSE, 2015).

Figure 7.7 Gonakier forests of the Senegal River floodplain showing dominance of *Acacia nilotica* with sparse understory (left) and aerial view of forest structure with Senegal River in the foreground (right) (Photos: Gray Tappan). Courtesy of the U. S. Geological Survey, EROS Center.

major sets of agricultural ecosystems, namely the agroforestry parks with species of the genus *Acacia*, *Faidherbia albida*, *Cordyla pinnata*, *Borassus akeassii*, and *Elaeis guineensis* (Figure 7.8). In the northern part of the country, there are *Balanites aegyptiaca* parks relatively located in the regions bordering the Senegal River Valley. These species are generally spared by

Figure 7.8 *Faidherbia albida* park in Ferlo (Photo credit, CSE, 2009).

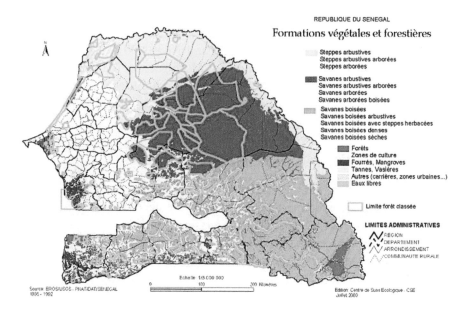

Figure 7.9   Plant and forest formations of Senegal.

farmers during clearing. These agroforestry parks are characterized by the integration of woody species and cultivated plants (Figure 7.8).

7.3   Senegalese Flora and Its Uses

7.3.1   Diversity

The flora of Senegal is estimated at about 3,600 species in 1,277 genera, dominated by vascular plants that would constitute 2,500 species (CSE, 2015; Ba and Noba 2001, Mugnier 2008, Berhaut, 1967). The southern and south-eastern parts have the greatest biodiversity diversity (CSE, 2015; Ba and Noba 2001; Traoré, 1997). Knowledge about algae, lichens, pteridophyta, bryophyta, bacteria and viruses is insufficient and fragmentary (Mingou and Guèye, 2017; CSE, 2015; Ba et al., 2006). The few data available estimate 1,141 species: 44 viruses, 19 Bryophyta, 38 pteridophyta, 7 lichens, 125 cyanophyta, 648 microalgae, and 260 macroalgae (CSE, 2015) (Figure 7.9).

The seed plants 2,500 species are well known and include 1,000 genera distributed in 165 families (CSE, 2015). Dicotyledons are largely in the majority with nearly 70%. Herbs make up more than 50% (MEDD, 2014).

The genera with the greatest species diversity are: *Cyperus* and *Indigofera* (44 each), *Ipomoea* (38), *Crotalaria* (33), *Ficus* (30), *Hibiscus* and *Tephrosia* (22 each), and *Euphorbia* (20) (Ba and Noba 2001). The least represented families (1 genus 1 species) are: Moringaceae, Medusandraceae, Linaceae, Opiliaceae, Surianaceae, Illecebraceae, Humiriaceae, Hernandiaceae, Balsaminaceae, Basellaceae, Droseraceae, Geraniaceae, Gesneriaceae, Goodeniaceae, and Holaragaceae.

7.3.1.1   Species Considered Endemic

An endemic plant is a plant whose range is restricted to a country or region. Species considered endemic to Senegal (33) are according Ba and Noba (2001): *Acalypha senegalensis, Alectra basserei, Andropogon gambiensis, Berhautia senegalensis, Cissus oukontensis, Cissus gambiana, Ceropegia senegalensis, Combretum trochainii, Salicornia senegalensis, Vernonia bambilorensis, Abutilon macropodum, Ceropegia practermissa, Crotalaria sphaerocarpa, Cyperus latericus, Digitaria aristulata, Eriocaulon inundatum, Ficus dicranostyla, Ilysanthes congesta, Indigofera leptoclada, Laurembergia villosa, Lipocarpa prieuriana, Nesaca dodecandra, Polycarpaea gamopetala, Polycarpaea linearifolia, Polycarpaea prostatum, Rhynchosia albiflora, Salicornia praecox, Scirpus grandicuspis, Solanum ceraxiferum, Spermacoce phyllocephala, Spermacoce galeopsidis,* and *Urginea salmonea.*

In view of this list and several studies, a meeting at the regional level is necessary to solve the problem of endemism in the area.

7.3.2   *Threatened Protected Plants*

The list of threatened and protected plants in Senegal presented here has been defined from several studies and legislative texts (CSE, 2015; CSE, 2013; CSE, 2009; Code Forestier, 1998).

7.3.2.1   Threatened Plants

Pterocarpus erinaceus, Bombax costatum, Landolphia heudelotii, Borassus akeassii, Parkia biglobosa, Oxythenanthera abyssinica, Raphia sudanica, Cordyla pinnata, Khaya senegalensis, Dalbergia melanoxylon. Sterculia setigera, Cyrtosperma senegalensis, Linaria sagitta, Rocella tinctoria, Anthocleista djalonensis, Mitragyna stipulosa, Pentaclethra macrophylla, Sterculia tragacantha, Cola laurifolia, Pandanus candelabrum, Raphia spp.

Calamus deeratus, Alstonia boonei, Pseudospondias microcarpa, Ficus abutilifolia, and *Pentaclethra macrophylla*.

7.3.2.2   Protected Plants

Albizzia sassa, Vitellaria paradoxa, Celtis integrifolia, Daniellia oliveri, Diospyros mespiliformis, Holarrhena africana, Mitragyna stipulosa, Piptadenia africana, Hyphaene thebaïca, Dalbergia melanoxylon, Faidherbia albida, Acacia senegal, Adansonia digitata, Afzelia africana, Borassus aethiopium, Ceiba pentandra, Cordyla pinnata, Khaya senegalensis, Prosopis africana, Pterocarpus erinaceus, Sclerocarya birrea, Tamarindus indica, Ziziphus mauritiana, Elaeis guineensis, Borassus akeassii, Raphia spp., *Chrysobalanus icaco, Parinari excelsa, Pentaclethra macrophylla, Sterculia tragacantha, Cola laurifolia, Ficus sagittifolia, Milicia regia, Carapa* spp., and *Alstonia boonei*.

7.3.3   *Use of Flora of Senegal*

7.3.3.1   Food

Various plant organs of local plants are consumed in Senegal. We will only present here fruits and leaves.

7.3.3.1.1   Fruits

Wild fruit species in Senegal represent about 2.8 to 3.5% of the flora. There are about 70 wild species whose fruits are eaten. They belong to 30 families, the most represented being the Moraceae, Fabaceae, Anacardiaceae, Annonaceae, Tiliaceae, Apocynaceae, Rhamnaceae, and Rubiaceae. This list does not include species whose fruits are potentially edible and are consumed more often in cases of famine or dearth. However, it remains to improve despite the reference works consulted. Consumption patterns are as diverse as they are varied. Species whose fleshy fruits are eaten raw are by far dominant. For some dry fruits are prized for their flesh, their seed or their aril. Others are used for their almond or seed. Note that there are fruits eaten raw or after preparation (fresh drink, cooking oil, spice, extraction of almond). Depending on their level of popularity wild fruits can be subdivided into popular fruits (known to the general public and are commercially traded), moderately known (not commercially traded or very rarely and locally) and little known (exploited at the local level) (Guèye and Samb, 2016; Guèye et al., 2014).

Among the   we can list (Plate 7.1): *Adansonia digitata, Detarium microcarpum, Detarium senegalense, Balanites aegyptiaca, Saba senegalensis, Tamarindus indica, Ziziphus mauritiana, Dialium guineense, Landolphia heudelotii, Parkia biglobosa,* etc.

In the second group (Plate 7.2) there is *Aphania senegalensis, Cordyla pinnata, Sclerocarya birrea, Spondias mombin, Diospyros mespiliformis, Neocarya macrophylla, Vitex doniana, Borassus akeassii,* etc.

In the third group (Plate 7.3) we have among others *Boscia senegalensis, Parinari excelsa, Landolphia dulcis, Ficus sycomorus* subsp. *gnaphalocarpa, Ficus sur, Hexalobus monopetalus, Cola cordifolia, Icacina senegalensis, Uvaria chamae, Sarcocephalus latifolius,* etc.

7.3.3.1.2   Leaves (Plate 7.4 A–D)

The various works carried out indicate about 40 local plant species whose leaves are traditionally consumed (Guèye and Diouf, 2007; Diouf et al., 1999). They are distributed in 21 families, of which the best represented are respectively Amaranthaceae, Malvaceae, Moraceae and finally Fabaceae and Tiliaceae (Guèye and Diouf, 2007; Diouf et al., 1999).

According to the subdivision of leafy vegetables according to their form of consumption (Stevel, 1990), all species except *Sesuvium portulacastrum* are eaten as a vegetable herb: cooked leaves mixed in a sauce or with other vegetables (Guèye and Diouf, 2007). All species of *Hibiscus* spp. are also consumed as spinach and condiment. *Tamarindus indica* is also used as a condiment, *Senna obtusifolia* as spinach while only *Sesuvium portulacatrum* is eaten as a salad (Guèye et Diouf, 2007).

7.3.3.1.3   Other Food Organs

In the southern and southeastern parts of the country tubers of various *Dioscorea* spp.and *Raphionacme* spp. (Plate 7.4 E) are locally consumed. Gum arabic (Plate 7.4 F) obtained by bleeding *Acacia senegal* is widely exploited. A few years ago Senegal was the world's largest producer (CSE, 2015; CSE, 2013; CSE, 2009). This gum is used in the preparation of many confectionery and other products at the international level. By bleeding *Sterculia setigera* we get an eraser used in the preparation of couscous. These two products are often exported (CSE, 2015; CSE, 2013; CSE, 2009).

Plate 7.1   Popular fruits: (A) *Detarium senegalense,* fruits for sale, (B) *Detarium microcarpum,* fruits in market, (C) *Parkia biglobosa* fruits, (D) Seeds of *P. biglobosa* transformed into a condiment in market, (E) *Ziziphus mauritiana* fruits, (F) *Saba senegalensis* fruits, and (G) *Tamarindus indica* fruits in market.

Plate 7.2 Less popular fruits: (A) Gathering of *Cordyla pinnata* fruits, (B) *Cordyla pinnata* fruits, (C) *Diospyros mespiliformis* fruits on sale, (D) *Neocarya macrophylla* fruits.

A

Plate 7.3   Little known fruits: (A) Various gathering fruits, (B) *Hexalobus monopetalus* fruits.

Plate 7.4   Food leaves and other food organs: (A) *Ceratotheca sésamoïdes* in bloom, (B) Drying leaves of *Ceratotheca sesamoïdes*, (C) gathering of *Senna obtusifolia* leaves, (D) *Sesuvium portulacastrum*, (E) Tuber of *Raphionacme splendens* next to an ungunged, (F) *Acacia senegal* gum after a few days of bleeding.

7.3.3.2 Medicine (Plate 7.5 A)

In Senegal, on a floristic potential of about 2,500 species of higher plants, 542 species used in the traditional pharmacopeia are mentioned (Kheraro and Adam, 1974). It has also been reported that medicinal plants represent 24% or about 589 species (Fall, 1996). Many of the medicinal plants are also known in the subregion and some at the international level. All plant organs are used in medicine but the most used are roots, barks, leaves, fruits, rhizomes or bulbs, leafy stems and flowers. The pathologies or symptoms regularly treated are gastroenterology, infectious diseases, gynecological problems, sexual asthenia, snakebites, fatigue, wounds, osteo-articular pain, bronchopulmonary diseases. Some emerging diseases such as diabetes, cardiovascular problems and renal insufficiency are also cared.

7.3.3.3 *Timber* (Plate 7.5 B and D)

Eastern Senegal and natural Casamance are the major wood providers at the national level. Local production comes mainly from the following species (CSE, 2013): *Khaya senegalensis, Cordyla pinnata, Pterocarpus erinaceus, Afzelia africana, Bombax costatum* and *Dalbergia melanoxylon.* The most popular and the most exploited species are *C. pinnata, P. erinaceus* and *K. senegalensis.* There are other species like: *Mitragyna stipulosa, Sclerocarya birrea, Balanites aegyptiaca, Antiaris africana, Ceiba pentandra, Detarium senegalense, Lophira lanceolata, Spondias mombin, Terminalia macroptera, Pseudocedrela kotschyi, Lonchocarpus sericeus, Pterocarpus lucens, Alstonia boonei, Milicia excelsa, Isoberlinia doka, Isoberlinia tomentosa,* and *Zanha golungensis* whose wood is locally exploited. Local production covers only about 5% of national requirements. Thus, Senegal imports a lot of logs.

7.3.3.4 Fuel Wood

Today firewood and charcoal come mainly from the regions of Tambacounda, Kédougou, and Kolda (CSE, 2013). Previously, the regions of Kaolack and Fatick were the big purveyors. The most exploited species: *Pterocarpus erinaceus, Cordyla pinnata, Afzelia africana, Acacia* spp., *Anogeissus leiocarpus, Burkea africana, Prosopis africana, Terminalia macroptera, Piliostigma reticulatum, Piliostigma thonningii, Albizia* spp. *Pseudocedrela kotschyi, Vitex doniana, Tamarindus indica, Detarium microcarpum, Detarium senegalense,* and *Parkia biglobosa.*

Plate 7.5 Other uses of flora: (A) Etale sale of medicinal plants, (B) Pirogue produced with *Khaya senegalensis* wood, (C) Woven palisade with stems of *Andropogon gayanus*, (D) Poutres de *Borassus akeasii* beams for construction, (E) Bunch of ropes of *Piliostigma reticulatum* bark.

In recent years, more and more private produce charcoal from alien species in reforestation programs. Species such as *Anacardium occidentale*, *Prosopis chilensis*, *Azadirachta indica*, *Casuarina equisetifolia*, *Eucalyptus* spp. are often used (Guèye, 2000).

7.3.3.5   Construction (Plate 7.5 C–E)

Local people use various forest products (wood, ropes, straw, leaves, etc.) for the preparation of their habitats (Guèye, 2012). The favorite woods are those of: *Borassus akeassii*, *Bombax costatum*, *Grewia* spp., *Loeseneriella africana*, etc.

The most used ropes are drawn from plant bark as: *Adansonia digitata*, *Piliostigma réticulatum*, *Piliostigma thonningii*, *Grewia* spp. (Piqué et al., 2016; Guèye, 2012),

The straws of *Pennisetum* spp., *Andropogon* spp., *Schizachyrium* spp. are

☐   ☐

7.3.3.6   Cultivated Species

The cultivated plants in Senegal are diverse. We will only present the winter crops and the most cultivated vegetables. The official catalog of species and varieties cultivated in Senegal lists 14 species and 174 varieties. This list only concerns the recommended varieties accompanied by their agronomic characteristics. This list consists according to the official catalog of species and varieties cultivated in Senegal by the Ministry of Agriculture and Rural Equipment in 2016:

- 69 varieties of straw cereals of 4 species (*Sorghum bicolor*, *Oryza sativa*, *Pennisetum glaucum*, *Zea mays*);
- 30 varieties of grain legumes and 2 varieties of *Vigna* or cowpea forage for 2 species (*Arachis hypogaea*, *Vigna unguiculata*);
- 73 varieties of vegetables of 8 species (*Alium cepa*, *Lycopersicon esculentum*, *Capsicum chinense*, *Solanum aethiopicum*, *Solanum tuberosum*, *Ipomoea batatas*, *Abelmoschus esculentus*, *Solanum melongena*).

Other species like *Colocassia esculenta*, *Vigna subterranea*, *Digitaria exilis*, *Manihot esculenta*, *Sesamum indicum*, *Gossypium bardadense*, *Citrullus lanatus*, *Hibiscus sabdariffa* are also widely cultivated in the rainy season.

7.4 Diversity of Wildlife in Senegal

7.4.1 The Birds

The birds recorded in Senegal total 623 species in 100 families. Seabirds are represented mainly by gulls, terns, cormorants, osprey, and oystercatchers.

It is also important to point out in land birds the presence of the Red-necked Ostrich, the Crowned Crane and the Great Bustard, which have a particular status. Poultry is also represented with 8 species and high numbers.

Palearctic migrants were mainly recorded in Anatidae, Apodidae, Ardeidae, Charadriidae, Laridae, and Scolopacidae.

Concerning waterbirds (165 species – divided into 35 families and 19 orders) were counted during the international enumeration of waterbirds of January 2017 in Senegal. The distribution by the family has shown that anatidae dominate largely with the following: laridae, scolopacidae, ardeidae, phalacrocoracidae, pelecanids, recurvirostrids, helicopterids, charadriids, and columbidae.

7.4.2 The Mammals

They are estimated to 192 species, distributed in 65 genera and 32 families. This group includes large wildland mammals, marine mammals and domestic mammals. Large wild mammals are found in national parks, mainly in Niokolo-Koba, Saloum Delta and Lower Casamance. They are also present in the area of cynegetic interest of Falémé. The large herbivorous fauna is represented by the Derby Eland (*Taurotragus derbianus*), the Roaring Horse (*Hippotragus equinus*), the Hartebeest (*Alcelaphus buselaphus major*), the Buffalo (*Syncerus caffer*), and the Kobe (*Kobus ellipsiprymnus defassa*). Megaherbivores are only represented by hippopotamus (*Hippopotamus amphibius*) and elephant (*Loxodonta africana*).

The average herbivore fauna is represented by bushbuck (*Tragelaphus scriptus*), reedbuck (*Redunca redunca*), Buffon's kob (*Kobus kob*), Grimm's duiker (*Sylvicarpa grimmia*), red-capped duiker (*Cephalophus rufilatus*) and the orbit (*Ourebia ourebi*). Suidae are represented by warthog (*Phacochoerus africanus*), and bush pig (*Potamochoerus porcus porcus*).

Among carnivores there are about 20 different species, including lion (*Panthera leo*), leopard (*Panthera pardus*), wild dog (*Lycaon pictus*), spotted hyena (*Crocuta crocuta*), jackal with striped   ⬚(*Canis adustus*), the serval (*Felis serval*), the caracal (*Felis caracal*), etc. Concerning primates Senegal is located at the northwestern limit of the range of nine African primate species: the chimpanzee: *Pan troglodytes*; the green monkey, *Cercopithecus aethiopssabaeus*, Campbell's moth, *Cercopithecus campbelli*, patas or red monkey, *Erythrocebus patas,* Guinea baboon *Papio papio,* smoky Mangabe, *Cercocebus atys*, West African bay colobus, *Colobus badius temmincki*, galago of Senegal, *Galago senegalensis* and Galago de Demidoff, *Galagoides demidovii.*

The chimpanzee (*Pan troglodytes*), has become a threatened species, its population is reduced to less than 500 individuals. It is the most northerly region of Africa and remains   ☐A

Other notable animals are present in the country such as the Aardvark (*Orycteropus afer*), the giant pangolin (*Manis gigantea*), the civet (*Viverra civetta*), many mongooses, the rock hyrax (*Procavia capensis*), etc.

Certain mammals such as the Derby Elk (*Taurotragus derbianus*) clas-
☐A
ing (IUCN, 2013). Indeed, its numbers are gradually decreasing from 1,000 individuals in 1990 (Sournia and Dupuy, 1990) to about 170 individuals (Hájek and Verne, 2000).

The same trend is also noted in the lion whose numbers have declined from the 50s to the present day. In the years 1950–1960, the lion population was estimated at more than 500 individuals. The study conducted in the National Park of Niokolo Koba (PNNK) by Henschel and Ndao in 2011 reported a population of about ten individuals.

Marine mammals inventoried include whales, dolphins, manatees, porpoises, and monk seals (Thiao, 2009).

Domestic mammals are mainly represented by cattle, sheep, goats, pigs, horses, insects, camels and small cats such as cats and dogs with several breeds.

7.4.3 The Reptiles

They are found with about 100 species, including crocodiles, snakes, lizards, and turtles. Marine turtles well present in Senegal with its 700 km of coast. The country has a very important coastal and marine biological diversity. An important host country for marine turtle species, it possesses with Mauritania, the most northern nesting sites of the Atlantic coast of Africa. The Senegalese coastline constitutes spawning sites, food and/or migration corridors for at least five of the six sea turtle species belonging to two families, the Cheloniidae (*Chelonia mydas* or green turtle, *Lepidochelys olivacea* or turtle), olive-tree, *Eretmochelys imbricata* or hawksbill turtle, *Caretta caretta* or loggerhead turtle) and Dermochelyidae with a unique genus (*Dermochelys coriacea*). The green turtle is the most abundant species. Leatherback turtles, nested, olive and loggerhead are common. For the Kemp turtle (*Lepidochelys kempii*) its presence is suspected (observation of a carapace belonging to the species Joal by Fretey in 2001) in our coasts.

7.5   State of Biodiversity Conservation in Senegal

Almost all of the country's development sectors rely on biodiversity. This is why the loss of biodiversity will inevitably affect the livelihoods of the people and may jeopardize the sustainable development that Senegal aspires to achieve. Fortunately, conservation of biodiversity has always been a national concern since the colonial period with the creation of protected areas (Niokolo Koba National Park, 1951, designated a Wildlife Reserve in 1953, and a National Park in 1954) and even earlier, with traditional conservation practices (sacred forests, totems, cemeteries.). In addition, the ratification of numerous international conventions, the development of a concerted management policy at the national level and with border countries, as well as the existence of national texts, documents and policies in favor of the conservation of biodiversity attest a clear and continuing desire to conserve biodiversity. Efforts have been made to strengthen the legal and regulatory framework with the ratification of a convention and protocol (Convention on Biological Diversity, Cartagena Protocol on Biosafety, Nagoya Protocol on Access and Benefit Sharing from conservation). It is the Ministry in charge of Environment that sets the guidelines through its sectoral policy letter.

However, because of the cross-cutting nature of biodiversity, several sectoral ministries intervene in its management. These include the Ministry of Fisheries, the Ministry of Agriculture, the Ministry of Livestock, the Ministry of   ⌐A
and so on. Coordination between these different sectoral entities is facilitated by the National Biodiversity Committee.

Other structures also facilitate synergy between sectors. This is particularly the case of the Higher Council of Natural Resources and the Environment (CONSERE), the National Commission for Sustainable Development (CNDD), the National Biosafety Authority (ANB), the National Biosafety Committee (CNB), etc.

As for the legislative and regulatory framework, the provisions that apply to biodiversity in Senegal are divided mainly between the various codes governing the management of natural resources (Forest Code, Hunting and

Wildlife Protection Code, Code of maritime   A ☐
Environment Code, General Code of Local Communities), certain laws (agro-forestry-pastoral orientation law, biosafety law, orientation law on the biofuels sector, law on bioethics), and the accompanying decrees of application.

In the interests of conserving biodiversity and combating the multiple aggressions on this biological diversity, Senegal, has made a lot of efforts in its preservation. Efforts to improve the conservation status have been noted in some species such as the Ostrich (*Struthio camelus camelus*) whose range that once covered much of the African continent, is currently very threatened. Apart from Chad and Senegal, where its presence in the wild with low numbers is   in Louguignibi locality by recent observations made in 2013, the species no longer exists in the wild state.

Its conservation and restoration is a priority for Senegal. For example, restoration initiatives involving the raising of ostriches in the katané enclosure have been undertaken since 2008. Today, these efforts are continuing and the species is regularly monitored.

The animal species that have disappeared from the country, such as scimitar-horned oryx (*Oryx dammah*), dama gazelle (*Gazella dama mhorr*) and dorcas gazelle (*Gazella dorcas neglecta*) are now reintroduced into the Gueumbeul Wildlife Reserve and the Reserve Fauna of Ferlo Nord. The giraffe (*Giraffa camelopardalis peralta*) is also one of the species that has disappeared from Senegal, but since 1996 another subspecies of Giraffe (*Giraffa camelopardalis giraffa*) from South Africa has become semicaptive in the animal reserves of Bandia and Fathala.

Private initiatives at the reserve level of Bandia and Fathala, have also made it possible to reintroduce species that had completely disappeared in Senegal, such as the Rhinoceros, the Derby eland, etc.

Parallelly, in terrestrial environments, in situ conservation focused on the establishment of a network of protected areas (Table 7.2 and Figure 7.10). This network has 6 national parks, 4 wildlife reserves and 3 special reserves, 213   ☐A
several community forests.

In the marine and coastal environment, in addition to the national parks with marine fringes, new *in situ* conservation actions are aimed at setting up a national network of Marine Protected Areas (MPAs) (Table 7.2 and Figure 7.11).

Thus, by their importance internationally recognized protected areas are on the list of Biosphere Reserves, Ramsar sites or natural sites of World Heritage.

Table 7.2 The Network of National Parks Wildlife Reserves, Marine Protected Areas, and Their Characteristics

Protected Area	Area (ha)	Biodiversity Interests/International Recognition
National Park of Niokolo-Koba (PNNK) created in 1954	913,000	Last refuge area of the great fauna in West Africa including lion, elephant, wild dog, Derby eland and chimpanzee.
		Biosphere Reserve, World Heritage Site.
National Park of Basse Casamance (PNBC) created in 1970	5,000	Last vestiges of the Guinean forest of Senegal (*Parinari excelsa, Treculla africana, Pithecelobium altissimum*, etc.). The park is also home to mammals such as the forest buffalo, panther, Campbell's Mone (*Cercopithecus campbelli*), Mangabey enfumé (*Galagoïdes demidoff*), colobe bai
National Birds Park of Djoudj (PNOD) created in 1971	16,000	One of the 3 sanctuaries of West Africa for Palearctic and Ethiopian migratory birds.
		teal, northern shoveler, wildlife and widowed whistling ducks, Gambian goose, crowned crane, etc. nest in the park.
		UNESCO World Heritage Site, Ramsar Site, Central Core of the Senegal River Delta Transboundary Biosphere Reserve
National Park of Saloum Delta (PNDS) created in 1976	76,000	Birdhouses of many species of birds: Lesser Flamingo, Pelican, Goliath Heron, Meandering Gull, Gray-headed Gull, Royal and Caspian Tern, Daisy Egret, Black-tailed Godwit, Avocet, many Palearctic Waders. Mammals present: warthog, bushbuck, grimm sylvicarp, reedbuck (rare), spotted hyena, bay colobus, green monkey, patas.
		Central core of the Saloum Delta Biosphere Reserve, transboundary Ramsar site (Niumi (The gambia)-Delta)
National Park of Langue de Barbarie (PNLB) created in 1976	2,000	Varied Avifauna: gray and white pelican, gray-headed gull, meager gull, and other laridae (royal tern, caspian, sooty), many migratory waders. Sea turtles (*Chelonias mydas, Caretta caretta, Dermochelys coriacea*, etc.).
		Central core of the Transboundary Biosphere Reserve of the Senegal River Delta

Protected Area	Area (ha)	Biodiversity Interests/International Recognition
National Park National of Madeleine Islands (PNIM) created in 1976	45	Presence of a single vegetative group (steppe with *Andropogon gayanus, Brachiaria distichophylla* and *Bothrichloa intermedia*), a large breeding colony of crow, black
Ornithological Reserve of Kalissaye (ROK) created in 1978	16	Colonies breeding Caspian Terns, Royal Terns, White Pelicans, etc. Breeding area for several sea turtle species including *Caretta caretta* and *Chelonia mydas*
Special Wildlife Reserve of Gueumbeul (RSFG) created in 1983	720	Wintering site for thousands of birds such as avocet, black-tailed godwit, silver plover dimorph, big banded plover, etc. Site for the breeding of a group of dama mhorr gazelle, Oryx algazelle and gazelle dorcas as part of the reintroduction policy for Sahel-Saharan species. Central core of the Transboundary Biosphere Reserve of the Senegal River Delta, Ramsar Site
Tocc Tocc Community Nature Reserve created in 2011	273	It is composed of a terrestrial zone (20% of the total area), with *Tamarix senegalensis* as the dominant species and constituting the favorite domain of some terrestrial mammals (Patas, Jackal, etc.) and an amphibious zone (80% of the total area) consisting mainly of a freshwater bowl. This basin is a site of great importance, not only by its extent but also by the wealth of species: the West African manatee (*Trichechus senegalensis*), the Adanson's Pelusius (*Pelusios adansonii*) and the First Ramsar Community Site in Senegal.
Popenguine Nature Reserve (RNP) created in 1986	1,009	Sudano-sahel savannah in the rehabilitation phase, refuge of wildlife species such as: guinea fowl, jackal, porcupine, etc. Birds: Bluebird, Swallow etc.

Table 7.2 (Continued)

Protected Area	Area (ha)	Biodiversity Interests/International Recognition
Ferlo Nord Wildlife Reserve (RFFN) created in 1996	487,000	Presence of a residual population of *Gazella rufifrons*. In addition to the Sulcata turtle, more than 180 bird species are listed, including the ostrich (*Struthio camelus*), the ground hornbill (*Bucorvus abyssinicus*), the great Arab bustard (*Otis arabs*). Site of repopulation of the disappeared Sahelo-Saharan fauna with notably the oryx, the gazelle dama mhorr and the gazelle dorcas. Central core of the Ferlo Biosphere Reserve.
Nature Reserve of Community Interest of Somone (RNICS) created in 2001	700	- Jackals, ichneumeun mongooses, very diverse avifauna including: spatula, pelican, cormorant, egret, curlew, knight. , temporary watercourse, permanent watercourse (perennial course of Somone and channels), lagoon (lake), The most represented groups of fauna are birds,
Palmarin Community Reserve created in 2003	10,450	- Reproduction site of sea turtles, striped hyena, jackals, monkeys, very important avifauna. Terrestrial, Mangrove and Marine Ecosystems, Migratory habitat and migration corridor, Sea turtle breeding site, Spawning grounds, nursery and habitat for marine and estuarine fauna, Conservation site for characteristic species such as the hyena, dolphin, etc.

Protected Area	Area (ha)	Biodiversity Interests/International Recognition
Marine Protected Area of Bamboung created in 2004	7,000	- 188 species of plants, dominated by Combretaceae, *Daniela oliveri*, *Parkia biglobosa*
		- 95 species of migratory birds from the Western Palearctic (Royal Tern, Caspian Tern, Dominican Gull and Mockingbird, Gray-headed Gull)
		- 36 species of mammals such as dolphin, manatee, striped hyena, spotted hyena, bushbuck, mongoose, porcupine, etc.
		Spawning and feeding area for ichthyofauna, manatee, dolphin and sea turtles, Bambong bolong is made up of a great diversity of habitats (sandy zone, gravelly, mangrove, tannes, seagrass, small bolongs, pit, shoals, etc.
		This diversity of habitats and the variability of their physicochemical and microclimatic characteristics over the seasons favor the presence of a great
		10 species of bats were observed on the MPA and only 7 were determined with certainty. According to IUCN, live bats on the MPA are not at risk and are included in the category "Least Concern" or "minor concern." In insects, three groups of
		inventoried species) and Orthoptera (locusts, grasshoppers, crickets for 30 species), or "minor concern." Terrestrial mammals: 18 species of terrestrial mammals are A. According to IUCN, all species have the status of "Least Concern" or "minor concern."
		Marine mammals: The Atlantic humpback dolphin and the Senegal manatee are considered as two "vulnerable" species by IUCN, i.e., the category just below the "Endangered."

Table 7.2 (Continued)

Protected Area	Area (ha)	Biodiversity Interests/International Recognition
Marine Protected Area of Saint-Louis created in 2004	49,600	of sandy-muddy nature that is distinguished by the presence of shells. resources consist of demersal species such as *Sciaenidae* (white carps, jowl, sole, captain, etc.), *Sparidae* represented by sardinella, while at the edge of the plateau meet especially white shrimp, lobsters. Marine mammals are represented by the whale, dolphin blower (*Turciops truncatus*) and the monk seal (*Monachus monachus*). These species are green turtle (*Chelonia mydas*), a herbivore species, is more common. While the other species are much rare it is the hawksbill turtle (*Erethmochelis imbricata*) which is generally found in shallow waters; Leatherback turtle (*Dermochelys coriacea*) is a species of high seas that does closer to the coast than every two years for the purpose of laying and *Lepidochelys olivacea* and *Lepidochelys kempii*.
Marine Protected Area of Kayar created in 2004	17,100	Area. grounds are grouped into 4 main areas with a certain spatial, physical and biological homogeneity. The different levels of habitat types provide a broad spectrum for a rich There is great species richness due to diverse habitat (sandy, sandy-sandy, and rocky bottoms and great depths at or near the edge of the pit).
Marine Protected Area of Joal Fadiouth created in 2004	17,400	Existence of a seagrass meadow that is a breeding area of many species and a zone of important nurseries throughout the West African region especially for the green turtle, *Chelonia mydas* called "Ndoumar" in the local language, the leatherback turtle (*Dermochelys coriacea*) also called Wagnor

Protected Area	Area (ha)	Biodiversity Interests/International Recognition
Marine Protected Area of Abéné created in 2004	11,900	- Spawning areas and feeding for ichthyo fauna - Sea turtles - Many birds: white and gray pelican; Goliath heron; heron guard beef heron ash, gray-headed gull, terns; dimorphic crest, cormorants; anhinga;, etc. and many - Green monkey, patas - Nile crocodiles
Marine Protected Area of Gandoul created in 2014	15,732	Multifunctional space with regard to its natural potentialities with a diversity of landscape units, an important biological diversity characterized by seabirds, marine turtles, *Crocodylus niloticus* Nile crocodile and Manatee *Trichechus senegalensis*. reptiles, etc.
Marine Protected Area of Sangomar created in 2014	87,437	In the marine part, it includes a spawning area (the Sangomar Pit or Bakina Pit).
Marine Protected Area of Niamone Kalounayes created in 2015	66,032	Its main biotopes consist of maritime and coastal facades and mangroves. The cetaceans, birds and marine turtles.
Marine Protected Area of Kassa Balantacouda	23,200	The interest of the area lies in the presence of mangrove fairly preserved. It is crossed by the Casamance River cetaceans, birds and sea turtles

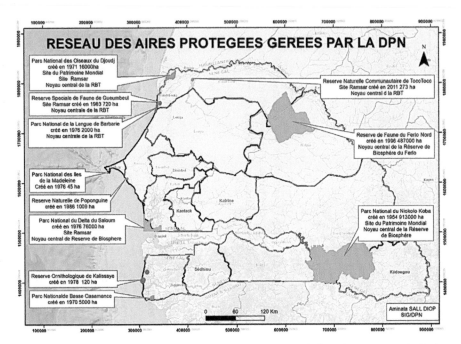

Figure 7.10 Network of protected areas managed by the National Parks management (Courtesy of the U.S. Geological Survey, EROS Center).

☐   ☐   ☐

- The Saloum Delta Biosphere Reserve with Saloum Delta National Park and the Bamboung, Gandoul, Sangomar and Palmarin Reserve Marine Protected Areas as the main centers.
- The Senegal River Delta Transboundary Biosphere Reserve with the Djoudj Bird National Park (PNOD) as its main points, the Barbarie Langue National Park (PNLB), the Guembeul Wildlife Reserve (RSFG), the marine protected area of Saint Louis and the Reserve of Birdlife Ndiaël (RSAN).
- The Niokolo Koba Biosphere Reserve with Niokolo Koba National Park as its central core.
- The Samba Dia Biosphere Reserve with a central part of the Samba
        ☐
- The Ferlo Biosphere Reserve which concerns the North and South Ferlo.

Senegal also has eight Ramsar sites (wetlands of international importance) and that are:

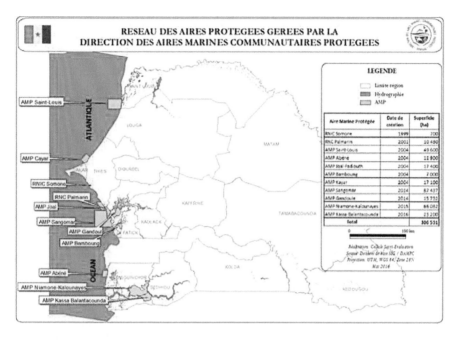

Figure 7.11   Protected areas network managed by the Protected Marine Community Areas Directorate.

- Djoudj National Bird Park.
- The Special Bird Sanctuary of Ndiael.
- Saloum Delta National Park.
- The Guembeul Wildlife Reserve.
- Tocc Tocc Community Nature Reserve.
- Community Nature Reserve of Somone.
- The Palmarin Community Nature Reserve.
- The Ornithological Reserve of Kalissaye.

7.6   Challenges of Biodiversity Conservation

As many Sahelian countries, Senegal has suffered for several decades, a decrease in rainfall and increasing pressure on natural resources (CSE, 2003). The combined effects of population growth and climatic disturbances have led to land degradation.

In Senegal, the degradation of ecosystems is linked to the combined action of several factors including the expansion of agricultural land, the overexploitation of biological resources, overgrazing,   A

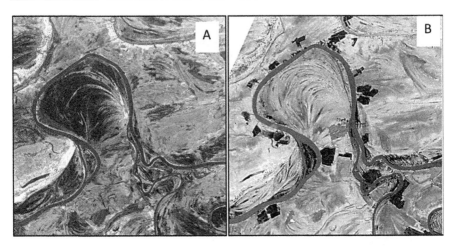

Figure 7.12   Gonakier forest in northern Senegal in 1965 (A) and 1990 (B) (*Source:* Tappan, 2000). Courtesy of the U. S. Geological Survey, EROS Center.

species, increasing urbanization, and climatic changes. This degradation has not spared plant and animal species, some of which have disappeared, and others are now threatened with extinction despite the efforts made by the country for the conservation of biodiversity (DPN, 2014).

The destruction and fragmentation of ecosystems is the leading cause of biodiversity loss in the world, and Senegal is no exception. In the country, this phenomenon is mainly related to human activities, including agriculture, urbanization, the construction of dams, etc. (Figure 7.12).

Bush   annually destroy vast areas of forest and result in the loss of many plant and animal species. The Ecological Monitoring Center (ESC) estimates that 847,600 ha of area burned by   in 2011 compared to 755,900 ha in 2010. This   increase in area burned is greater in the southern regions and the eastern part of the country. The   A◻ amounted to 5,741,810 tons in 2011. In 2012, the area burned is estimated at 89,824 ha with a total of 393 reported   (DEFCCS, 2013). Overexploitation of natural resources is also a source of biodiversity decline. This abusive and fraudulent exploitation not only takes into account the species' renewal capacities, but also uses very destructive techniques and practices. In Senegal, various factors contribute to this phenomenon. These include over-
        ◻A
(exotic) invasive plants are the third leading cause of biodiversity loss in the world. In Senegal, native species such as *Typha domingensis* and introduced species such as *Hyptis suaveolens, Salvinia molesta, Mimosa pigra*

(Hutchinson and Dalziel, 1958) have invaded most of the country's freshwater ecosystems. The invasion is more marked in the Senegal River and in the Niokolo Koba National Park where *Mitragyna inermis* and *Mimosa pigra* have colonized most of the pools.

Water, soil and air pollution and land-based pollution directly or indirectly affect certain terrestrial and aquatic ecosystems and species. Pollutants from industrial, agricultural and domestic wastes seriously threaten the survival of many species and cause ecosystem degradation.

Climate change, the main effect of which is global warming, will lead to sea level rise of 20 cm by 2030 and 80 cm in 2080 in Senegal. The main climatic risks are mainly the rainfall   A resources, the advancement of the sea and coastal erosion, the recurrence of torrential rains and the off-season rains, the   the salinization of the lands and water, extreme temperatures and heat waves, cold spells, shrinking areas of major natural wetlands, etc. (Republic of Senegal, 2006, 2010). The impact of climate change is perceptible on biodiversity. According to the IPCC (2014), climate change will cause physiological disturbances in some species, but also changes in the functioning of ecosystems. Species with limited adaptive capacity are at very high risk if the average global temperature increases by 2°C.

Coastal erosion poses a real threat to the Senegalese coastline and causes a degradation of biodiversity and numerous socioeconomic impacts. Salinization is also one of the main factors of ecosystem degradation. Soil salinization is known in two forms: primary salinization from the source rock and secondary salinization that depends on a whole set of processes and environmental factors.

The conversion of ecosystems to farmland model is shown in Figure 7.13. Land-use statistics between 1975 and 2010 revealed an increase in

Figure 7.13 Woodland savannah in 1983 (A) converted to agricultural land in 1988 (B) (Tappan, 2000).

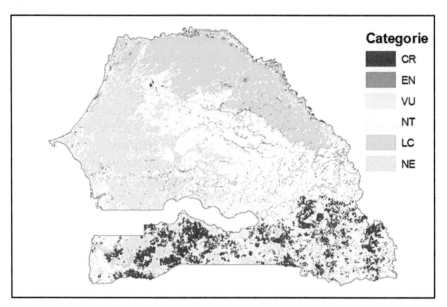

Figure 7.14   Draft of ecosystem status of Senegal (draft) according to the criteria of the Red List of Ecosystems (Rodriguez et al., 2010) (*Source: CSE, 2012*).

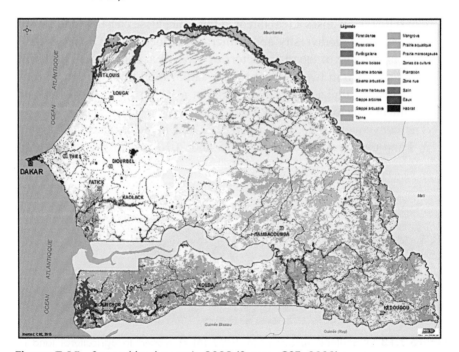

Figure 7.15   Senegal land cover in 2008 (*Source: CSE, 2008*).

Table 7.3   Evolution of the Areas of the Ecosystems in Senegal Between 1990 and 2008 (CSE, 2012)

Land cover class	Area (sqm)_1990	Area (sqm)_2008
Dense forest	6354.25	5948.29
Open forest	17270.23	11906.28
Gallery forest	6413.22	6305.67
Gonakier relict forest	295.70	244.89
Tree Savannah	14646.09	18802.05
Savane arbustive	65191.94	62403.08
Thicket	165.75	162.00
Tree to shrub steppe	3465.53	45.13
Shrub Steppe	24329.41	32399.36
Mangrove	2306.10	1760.05
	2231.31	2400.02
Aquatic vegetation (Typha)	507.65	539.22
Swampy meadow	1498.25	1472.39
Mud Flat	333.83	525.72
Flooded bare soil	1047.55	1444.35
Bright dune	39.51	25.03
Sandy beach	45.16	42.87
Waterbody	1932.62	1962.65
Lake and pond	385.53	410.34
Recession crop	88.36	88.37
Irrigated farming	1140.13	2111.35
Rain crop	36539.58	43456.74
Fruit tree plantation / Orchard	328.58	360.90
Forest plantation	373.07	429.50
Inhabited area	1008.00	1221.85
☐ habitat	90.01	119.36

areas of higher cultivation in the southern forest zone of Casamance (2,480 km²) followed by that in the agro-silvo-pastoral zone (1,576 km²), the groundnut basin (1,056 km²), the sylvo-pastoral zone of Ferlo (816 km²) and the Senegal River valley area (616 km²). In the Niayes area, the area under cultivation decreased by 132 km². It should also be noted that when growing areas increased by 6,412 km², savannah areas decreased by 8,708 km² (CSE, 2015).

Fraudulent carbonization is practiced in managed forests (quota over-runs) as well as in other types of forest, with the result that deforestation is accentuated, especially in areas where carbonization has been totally prohibited to allow natural regeneration of vegetation.

In general, the trend in the physiognomy of ecosystems gives the appearance of a deteriorating environment with   consequences for biodiversity.

This observation motivated the process of drawing up Senegal's Red List of Ecosystems (Figure 7.14), which should ultimately lead to a clear idea of their health in order to establish sustainable regeneration or restoration strategies (Figure 7.15 and Table 7.3).

Keywords

- challenges
- ecosystems
- fauna
- flora

References

ANSD, (2014). General Census of Population and Housing, Agriculture and Livestock.

Ba, A. T., & Noba, K., (2001). Flora and plant biodiversity in Senegal. *Drought, 12*(3), 149–155.

Ba, A. T., & Noba, K., (2001). Flore et biodiversité végétale au Sénégal. *Sécheresse, 12*(3), 149–155.

Berhaut, J., (1967). *Flora of Senegal more complete with the humid forests of Casamance.* ed. Clear Africa.

CSE, (2002). *Rapport sur l'Etat de l'Environnement.*

CSE, (2003). Land Degradation Assessment in Senegal. *FAO Land Degradation Assessment Project* (LADA): Preliminary Report.

CSE, (2008). Mapping report of the land use of Senegal.

CSE, (2009). *Directory on the environment and natural resources of Senegal.* Ed. CSE, 2nd edition.

CSE, (2012). *Report of the Regional Workshop on the Red List of Ecosystems Development Process.*

CSE, (2013). *Directory on the environment and natural resources of Senegal. Ed. CSE, 3rd edition.*

CSE, (2015). *Update of the cutting and characterization of Ecogeographical zones of Senegal.*

DEFCCS, (1993). *Forest Action Plan of Senegal. flight. 1, II & III* - Dakar.

DEFCCS, (2013). *National Report on Bushfires.*

Diouf, M., Diop, M., Lô, C., Drame, K. A., Sene, E., Ba, C. O., Guèye, M., & Faye, B., (1999). Prospecting of traditional African leafy vegetables in Senegal. In: Chweya, J. A., & Eyzaguirre, P. B., (eds.). *The Biodiversity of Traditional Leafy Vegetables*, IPGRI, pp. 111–154.

DPN, (2014). *Fifth National Report on the Implementation of the Convention on Biological Diversity.*

Fall, A. D., (1996). *Protection of natural ecosystems in Senegal: example of medicinal plants. Surveys conducted at the level of Dakar and Thiès regions.* Thesis of Doct. Pharmacy, No. 3, UCAD.

Forest Code, (1998)., Law 98/03 of 08/01 /, (1998). Decree no. 98/164 of 20/02. CSE, (2002). State of the Environment Report.

Guèye, M., & Diouf, M., (2007). Traditional leafy vegetables in Senegal: Diversity and medicinal uses. *African J. Traditional, Complementary and Alternative Medicine, 4*(4), 469–475.

Guèye, M., & Samb, P. I., (2016). Wild edible plant use among the people of Tomboronkoto, Kédougou region, Senegal. In: Hardy, K., & Kubiak-Martens, L., (eds.). *Wild Harvest: Plant in the Hominin and Pre-Agrarian Human Worlds*, Oxbow Books, Oxford & Philadelphia, 341–360.

Guèye, M., (2012). *Ethnobotanical study among the Malinké of the Rural Community of Tomboronkoto (Kedougou Region) and enhancement of the historical collections of the Herbarium of the Fundamental Institute of Black Africa* (IFAN) A. DIOP / UCAD Ch. PhD thesis in Natural Sciences, UCAD.

Guèye, M., Ayessou, C. N., Koma, S., Diop, S., Akpo, L. E., & Samb, P. I., (2014). Wild fruits traditionally gathered by the Malinke ethnic group in the edge of Niokolo Koba park (Senegal). *American J. Plant Sci., 5*(9), 1306–1317.

Guéye, N. S., (2003). Rural women's access to land: a key factor in Food Security. *International Workshop "Rural Women and Land"–Thiès*, 25–27 February, Senegal..

Guéye, S., (2000). *Study on forest resources and forest plantations in Senegal*, Period: 1992–1999. Data collection and analysis for sustainable forest management - joint national and international efforts. EC-FAO Partnership Program (1998–2002). Technical Report FDCA / TR / 15.

Hájek, I., & Verne, P. H., (2000). Aerial Census of Big Game in Niokolo-National Park and Falemé Region in Eastern Senegal. *Proc. All Africa Conference on Animal Production, Alexandria, 3*, 5–9.

Hutchinson, J., & Dalziel, J. M., (1958). *The Flora of West Tropical Africa, 2nd ed.* (Revised by Keay, R. W. J.). Crown agents, London.

Kheraro, J., & Adam, G., (1974). *The Senegalese PharmacopoeiaTraditional: Medicinal and toxic plants.* ed. Vigot brothers.

MEDD, (2014). (Ministry of Environment and Sustainable Development), Nationally Determined Intended Contribution (INDC) of Senegal.

Mingou, P., & Guèye, M., (2017). The Pteridological Flora of Some Wet Places in the Kedougou Region (Senegal). *European Scientific Journal*, 13 (12), 127-144. URL: http://dx.doi.org/10.19044/esj.2017.v13n12p127.

Ministry of Agriculture and Rural Equipment, (2012). *Official catalog of species and varieties cultivated in Senegal*, 1st edition, Dakar, Senegal.

Mobius, Die Auster und die austernwitschaft, (1877). *Berlin: Vrelag von Wiegandt, Hempel and Parey*. Trans, by H J Rice: in report of commissionner of 1880, part III, U.S commission of fish and fisheries, pp. 683–751.

Mugnier, J., (2008). *New flora of Senegal and neighboring regions.*

Piqué, R., Guèye, M., Hardy, K., Camara, A., & Dioh, E., (2016). *Not Just Fuel: Food and Technology from Trees and Shrubs in Falia*, Saloum Delta (Senegal). In: Biagetti, S., & Lugli, F., (eds.). *The Intangible Elements of Culture in Ethnoarchaeological Research.* Barcelona (Spain) and Rome (Italy), 217–230. doi: 10.1007/978-3-319-23153-2_17.

PRSP, (2002). *Strategic Paper for Poverty Reduction in Senegal.* Report.

Rodriguez, J. P., Kathryn, M., Rodriguez, C., Baillie, J. E. M., Neville, A., Benson, J., et al., (2011). Elaboration of the IUCN Criteria for the Red List of Threatened Ecosystems. *Conservation Biology. 25*, 21–29.

Sournia, G., & Dupy, A., (1990). Senegal. In: East, R., (ed.). *Antilopes. Global Survey and Regional Action Plans, Pt 3: West and Central Africa.* IUCN Gland.

Stevels, J. M. C., (1990). *Traditional vegetables from Cameroon, an agro-botanical study.* Wageningen Agricultural University (WAU), pp. 90–91.

Study report. CSE, (2015). *Report on the state of the environment in Senegal.* Ed. CSE, Dakar.

Tansley, A. G., (1935). The use and abuse of vegetational concepts and terms. *Ecology, 16*(3), 284–307.

Tappan, G. G., Hadj, A., Wood, E., & Lietzow, R. W., (2000). Use of Argon, Corona, and Landsat imagery to assess 30 years of land resource changes in Western Senegal. Photogrammetric Engineering & Remote Sensing. *American Society for Photogrammetry and Remote Sensing, 66*(6), 727–735.

Traore, S. A., (1997). *Analysis of the woody flora and vegetation of the Simenti area (Niokolo Koba National Park), Eastern Senegal.* Postgraduate thesis, Faculty of Sciences, Dakar Cheikh Anta Diop University.

CHAPTER 8

Biodiversity in Sierra Leone

PRINCE E. NORMAN, JENNEH F. BEBELEY, JANATU V. SESAY, and YVONNE S. NORMAN

Sierra Leone Agricultural Research Institute (SLARI), PMB 1313, Tower Hill, Freetown, Sierra Leone, West Africa, E-mail: penorman2008@yahoo.com/p.norman@slari.gov.sl

8.1 Introduction

Sierra Leone is situated along the Atlantic Ocean in West Africa between latitudes 6° 55' N and 10°N and longitude 10° 14' W and 13° 17' W. The country has a total land surface area of 71,740 km² (27,699 sq. miles) with an arable land of 6,026 km² or 0.6 million ha (representing 8.4% of the total surface area); agricultural land is 28,839.5 km² or 2.9 million ha (40.2%); permanent cropland is 789.14 km² (1.1% of the total land surface area); irrigated area is 37.1 km² (4.7% of permanent cropland); forest land is 27,620 km² (38.5% of total land surface); nationally protected land is 2,941 km² (4.1% of the total surface area); while other land use accounts for 7.7% of the total surface land area (GOSL, 2014).

According to the country environmental   A
the Upper Guinean Lowland Forest Ecosystem with an abundant richness in ecosystem and species biodiversity (lowland rainforests, mountain forests, savannah woodlands, agricultural, freshwater and wetlands). There are 48 forest reserves and conservation areas, representing about 4% of the land area. The total area of government wildlife reserves is estimated at 173,000 ha. There are over 2,000 species of plants including 74 endemic species
        A
antelopes and duikers, 9 bat species and over 500 bird species have been recorded in the country. An estimated 4,837.8 km² of Sierra Leone is covered by wetlands with vegetation, which is typically of freshwater swamp forests, riparian and mangroves. There are essentially two seasons; wet (May–October) and dry (November–April) seasons, each lasting approximately 6 months. The annual rainfall varies from about 1,800 mm in the Northeast of the country to about 5000 mm in the Freetown Peninsula. The average monthly temperatures are around 26°C. The heavy rains and maritime   A ence lead to humidity values of up to 92% in the wet season and 45% inland in the dry season (GOSL, 2014).

The vegetation of Sierra Leone is determined by natural features such as geography, geology, and topography. The country is divided into four topographic regions, namely, the coastal lowland, the interior plains, the interior plateau and scattered mountains and hills. The coastal lowlands occupy the West to Southwest of the country, and consist of a narrow strip about 40 km wide that lies below 7 m a.s.l.. The interior plains extend from 50 to 130 km inland, rises to 200 m in the East and cover 43% of the land. The interior plateau emerges as an abrupt escarpment that runs almost parallel to the interior plains, giving rise to higher plateaus in the Eastern sector, which is topped by two mountain ranges, the Loma Mountains (that peak at Mount Bintumani, 1945 m a.s.l.) and Tingi Hills (that peaks at SankanBiriwa, 1709 m a.s.l.). Mount Bintimani is the highest peak in West Africa, west of Mount Cameroon). Other mountain ranges include the Western Area Peninsula, the Kangari Hills on the central region and the Kambui Hills on the south of the country. The Freetown Peninsula is made up of dissected mountainous Peaks with Sugar Loaf and Picket Hills being the highest. Sierra Leone's rich biological diversity is   A ☐
fauna are impressive and consists of wildlife and domesticated species and their richness and diversity had been recognized since colonial times. There are approximately 48 forest reserves and conservation areas in Sierra Leone.

8.2   Importance of Biodiversity

Biodiversity is noted as the foundation for human health (GOSL, 2014). By securing the life-sustaining goods and services which biodiversity provides, the conservation and sustainable use of biodiversity can provide significant benefits to humans. The value of the biodiversity of Sierra Leone is a reflection of the uses by the man of the diverse resources found in various ecosystem types. In Sierra Leone, natural resources of all categories determine the pattern of economic growth, depending mostly on how they are used, valued and managed, and on the economic policies and institutions put in place (Blinker, 2006); hence, the need to focus on the institutions managing these resources. With its rich biodiversity, the country has potentials to raise funds through sales of biodiversity resources, ecotourism, genetic modification, and other ecosystem services. The nation's biodiversity can also contribute to and promote its health through its diverse natural capacity to mitigate both natural and anthropogenic stresses.

Sierra Leone is a predominantly agricultural country with agriculture sector contributing 31% GDP. The forest resource is a   contributor to the national economy, both in monetary terms and indirect   The country's forestry resources are derived from the current forest cover representing about 5% of total area. Moreover, a number of forest reserves are being subjected to timber exploitation. A large part of the rural population depends on wildlife cropping targeting animals including mammals, birds, reptiles, frogs, and insects. Besides the direct consumptive value of animals to man, the bushmeat trade is widespread in the country (GOSL, 2014).

Local communities also forage for a variety of products for use as food, fuel (fuelwood and charcoal), construction materials, thatching and materials, ropes, crafts, medicinal plants, fodder, recreational materials (raf-

A

ticides. The reptile, amphibian, and manatee biodiversity is a category of biodiversity in Sierra Leone, and they provide pest control services in the food productive cycle of the country. They are part of the culture of Sierra Leoneans, who used them for different life-sustaining values such as food and medicine (GOSL, 2014).

The wildlife parks and reserves could serve to increase the ecotourism potential and the rich biodiversity affords an opportunity for research. There are cultural and spiritual values attached to certain elements of biodiversity such as the existence of sacred groves for initiations into secret societies. Traditional healers depend on medicinal plants to cure various ailments at the community level. The forest resources, though heavily depleted during the war, contribute   to the economic development of the country. Forestry contributes about 9.3% to the GDP, but this calculation excludes nonmonetary   A

forest product such as medicines. Tourism from National Parks and Game Reserves are a latent demand yet having great potentials. When all these are compiled the contribution of the forestry sector could be as high as 15% (GOSL, 2014).

The livelihood of the rural population in Sierra Leone, as in most countries in Africa and Asia, incorporate natural resources and high diversity, regardless of whether the agro-ecosystems are based on permanent cropping, predominantly pastoral or mixed. This helps to provide resilience in the face of adverse trends or shocks, and offers a greater choice of livelihood options. Traditional medicine, which relies on species of wild and cultivated plants, is the basis of primary health care for the majority of people in developing countries like Sierra Leone. Recreational opportunities and aesthetic value associated with wild birds, salt-extraction, water/freshwater recreational   and parks bring in much-needed revenue. Biodiversity, from which all these   A

economic and cultural development.

From an agriculture standpoint, different types of biodiversity are used at different times and in different parts of the country; thereby contributing to livelihood strategies in a complex fashion. For example, wild resources are particularly important for food and livelihood security of the rural poor, women and children, especially in times of stress, such as the hungry period, when food stocks are low or in a period of drought. These groups generally have less access to land, labor, and capital and thus need to rely more on the wild diversity available. At least 70% of the country's population depends on agricultural biodiversity for livelihood. The sector continues to be the main contributor to growth in terms of share (45% of value-added) and of added GDP (just under half of the real GDP growth) (GOSL, 2014).

The Upper Guinea forests of West Africa encompass a belt of lowland rainforest stretching from Guinea and Sierra Leone eastwards to Ghana. In some places, this rainforest belt is up to 350 km wide, but becomes as narrow as 100 km in Cote d'Ivoire. It is   ⌐A
biodiversity hotspots in the World, with a total of 2,800 species of vascular plants of which about 650 species (23%) are endemic to the region. Some 15 endemics such as the White-necked Picathartes and White-breasted Guinea fowl are among a large number of bird species present. The Gola forests of Sierra Leone have been described as the region's center of diversity and endemism (Klop et al., 2008). There are lots of other species of value in conservation. The Outamba Kilimi National Park (OKNP), the only extensive area of savanna woodlands and grassland savanna with protection in Sierra Leone, and with a high primate population, especially chimpanzees have a potential for education and ecotourism development.

8.3   The Concept of Genetic Diversity

Genetic diversity is defined as the heritable variation created, enhanced or maintained in populations of living organisms (FAO, 1999). Such diversity is often reflected in the allelic and genotypic frequencies of populations of living organisms. It plays a great role in the adaptation of living organisms to changing environments. The continuity of generations of living organisms depends on the suitability of allelic variations and the successful survival capability of their offspring to changing environments (NBII, 2011).

Genetic diversity and species diversity are interdependent. Species diversity is a measure of within an ecological community. It involves both species richness and evenness. Species richness refers to the total number of different species in a community, whereas species evenness refers to the variation of abundance within individuals per species in a community. An alteration in species diversity leads to changes in the environment, leading to adaptation of the remaining species. Loss of species is caused by alterations in genetic diversity leading to loss of biological diversity (NBII, 2011). For instance, loss of genetic diversity within populations of livestock populations is caused by the extension of markets and economic globalization (Groom et al., 2006; Tisdell, 2003).

8.4 Diversity of Plants

According to *Earth Trends* (2003), there are about 2090 higher plant species. However, the number of known plant species in Sierra Leone is estimated at approximately 1,576 belonging to 757 genera and 135 families. Of these, about 7 plants are endemic including *Triphyophyllum peltatum* Hutch. & Dalziel, *Octoknema borealis* Hutch. & Dalziel, *Napoleona leonesis* Hutch. & Dalziel., *Afrotrilepis jaegeri* J.Raynal., crabgrass or finger-grass (*Digitaria phaeotricha* (Chiov.) Robyns) and *Loxodera strigosa* (Gledhil) Clayton. In the closed forest and savannah communities, *Uapaca togoensis* L., Red Monkey kola (*Cola lateritia* K.Schum), Guinea Plum (*Parinari excelsa* Sab.), Kpindii or ishin odu (*Ochna membranacea* Oliv.), Gorli shrub (*Caloncoba echinata* (Oliv.) Gilg) and Dwarf Red Ironwood (*Lophira lanceolata* Tiegh. ex Keay) are more predominant compared to the Sub-montane flora, which comprises of Guinea Plum (*Parinari excelsa* Sab.), African rosewood (*Anthonotha macrophylla* P. Beauv.), *Aphimas pterocarpoides*, Ogea or Shedua (*Daniella thurifera* (J. J. Bennett), L.), *Dissotis elliotii* Gilg. or *Dissotis thollonii* Cogn., ke-rin-ke-ra-lal or bulom (*Ctenium newtonii* Hack.), *Londettia kagerensis* (K. Schum.) Hutch. and *Cyanotis longiflora* Wight (Cole, 1974). In the savannah ecosystem, the dominant tree species include *Daniella oliveri*, African locust bean (*Parkia biglobosa*(Jacq.) R. Br. ex G. Don f.), drumstick tree (*Cassia sieberiana* DC.), Dwarf Red Ironwood (*Lophira lanceolata* van Tiegh. ex Keay), African fan palm (*Borassus aethiopum* Mart.), clusterleaf or silver cluster-leaf or silver terminalia (*Terminalia albida* Scott-Elliot) and black plum (*Vitex cuneata* Schumach. & Thonn.), and the dominant grasses are blue stem (*Andropogon gabonensis* Stapf), World Grasses (*Andropogon tectorum* Schumach. & Thonn.), English cane grass (*Chasmopodium caudatum* (Hack.) Stapf) and jaragua grass or giant thatching grass (*Hyparrhenia rufa* (Nees) Stapf).

8.4.1 Diversity Within Crop Species

The diversity within crop species grown in Sierra Leone is variable. A total of 142 economic crop species were recorded in the indigenous farming system.

These crops were grouped into eight and their various proportions include edible fruits (29.6%), vegetables (15.5%), potherbs (14.8%), beverages (12.7%), root and tuber crops (9.9%), cereals (6.3%), nuts and oils (6.3%), and legumes (4.9%) (FAO, 1996). About 55–65% of the crop species including edible fruits, legumes, cereals and nuts were underexploited and grown in the wild. Beverages, vegetables, and root and tuber crop exhibited lesser (35–45%) wild species, whereas pot-herbs or leafy vegetables formed 9.5% (i.e., 19 out of 21 species) wild species. Pot-herbs are mostly cultivated in backyard gardens for daily use (FAO, 1996).

Wild edible fruits constitute 42 species of tropical fruits, of which mangoes and bananas are eaten raw. About 64% of these species are considered mainly wild in closed forests, forest regrowth, and savannah woodlands. Some of the wild fruits include sour tumbler (*Tamarindus indica* L.), African almond (*Tarrietia utilis* Sprague), dita (*Detarium senegalense* Gmelin.), damson (*Dacryodes trimera* (Oliv.) H.J. Lam), black plum (*Vitex doniana* Sweet.), monkey apple (*Anisophyllea laurinia* R.Br. ex Sabine), A *Spondias mombin* L.), blackberry (*Flacourtia vogelii* Hook. f.), botlem (*Omphalocarpum pachysteloides* Mildbr. ex & Dalziel), bobby water (*Chrysophyllum cainito* L.), bush lime (*Erythrococca anomala* Benth.), mammy supporter (*Mammea americana* L.), passion fruit (*Passiflora foetida* L.), A *Uvaria chamae* P. Beauv.), malombo (*Salacia senegalensis* Lam.), monkey *Uvaria afzelii* Sc. Elliot) and wild citrus (*Citropsis articulata* (Willd. ex Spreng.) Swingle & Kellerm.) (FAO, 1996).

Some of the wild vegetables reported in Sierra Leone include white pumpkin, bottle gourd (*Lagenaria siceraria* (Mol.) Stand.), wild jakato or Gboma eggplant (*Solanum macrocarpon* L.) and *S. dasyphyllum* Schum. & Thonn. These are both wild relatives of tomato, squash and jakato. Snake tomato (*Trichosanthes cucumerina* L.) serves as substitute for tomato, whereas tola (*Beilshmielda mania*) substitutes dried powdered okra, and A *Luffa acutangula* Roxb.). Watermelon (*Citrullus lanatus* (Thunb.) Mansf.) and pumpkin (*Cucurbito pepo* L.) seeds are ground into egusi powder to thicken leafy vegetable soups. These are sometimes substituted by castor oil seeds (*Ricinus communis* L.) and groundnut (*Arachis hypogaea* L.) (FAO, 1996). Some of the cultivated vegetable species include cucumber (*Cucumis sativus* L.), pumpkin, e.g., summer squash (*Curcubita pepo* L., *C. maxima* Duch., *C. moschata* Duch. ex Lam.), tomato (*Lycopersicon esculentum* Mill), garden eggs (*Solanum melongena* L.), cowpea (*Vigna unguiculata* (L.) Walp.), broad bean (*Vicia faba* L.), pepper (*Capsicum annuum* L.), okra (*Abelmoschus esculentus* (L.) Moench).

Pot-herbs (leafy vegetables) are widely used in West Africa, including Sierra Leone, as a cheap source of nutrient supplements in starchy staple

food diets such as rice, yam, cassava, fundi, etc. About 21 potherb species are domesticated. However, over 70% potherb species are found in farm bush growing as weeds. Some of them include celosia or shokotoh-yokotoh (*Celosia argentea* L.), ajefawo (*C. leptostachya* Benth.), pitch-forks or bush needle (*Bidens pilosa* L.), rat's ear (*Portulaca oleracea* L.), purslane or ogamoh and efo-odu (*Solanum nigrum* L.), bologi or woro (*Crassocephalum biafrae* (Oliv. & Hiern) S. Moore), broad bologi (*Basella alba* L.), Ceylon spinach or Lagos bologi (*Talinum triangulare* (Jacq.) Willd), scarlet
□A   *Emilia coccinea* (Sims) G. Don) and efonyori (*E. praetermissa* Milne-Redh), patmenji (*Ocimum basilicum* L.), and water bitters (*Struchium sparganophora* L.).

Beverage plant species are important in Sierra Leone. Two indigenous wild coffee species include *Coffea stenophylla* and *C. liberica* discovered in the wild Gola-Gofa rainforest. Fragrant teas brewed from dried leaves of lemon grass (*Cymbopogon citratus* Stapf), Gambia tea bush (*Lippia chevalieri* Moldenke), tea bush (*Ocimum gratissimum* L.) or bush tea bush, pignut or chan (*Hyptis suaveolens* (L.) Poit.). Ketenfe (*Thaumatococcus daniellii* (Benn.) Benth.) and serendipity berry (*Dioscoreophyllum cumminsii* (Stapf) Diels) are the two natural beverage protein sweetener species found in Sierra Leone. The major cultivated beverage crops are oil palm (*Elaeis guineensis* Jacq.), Cacao (*Theobroma cacao* L.), Coffee [Arabian coffee or coffee (*Coffea arabica* L.); robusta coffee (*C. canephora* L. syn *C.robusta* (L.) Linden); Liberian coffee or Kafeng barako (*C. liberica* W. Bull ex Hiern); Coffee excels or Kafeng barako (*C. excelsa* A. Chev.); narrow leaf coffee or highland coffee (*C. stenophylla* G. Don)]. Coconut palm (*Cocos nucifera* L.). Rubber (*Hevea brasiliensis* Willd. ex Adr. Juss.), cashew (*Anacardium occidentale* L.) and citrus including sweet orange (*Citrus sinensis* L.), lime (*Citrus aurantifolia* (Cristm.) Swingle.), lemon (*C. limon* L.), and grapefruit (*Citrus paradisi* Macf) are also important economic tree crops grown in the country. Bitter Kola (*Garcinia kola* Heckel.) is sometimes used as a substitute for kola nut (*Cola nitida* (Vent.) Schott & Endl.), due to its medicinal attributes including anticongestant, diuretic and antipyretic. Common wild relatives recorded of tuber crops so far is on the yam including *Dioscorea bulbifera*, *D. hirtiflora* Benth., *D. minutiflora* Engl., and cocoyam (*Alocasia macrorrhizos* (L.) G. Don). About   A]
Dioscorea alata L., *D. rotundata* Poir., *D. cayennensis* Lam., *D. bulbifera* L., *D. esculenta* L., *D. prahensilis* Benth), cocoyam (*Xanthosoma sagittifolium* (L.) Schott, *Colocasia esculenta* (L.) Schott). Root crops cultivated in the country include cassava (*Manihot esculenta* Crantz), sweet potato (*Ipomoea batatas* (L.) Lam.), and potato (*Solanum tuberosum* L.).

About nine cereal species are commonly cultivated in Sierra Leone. Rice is the staple food of the country and about two species including Asian or white rice (*Oryza sativa* L.) and local or brown rice (*O. glaberrima* Steud) are cultivated. The common wild relatives of *O. glaberrima* include *O. brachyantha* of upland rice farms, *O. barthii* and the perennial wild rice (*O. longistaminata* A. Chev. & Roehr.) of the riverain grassland ecology. Some of the cultivated cereal crop species include sorghum (*Sorghum margaritiferum* Stapf), sorghum (*Sorghum bicolor* (L.), pearl millet (*Pennisetum glaucum* [L.] R. Br.), maize (*Zea mays* L.), bulrush millet (*Pennisetum americanum* (L.) Leeke), fundi (*Digitaria exilis* Stapf).

Different wild species and relatives of the tropical spices, legumes, and nuts producing have been noted. The common spicy wild species include spearmint (*Mentha spicata* L.), cinamon (*Cinnamomum zeylanicum* Breyne.), symingi (*Xylopia aethiopica* (Dunal) A. Rich) and the local farmer's tobacco (*Nicotiana rustica* L.). Wild nut species used for edible oil extraction include bush palm kernels (*Diospyros heudelotii* Baill.), oyster nuts (*Telfairia occidentalis* Hook.f.), fula butter (*Cassia bicapsularis* (L.) Roxb.), shea butter (*Butyrospermum paradoxum* J. Gaertn.), fawei (*Pentaclethra macrophylla* Benth.) and Sodei (*Pentadesma butyracea* Sabine). Wild legumes noted include calabar beans (*Physostigma venenosum*Balf. f.), kenda or locust beans (*Parkia biglobosa* (Jacq.) Benth.) and bambara groundnut (*Voandzeia subterranea* L.). The commonly cultivated legume species include cowpea (*Vigna unguiculata* (L.) Walp.), broad bean (*Vicia faba* L.), groundnuts (*Arachis hypogaea* L.), pigeon pea (*Cajanus cajan* (L) Millsp.) and bambara groundnut (*Vigna subterranea* (L.) Verdc.).

8.4.2 Medicinal Plants

Indigenous medicinal plant species used in conventional medicine are curative or poisonous. Most (70%) of the wild-type medicinal plant species in West Africa have been reported (Oliver-Bever, 1983; Burkill, 1985, 1994). A comprehensive list of these species is reported by Cole (1994). Findings revealed that about 30% of the alien drug plants are now locally cultivated as food crops or ornamentals, whereas 40% are found in the closed forests, 20% in savannah woodlands, 7% in the arid lands and 3% in wetlands. Some of the poisonous medicinal plants include *Dichapetalum toxicarium* (G. Don) Baill. (broke-back or West African rat's bane), English ordeal tree; red water tree (*Erythrophleum guineense* (Guill. & Perr.) Brenan), English 'arrow poison' or brown Strophanthus (*Strophanthus hispidus* DC.), English swizzle-stick (*Rauvolfia vomitoria* Afzel,), Voacanga

(*Voacanga africana* Stapf), African allophylus (*Allophylus africanus* P Beauv.), Tietie (*Paullina pinnata* L.), monkey rope or snake climber (*Adenia cissampeloides* (Planch. ex Hook.) Harms), blue bitter-berry (*Strychnos usambarensis* Gilg), hairy thornapple (*Cnestis ferruginea* DC.), *Alafia multiflora* (Stapf) Stapf, Chinese lantern (*Dichrostachys glomerata* (Forssk.) Chiov.), gobi or kowi (*Carapa procera* DC.), Christmas bush (*Alchornea cordifolia* (Schum. & Thonn.) Müll.-Arg.), *Craterispernum laurimum*, ackee (*Blighia sapida* K.D. Koenig), African nut tree (*Mareya micrantha* (Benth) Mull-Arg.), cowitch (*Mucuna pruriens* (*L.*) DC.) and fish poison (*Adenia lobata* Forssk. and *Tephrosia vogelii* Hook.f.). Common medicinal plant species of the savannah woodlands and grasslands are drumstick tree (*Cassia sieberiana* DC.), bushveld peacock-berry (*Phyllanthus discoideus* (Baill.) Müll.Arg.), kwandari (*Terminalia macroptera* Guill. & Perr.), shea butter tree (*Butyrospermum paradoxum* (C.F. Gaertn.) Hepperor, sheanut tree (*Vitellaria paradoxa* C.F.Gaertn.), large red-heart (*Hymenocardia acida* Tul.), cabbage tree (*Anthocleista procera* Lepr. ex Bureau.), wild plum (*Grewia carpinifolia* Juss.), bitter or trifoliate wild-yam (*Dioscorea dumetorum* (Kunth) Pax) and the lily (*Urginea indica* (Roxb.) (Kunth). Some of the weedy medicinal plants noted in wasteland communities include goat weed (*Ageratum conyzoides* L.), four-leaved senna or black grain (*Cassia absus* L.), wonder berry or black nightshade (*Solanum nigrum* L.), asthma weed (*Euphorbia hirta* L.), stone breaker (*Phyllanthus amarus* Schum. & Thonn.), gotu kola (*Centella asiatica* L.), shrubby false buttonwood (*Borreria verticillata* (L.) G. Mey.), blackjack or hairy beggar (*Biden pilosa* L.), wormgrass (*Spigelia anthelmia* L.), sal leaved Desmodium (*Desmodium gangeticum* (L.) DC.), red hogweed (*Boerhavia diffusa* L.) and spider flower (*Gynandropsis gynandra* (L.) Briq.). Farm bush shrubs considered as medicinal plant species are candle bush (*Cassia alata* L.), sickle pod or coffee pod (*C. tora* L.), *C. podocarpa* (Guill. & Perr.) Lock, stinking weed or coffee weed (*C. occidentalis* L.) and pot-herb kren-kren or nalta jute (*Corchorus olitorius* L.). Ginger (*Zingiber officinale* Rosc.) is an important medicinal and economic spicy crop grown in Sierra Leone. However, the dearth of information exists on the extent of diversity of this crop (Figure 8.1).

8.5 Diversity of Animal Species

Sierra Leone exhibits reasonable diversity within existing animal species. The animal diversity can be broadly categorized into invertebrate and vertebrate animals. Reports on diversity within existing invertebrate species

| Coffee | Cocoa | Cocoyam |
| Yam | Pigeon pea | Patmenji |

Figure 8.1   Photographic Display of Selected Crops Grown in Sierra Leone.

are scanty and variable. A total of 600 species of butterflies and Odonata was proposed based on analysis of biogeographical and ecological patterns in West Africa and the Gola Forest of Sierra Leone (Klop et al., 2008). Of this amount, 44 species are endemic to Upper Guinea, and seven species including *Ornipholidotos nympha*, *Eresiomera petersi*, *Hewitsonia occidentalis*, *Spindasisiza*, *Anthene mahota*, *Bicyclus dekeyseri* and *Caenides dacenilla* were new records for Sierra Leone. The extremely rare and highly fragmented range species, *Euriphene lomaensis* (Loma nymph) and another species firstly discovered on Tiwai Island, *Cymothoe hartigi* (Hartig's Red Glider) were also noted. However, the previous report on the total number of butterflies and Odonata (dragonflies and damselflies) in Serra Leone was 363 species, giving a total of 370 species (Klop et al., 2008). According to GOSL (2014), the number of species of butterflies and moths is 108, and together with several species of bees and beetles, serve as pollinators of both cultivated and wild plants.

Comparatively, there are more research reports on vertebrates. There are about 178 species of mammals   of which, 15 species are primates. Of the 15 species, six species are threatened including the black and white Colobus Monkey (*Procolobus polykomus*), Red Colobus Monkey (*Colobus badiuspolykomos*), Diana Monkey (*Cercopithecus diana*), the Western Chimpanzee (*Pantroglodytes verus*), Olive colobus monkey (*Procolobus verus*) and Upper Guinea Red *Colobus* (*Piliocolobus badius*). Antelopes constitute

18 species, of which, nine are threatened and six endangered. These include Jentinck (*Cercopithecus jentinki*) and Zebra (*Cephalophus zebra* duikers). Other mammals under threat are the Forest elephant (*Loxodonta africana cyclotis*), the West African Manatee (*Trichechus senegalensis*), the pigmy Hippopotamus (*Hexaprotodon liberiensis*) and the Leopard (*Pantera pardus*), etc. Fruit bats comprise of nine species, while crocodiles exhibit three species including Nile, Slender-snouted and Dwarf (GOSL, 2014).

The amphibians and reptiles comprise of 55 and 67 species, respectively (wikipedia.org). Of the 67 reptile species,   A
green turtles (*Chelonia mydas*), hawksbill (*Eretmochelys imbricata*), olive ridley (*Lepidochelys olivacea*), leatherback (*Dermochelys coriacea*) and loggerhead (*Caretta caretta*) (GOSL, 2014). In Sierra Leone, the sea turtle species nest on beaches along the coast. The population of sea turtles is believed to be increasing, whereas the number of amphibians is decreasing due to predation, habitat   environmental acidity and toxicants, diseases, climate change, and interactions among these factors. Moreover, there is lack of information on primary   A
ians and reptiles in the country. Two important amphibian species   A
cally noted are the endemic frog found (*Bufo cristiglands*) and an endemic toad (*Cardioglossus aureolli*) noted in the Tingi Hills and the Western Area Peninsular, respectively.

Manatees are among the data   species in Sierra Leone. Only one species of manatees, *Trichechus senegalensis* (Seacow) has been reported. This species inhabits solely wetlands. Their activities are limited by natural and anthropogenic threats such as mining, logging, unsustainable farming, erosion and climate change (GOSL, 2014).

The diversity of birds in Sierra Leone is estimated at 642 species, of which 632 are regular, while 10 are vagrants. This presents a fairly reasonable level of diversity within the avian population possibly due to the two biogeographic vegetation including the Guinea-Congo forest biome and the Sudan-Guinea savannah biome. Moreover, 27 species of birds arouse global conservation concern (IUCN, 2013; BLI, 2013a), since two of them are endangered, 10 are vulnerable, 11 are near threatened and two are data   A
Eagle, Brown-cheeked Hornbill and Yellow-casqued Hornbill were upgraded to vulnerable status, whereas Bateleur, Martial Eagle and Blue-mustached Bee-eater, which had least conservation concern, were upgraded to near-threatened status. Nevertheless, the endangered Rufous   Owl was downgraded to vulnerable, and the near-threatened Turatis Boubou was also downgraded to least concern (IUCN, 2012; BLI, 2013b).

There are 489 resident species, accounting for 76.1% of the total avian population, of which 307 (47.8%) is a breeding population. Of the 307 resident species, 274 are Guinea-Congo Forest Biome (GCFB) species and 28 Sudan-Guinea Savanna biome (SGSB) species. About 15 GCFB species are endemic to the Upper Guinea forest with 13 species exhibiting global conservation concern (GOSL, 2014).

About 28 species are restricted to GCFB, accounting for about 4.4% of the total avian species. Only Emerald Starling is categorized data - cient, which attract research by many local and international bird enthusiasts. Many of the SGSB species are found in the Outamba-Kilimi National Park (D01) and the Lake Sonfon and environs (D02) (Okoni-Williams et al., 2005). However, Emerald Starling, Dybowski Twinspoti, Togo Paradise Whydah, Splendid Sunbird, Oriole Warbler and Piapiac have also been noted in the Ferrengbaia Hills of the African Minerals Ltd and the Bumbuna Hydroelectric Project areas (GOSL, 2014).

Most of the coastal and marine avifauna is migratory species. Of the 642 avian species, 146 species are migratory. Their migration is dependent on seasonal and moisture variability between the two locations they stay. The migratory birds are broadly [A and Afrotropical migrants, constituting 98 and 48 species, respectively. The Palearctic migrants migrate from Europe to the Mediterranean region in summer and to Africa in winter, contributing 98 (15.3%) of the total bird species in Sierra Leone. According to data collected during the IBA and other surveys, About 10 species of Palearctic migrants enter the country with 1% of their biogeographic population. The Afro-tropical migrants (intra-African migrants) often migrate in response to moisture rhythm, depending on the species, contributing 48 (7.5%) of the total bird species in the country. The Afrotropical migrants migrate more in the wet season (June to September). The only two species of Afrotropical migrants are Great White Egret and African Spoonbill, accounting for 1% of their biogeographic population (GOSL, 2014).

Livestock is another component of animal diversity covered in this review. Knowledge of livestock genetic diversity permits practicability of animal production across environments addressing various objectives. This serves as a good guide for selective breeding programs that permits adaptability of livestock populations to erratic environments (FAO, 2015). Genetic erosion and breed extinction account for the loss of livestock biodiversity. According to the 2014 Domestic Animal Diversity Information System (DAS-IS) report, 8774 breeds of livestock were recorded, of which, 7% are extinct and 17% is at the risk of extinction (FAO, 2015).

In Sierra Leone, various livestock farming activities are done. The most common ones include large ruminants such as cattle (N'Dama cattle, a West African breed); small ruminants such as goats (predominantly the West African dwarf goat); sheep (predominantly the indigenous Djallonké breed); pig (including large white, landrace, Duroc and indigenous breeds); poultry including local and exotic breeds (white leghorn and Rhode Island Red types); ducks, pigeons, guinea fowls and rabbits including local and exotic breeds (the New Zealand White and California White) (FAO/ LPVDMAFFS, 2007). Examples of some of the animals in Sierra Leone are shown in Figure 8.2.

8.6 Diversity of Microbes

Various microorganisms such as fungi, bacteria, protozoa, and viruses are noted for their infectious activities in higher plants and animals. A total of 339 fungi were recorded in mangroves, of which, 151 species in 84 genera, are ascomycetes (Alias, 1996). The mitosporic fungi comprise of 37 species in 29 genera, whereas basidiomycetes are the least members with 3 genera. The activities of 191 higher marine fungi are purportedly supported by 55 mangrove tree plants and their associates (Alias, 1996). In Sierra Leone,

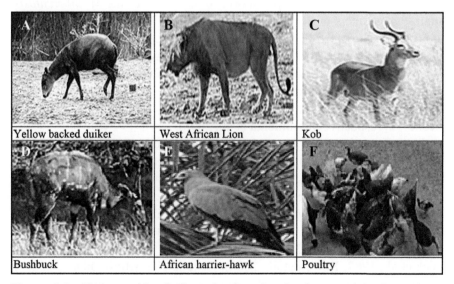

Figure 8.2 Photographic display of selected animals reared in Sierra Leon https://en.wikipedia.org/wiki/Wildlife_of_Sierra_Leone#/media/ File:Polyboroides_typus_0005.jpg (Photos via Wikipedia, Creative CommonsAttribution-Share Alike 3.0 and Public Domain: A: Raul654; B: Jonas Van de Voorde; C: Frank Dickert; D: Hans Hillewaert; E: MarioM)

Ascomycetes; *Halosphaeria viscidula, Rosellinia* sp. and *Torpedospora radiata* were frequent on mangroves, whereas the Mitosporic taxa. *Cirrenaliama acrocephala, C. pygmea, C. tropicalis, Periconia* and *Zalerion* spp. were copiously present on mangrove wood (Aleem, 1980). The distribution of marine fungi was studied based on hydrographic and climatic parameters. About 27 spp., consisting of 3 Phycomycetes, 14 Ascomycetes and 10 Deuteromycetes, of which, 50% inhabits tidally influenced mangrove swamps. The growth intensity and the number of Mangrove mycota increased in the wet season (May–November) than in the drier (December–April). The wet season favors more degradation of wood by members of Deuteromycetes including *Cirrenalia* spp., *Periconia prolifica* and *Zalerion* spp. than by Ascomycetes. During the wet season, sea foam is a richer source of fungal ascospores of 6 species of *Corollospora* and conidia of Deuteromycetes than in the dries. Whereas estuarine foam exhibits various Hyphomycetes (e.g., *Alternaria, Curvularia, Drechslera*, etc.) from freshwater or terrestrial ecosystems through runoff in the rainy season (Aleem, 1980).

High genetic diversity among *Mycobacterium tuberculosis* complex strains from Sierra Leone has been reported (Homolka et al., 2008). A total of 97 strains were studied in 2003/2004. Strain   A bp deletion in pks 1/15 grouped most (66 out of 74) of the *M. tuberculosis* strains as members of the Euro American lineage. About 23 strains belong to *M. africanum* and 74 strains to *M. tuberculosis*. Genotype *M. africanum* West African – 2 was noted as the predominant genotype followed by *M. tuberculosis* LAM and Haarlem (Homolka et al., 2008). Moreover, the two new strains with closely related IS*6110* DNA   and MIRU-VNTR patterns   ▢A Leone – 2, respectively (Homolka et al., 2008).

Lassa fever is a viral hemorrhagic fever caused by Lassa virus or Old World arenavirus (*Mastomys* sp.) found in rodents and endemic in Nigeria, Sierra Leone, Liberia, and Guinea (Granjon et al., 1997). Human infection is often through the excreta, tissues, or blood contact of infected rodent (McCormick, 1987; Ter Meulen et al., 1996); and also through from infected person-to-healthy person transmission (Bowen et al., 2000). About 5,000 deaths from 100,000 to 300,000 infected people occur annually (McCormick et al., 1987; Bowen et al., 2000). Genetic diversity studies among 54 Lassa virus strains revealed that Lassa viruses consist of four lineages, of which, three are in Nigeria and the fourth in Guinea, Liberia and Sierra Leone (Bowen et al., 2000).

Ebola hemorrhagic fever (EHF) was   reported in Zaire, now called Democratic Republic of the Congo (DRC) in 1976. About 21 Ebola virus

disease (EVD) outbreaks have been noted in humans in the tropics of sub-Saharan Africa. Ebola virus belongs to the genus Ebola virus, which is categorized into   ⌐A
virus, EBOV, and Bundibugyo virus. Of these, EBOV (57–90%) causes the highest fatality rate in humans, followed by SUDV (41–65%) and Bundibugyo virus (40%). The EBOLA virus strains of Sierra Leone are similar to those of the DRC. Unlike EBOLA virus, Tai Forest virus is noted to cause two nonfatal human infections, whereas Reston virus accounts for asymptomatic infection in humans (Geisbert and Feldmann, 2011; Jones et al., 2005). The viral hemorrhagic fevers (VHFs) are diverse animal and human diseases caused by RNA viruses. These viruses belong to four distinct families such as Arenaviridae, Filoviridae, Bunyaviridae, and Flaviviridae. The symptoms and severity of VHFs are   ☐ A
agent, and the epidemiological and clinical features of host.

8.7 Ecosystem Diversity

The main ecozones in Sierra Leone can be divided into two broad categories: terrestrial and aquatic. Forest, Montane, Savannah, and Agricultural ecosystems fall under terrestrial; while, wetlands, freshwater, coastal and marine ecosystems are classified under aquatic systems. Of all these ecosystems, there is evidence that the lowland rainforest ecosystem is the most endowed in species richness, diversity and endemism.

There are six main ecosystems in Sierra Leone; namely: Forest, Montane, Savannah, Agricultural, Wetlands and Freshwater, and Coastal and Marine Ecosystems (NBSAP, 2002).

8.7.1 Forest Ecosystem Diversity

Rapid deforestation rates in the country have drastically reduced the forest cover and forest regeneration has not kept pace with the rates of deforestation. The vegetation is now a mixture of various types of plant communities ranging from forest to open grassland savannah (Panagos et al., 2011).

There are two types of forests in Sierra Leone: Tropical moist evergreen forest and moist semi-deciduous forest and are found in the southeast and north of the country, respectively. The former can be further divided into lowland rainforest ecosystem and montane (Appendix 1).

A large proportion of the country's land surface (over 50%) is now occupied by farm bush and forest regrowth at various stages of succession. The changes in forest cover have been under a dynamic state characterized by clearing, cultivation, and regeneration. This cycle, which is basically a cultivation-fallowing ecological succession, is the most common agricultural practice in the country. The nation's forest resources have come under serious pressure from urbanization and sprawling, commercial logging and timber production, quest for additional farmlands, fuelwood and charcoal production, mining and monoculture cultivation. Bush fallowing provides an opportunity for forest regeneration, but the period of fallow, which determines the quantity of forest cover and soil nutrient replenishment within such agro-ecosystem has declined considerably over the last couple of generation. The average fallow periods have dropped from 10–15 years to 6–8 years in about two generations (Gleave, 1996; Okoni-Williams, 2013).

The national forest estate was estimated in the 1990s to have an area of 610,122 ha (8.4% of the total land area of Sierra Leone) comprising gazette and proposed forest reserves: proposed (360,622 ha); strict nature reserves (7,500 ha), proposed game reserves and game sanctuaries (60,100 ha) and gazette and proposed national parks (181,900 ha), (Allan, 1990; Mnzava, 1992). In 2006, the forest reserves (48 reserves in total) and conservation areas were estimated at 4% of the land area or 180,250 ha (Blinker, 2006). The Gola forest is noted as the largest forest area in Sierra Leone measuring about 77,000 ha (USAID, 2007). Much of the original forest in the country has been replaced by secondary farm bush and forest regrowth, mainly as a result of unplanned agricultural activities. Moreover, considering the fact that the country has lost 70% of its forest cover, with less than 5% of the original forest remaining in isolated forest reserves on tops of mountains and hillsides (Table 8.1). The shifting cultivation (slash and burn) farming system, mining activities and the heavy dependence on fuelwood present huge pressure on forest resources.

Sierra Leone has about 48 forest reserve and conservation areas (Table 8.2, Appendix 2). Most of these areas lack adequate protection and management. The total area of the forest reserves is estimated at 284,591 ha, whereas about 36,360 ha additional forestland has been proposed as forest reserves cover. The total protected wetland and marine ecosystems are estimated at 350,677 and 300,000 ha, respectively. Two of the reserved areas, Outamba Kilimi National Park (OKNP) and Tiwai Wildlife Sanctuary (TWS) have been upgraded to the national park and wildlife sanctuary, having   the criteria of the IUCN   A ☐ protected areas including the Gola forests have been proposed as national parks or game reserves.

Lowland rainforest although found in the rest of the country is more skewed to the East and South. The dominant tree species include *Heritiera*

Table 8.1 Reserves and Corresponding Areas in Sierra Leone

Reserve	Area (ha)	Reserve	Area (ha)
Gola	77,044	Nimini	15,557
Tonkoli	47,656	Freetown Peninsula (Western Area Peninsula Forest)	14,089
Loma	33,200	Sanka Biriwa	11,885
Kambui	21,213	Kangari Hills	8,573
Dodo Hills	21,185	Kuru Hills	7,001
Loma	17,094	Kasewe	2,333

Adapted from United States Agency for International Development (USAID) (2007).

utilis, Cryptosepalum tetraphyllum, Lophira alata, and *Erythrophleum ivo-rense.* Moist semi-deciduous forest is found mostly in the north of the country and the dominant plant species include *Terminalia ivorensis, Terminalia superba, Daniella thurifera, Parkia bicolor,* and *Parinari excelsa.* Where secondary forests are found the plant species are dominated by *Funtumia africana, Musanga cecropioides, Trema guineensis, Carapa procera, Elaeis guineensis,* and *Spondias mombin.* There are 11 protected areas within the forests (moist forest and semi-deciduous) including: Gola North, Gola East,

Table 8.2 Ecosystem Type, Number of Reserves, Areas, and Categories in Sierra Leone

Ecosystem type	Number of reserves	Total land area (ha)	Categories represented
Montane	2	43,720	National Park, Game Reserve
Rainforest	27	124,789	Forest Reserve, National Park, Game Reserve, Game Sanctuary
Savannah	3	113,500	National Park, Game Reserve, Game Sanctuary
Wetland	13	350,677	Strict Nature Reserve, National Park, Game Reserve, Game Sanctuary, Important Bank Area
Marine	1	300,000	Inshore Exclusion Zone (IEZ)

Adapted from United States Agency for International Development (USAID) (2007).

Western Area, Lake Sonfon, Loma Mountains, Yawri Bay, and Tingi Hills (Harcourt et al., 1992; Bomah, 2002; Lebbie, 2002).

8.7.2 Montane Ecosystem Diversity

The two mountain chains (Loma Mountains and Tingi Hills) found in the North and East occupy some 451 km². The highest peaks are Bintumani (1,947 m) and Sanka Biriwa (1,860 m) in the Loma and Tingi Hills, respectively. The annual rainfall varies between 1,600 and 2,400 mm. The soil is largely infertile. The mountain range is also the source of some major rivers. Plant associations encountered are dependent on the height a.s.l. Four identifiable plant associations are closed forest and Guinean Savanna (460–915 m), and submontane gallery forest and submontane savanna (915–1,700 m). About 1,576 plant species in 757 genera and 135 families have been identified. There are four endemic plants including *Triphyophyllum peltatum, Octoknena borealis* and *Napoleona leonensis*. Other endemic species include *Afrotrilepis jaegeri, Digitaria phaeotricha* and *Loxodera strigosa*. The dominant plant species in the closed forest and savannah are *Uapaca togoensis, Cola lateritia, Parinari excelsa, Ochna membranacea, Caloncoba echinata* and *Lophira lanceolata*. There are two protected areas within the montane ecosystem and the wildlife is unique consisting of leopards, monkeys, chimpanzees, baboons, buffalo, pygmy hippopotamus, duikers, elephants, and birds (Lebbie, 2002).

8.7.3 Savannah Biodiversity

Savannah ecosystems are found mostly in the North and Northeast of the country and they occupy about one-third of the country. Savannahs include forest savannah, mixed tree savannah and grassland savannah. Wildlife in the Savannah is characterized by elephants, leopards, hyenas, duikers, genets, civets, warthogs, aardvarks, chimpanzees, baboons, and monkeys; six species are recorded as endangered (NBSAP, 2002).

The savannah woodlands support widely spaced trees and tall grass. This habitat supports a more limited variety of wildlife than the forests in Sierra Leone. Common trees in this grassland are *Lophira, Parha biglobosa* (Locust beans), and *Piliostigma thennigir* (cow foot). Wild animals found are bush pigs (red rice hog), bush cats, and leopards. Other fauna are millipedes, snails, earthworms, millions of termites, army ants, and other species of insects. There are three protected areas within the Savannah ecosystem. These are the OutambaKilimi National Park with an area of 984 km², Bo plains and Port Loko plains both covering a total area of approximately 26 m².

8.7.4 Agricultural Ecosystem Diversity

Sierra Leone, situated at the westernmost tip of the upper Guinea forest, is a hotspot with abundant Agricultural biodiversity. There are 2,000 plant species reportedly recorded; of which, 74 species and one Genus are endemic to West Africa. Agricultural biodiversity in Sierra Leone, as in the African region, has had a history of unique challenges. The crop cultivars that are currently grown are largely introduced varieties compared to the livestock species and medicinal plants, which are largely indigenous and well-adapted to the agricultural biodiversity ecological systems. Some of the introduced crops include rice, mainly the *sativa* species and the interspecific breeds (NERICA and ROK varieties), cassava, sweet potatoes, yam and maize. Sorghum and millet are little-used cereal staples which together with cowpea, pigeon pea, Bambara groundnut, sesame and some vegetables (landraces) are indigenous to Sierra Leone. Fruit trees, namely citrus, mango, cashew and avocado, plantain, pineapple, and banana are among recent introductions to the country. Some introduced commercial perennial crops are coffee, cacao, cassava, rubber, coconut and oil palm. Several indigenous wild tree species producing edible fruits and nuts exist in the agricultural biodiversity system. Also of importance are the numerous medicinal plant species.

The Bolilands and Riparian grasslands comprise two important grassland agro-ecosystems in Sierra Leone. The Bolilands are saucer-shaped depressions often located on old riverbeds, which are   to varying depth in the rainy season. Drainage is poor and the soils are infertile. The native plant is mostly the coarse elephant grass. Rice is the main crop grown during the Rainy Season, but invariably subjected to high weed infestations of grasses and sedges. In the dry season, Bolilands may be left to fallow and serve as grazing   ⌐A
potato or to a limited extent groundnuts, vegetables, and maize. Migratory waterfowls and other tiny invertebrates, snails and worms that birds eat are common occurrences in the dry season.

In the livestock sector very little success has been recorded in terms of genetic improvement for beef and milk production. Most high yielding introduced breeds of animals intended for use in hybridization programs across the country perished as a result of disease infestations. The indigenous livestock are N'dama cattle, West African dwarf and Djalonke sheep and Goats

that are tolerant /resistant to both Trypanosomiasis and Streptothricosis. Other livestock are pigs (local and exotic), poultry (local and exotic), rabbits and Guinea pigs. It should be noted, however, that crop plants that are propagated from seeds tend to increase in genetic diversity. This is a result of many factors including the genetic mutations that normally occurs during
        □A
ing longer distance movements and some level of cultivar selection and breeding. Examples are the rice ROK series and NERICA cultivars, SLICASS, SLIPOT, SLINUT, and SLIPEA the cassava, sweet potato, groundnut, and cowpea cultivars.

8.7.5 Wetland and Freshwater Ecosystem Diversity

About 4,837.8 km² of the surface area of Sierra Leone is covered by wetlands with vegetation that is typically of freshwater swamp forests, riparian and mangroves (Blinker, 2006). Along the coast where the major river channels meet the sea are usually found thick deposits of clays and silt. They also occur at the foot of coastal terraces in the vicinities of the estuaries of the large rivers and intertidal creeks. The vegetation consists typically of Freshwater swamp forests, riparian type and mangroves. Freshwater swamp forests are present all over the country and the typical tree species include *Mitragyna stipulosa, Raphia palma-pinnus, Calamus deeratus, Heritiera utilis,* and *Rhychospora corymbosa.* The riparian or gallery forest vegetation includes species such as *Piptadeniastrum africanum, Uapaca togoensis, Pterocarpus santalinoides, Brachystegia leonensis, Anadelphia leptocoma, Panicum congoensis,* and *Cyperus pustulatus.*

Mangrove vegetation   ⌐A
are subject to tidal   In places, especially along the Scarcies Rivers, the mangrove has been extensively cleared. It is used as a source of energy. Some 500,000 ha of mangrove swamps fringe the coastline (Fomba, 1994). The drainage system consists of a series of rivers from North to South including the following: Great Scarcies, Little Scarcies, Rokel, Jong, Sewa, Moa and Mono. Other streams include Ribi, Gbangbaia, and Wanji rivers. There are in addition to the four main Estuaries (Scarcies, Rokel, Yawri, and Sherbro) numerous small Estuaries and lagoons.

The soil association varies in morphology depending on the degree of tidal   The saline content is likewise controlled by seasonality. In the rainy season when there is an abundance of fresh water the saline effect is distinctly less. However, in the dry season, the water is   saltier. In lower tidal   in and around Mambolo and Kobia in the Kambia, only

remnants of mangroves still exist. The rest have been cleared to make way for rice cultivation. The upper tidal   are covered by sedges such as *Sesuvium* sp., ferns and salt-tolerant grasses. Generally, *Rhizophora racemosa* is the species commonly present at the edge of the water while *Rhizophora mangle* and *Rhizophora harrisonii* are dominant upstream at the tidal limits where *Avicennia nitida* is also likely to be found. The dominant fauna in the lower tidal   are frogs, mudskippers, molluscs and oyesters. 29 species of
☐   ☐

The wetlands are very rich in animal life. An estimated 240 species of birds have been   About 200,000 migrant birds visit the wetlands annually (Becker, 1994). Three species of crocodiles (*Crocodylus niloticus, C. cataphractus* and *Osteolaemus tetrapsis*) are known to exist in these areas. The monitor lizard *Varamus* sp. and pythons (*Phthon sebae, Python regius*) are well represented. The most important mammals are the otter (*Aonyx capensis*), the carnivores (*Potamoga levelox, Atilax paludinosus*), herbivores (*Trichechus senegalensis*) and the pygmy hippopotamus (*Hexaprotodon liberiensis*). Mangrove and estuarine sediments are rich in invertebrate fauna including snails, bivalves, polychaetes, protochordate and Echinoderms (Ndomahina, 2002). The major   ⌐h⌐
the freshwater include *Alextesb longipinnus, Epiplatys fasciolatus, Hepsetum odoe, Sarotheodon kingsleyi, Ctenopoma kingsleyi, Polypterus palmos, Hemichromis fasciatus, Tilapia* sp., *Clarias lazera, Clarias laeviceps,* and *Mormyrus macrophaalus.* There are also several species of   (*Bagrus bayad, Synodontis nigrita, Clarias platycephalas, Clarias lazera,* and *Chysichthys furcatus*) found in lakes, rivers and Lagoons (NBSAP, 2002).

8.7.6 *Coastal and Marine Ecosystem Diversity*

Sierra Leone's coastline is 560 km long and the shelf covers an area of 30,000 km² (to 200 m depth). The Exclusive Economic Zone (EEZ) is 155,700 km². Plankton studies indicate that there are 5 genera of dinoflagellates, 14 genera of diatoms, 2 genera of Chlorophyta (Aleem, 1979). Bainbridge (1972) recorded 26 species of copepods, 9 species of Chaetognatha, 3 species of Protochordata, 2 species of Pteropods and 2 species Coelenterata, polychaetes, and protozoa. About 62 species of gastropods and 30 bivalves have been also identified (IMBO, 1996).

Other benthic fauna consists of a wide range of animals including Echinoderms, Gastropods, Bivalves, Crustacea, Polychaetes, and Protochordata. Seaweeds include *Caulerpa racemosa, Chaetomorpha pachynema, Lyngbya confervoides* and *Calothrix scopulorum.* Marine   A⌐

ics, Demersals and   (Edwards et al., 2001). Pelagic   ⌐A

sist of small pelagics (*Ethmalosa fimbriata, Sardinella maderensis, Caranx hippos, Scomber scombrus* and *Albula vulpes*). Large pelagics include *Xiphus gladius, Thunnus albacores* and *Auxis thazard.* Semi-pelagics consist of *Brachydeuterus auritus, Balistes capricus, Myctophum asperum, Diaphus dumerilli* and *Lepidophanes guassi.* The demersal stocks include *Pseudotolithus elongatus. P. senegalensis, Galeiodes decadactylus, Lutjanus agennes, Pagellus coupei* and *Dentex canariensis.*   consists of crustacea and molluscs. The crustaceans include *Penaeus duorarum, Parapenaeus longirostris, Panulirium regius, Callinectes pallidus* and *Cardiosoma armatum.* The molluscs include the   ⌐A*Sepia officinalis, Sepia berthelotti*) and Molluscs (*Pachymelina, Crassostrea tulipa* and *Iphegenia laeviagatum*).

Over 200 species of   have been documented in Sierra Leone (FAO, 1990). There are two types of   in the country: the artisanal and industrial   ⌐A

of different sizes and landing   ⌐A

variety of gear type in use include ring nets, beach seines, cast nets, traps, hooks, and lines. The industrial sector operates a number of vessels (demersal trawlers, shrimpers, canoe support vessels, purse seiners and longliners) and tar   ☐   ☐   ☐

In 1990, 22% of the   ⌐A

2002, the artisanal sector accounted for 75% of all   ⌐A

46,000 mt, with about 16,000 mt obtained from the freshwater systems every year. Aquaculture is rudimentary and there are less than 20 ponds nationwide growing Tilapia and   A ☐

among some coastal species including   ⌐A

exploitation and bad   practices are responsible for the present levels of threats.

8.8 Major Threats to Biodiversity

The major threats to biodiversity in Sierra Leone are mostly due to anthropogenic activities. These threats were exacerbated by the 10-year civil conflict (Koker and Kamara, 2002). Biological diversity in Sierra Leone is faced with diverse threats and the range of activities constituting major threats to biodiversity include poaching of fauna and inadequate valuation of natural resources, forest exploitation, agriculture, fishing, energy exploitation, mining, urbanization (infrastructure development), and waste disposal.

8.8.1 Threats to Natural Resources and Forests

Poaching robs the country of biological resources from all ecosystems. These stolen resources deprive the country of the much-needed revenue for economic development and the share of financial benefits belonging to the communities. Unfortunately, this activity is a source of major biodiversity loss, which is still difficult to contain due to limited staff strength and capacity. According to reports from the OKNP and the Gola National Park Management, poaching is happening between the OKNP (Sierra Leone) and MadinaOula and Fore Kaba axis (Guinea) and between the Gola National Forests (Liberia) and the Gola National Park (Sierra Leone). The poachers normally target elephants for the ivory, chimpanzees for the pet trade, buffalos and other animals for meat. It is difficult to assess the poaching loss due partly to lack of access, and fear for security reasons, as the poachers are often heavily armed in these remote areas. Resource-poor communities that provide biological resources at village level tend to undervalue these resources (GOSL, 2014).

Logging for timber remains in the interior, and is under severe assault. Logging companies often push further into areas with no proper management after felling the timber. Most of these sites receive little or no attention in terms of replanting or engaging in regeneration activities. Since forest reserves offer limited protection to most wildlife, logging activities coupled with hunting is a potentially devastating combination for forest biological diversity.

The lack of cheap and affordable electricity and fuel (kerosene) in the Urban and rural areas, mean that energy needs have to be met from alternative sources. The most common and frequently utilized energy sources are fuelwood and charcoal and the bulk of these come from the exploitation of preferred species from lowland rainforests, mangrove swamp forests and the Lophira savannah in the North of the country. An estimated 85% of the Sierra Leonean population is dependent on the use of fuelwood and charcoal for domestic heating and cooking. This percentage is expected to rise as the population increases and no investment is made in the production of modern electricity needs.

8.8.2 Threat to Agriculture

The agricultural biodiversity of Sierra Leone continues to face stiff and increased challenges from natural and man-made threats. Natural threats include climate change and related threats. Agrobiodiversity is among the thematic areas vaguely defined in the NBSAP as becoming vulnerable to the impacts of climate change. However, the occurrences of natural disasters such as flooding, high wind speed, erratic weather patterns and droughts are directly and indirectly considered as natural threats to the agricultural biodiversity ecosystems in many ways:

- Climate Change contributes towards loss of agricultural biodiversity in crops, resulting from adverse climatic effects, such as drought (water □   □   □ □   □
- Decline in crop and livestock yields due to   of water distribution patterns, the decline in soil nutrients and lack of fodder, etc.
- Decline in soil fertility conditions and structure resulting from heavy use of farm machinery in mono-cropping, unsustainable fallow systems and repeated cropping, etc. The impact of climate change on crops and livestock is most severe when several climatic factors occur simultaneously. For instance, prolonged drought coupled with heavy wind causes severe crop losses in cereals as a result of lodging particularly on mono-cultured commercial farms.

The agricultural practice of shifting cultivation involving slash and burn method and forest exploitation for fuel (fuelwood, charcoal) and timber production has reduced the forest cover from 70% in 1900 to merely 5% in 1990 (Grubb et al., 1998). Slash-and-burn agriculture has been blamed for the large-scale deforestation of Sierra Leone's forests and continues to degrade the remaining forest as fallow periods fall with increasing human population. On some of the most   terrains (steep slopes), farmers perilously stake claims to land for the cultivation of crops. Such sites are prone to erosion and are known to lead to the impoverishment of biological diversity. Most farming activities nowadays, extend very close to the riverbanks, and potentially result in siltation of freshwater streams and rivers. The by-product of slash and- burn agriculture is farm bush and is increasingly becoming the dominant vegetation in most areas in the country. This is occurring at the detriment of species dependent on high forest.

Pest and disease infestations of crops are exacerbated by a combination of poor farming practices and increased adverse effects of climate change. For instance, mono-culture coupled with poor rainfall (drought) contributes towards substandard plant growth, thereby increasing vulnerability to pests and diseases attacks. The increased number of pests and diseases is closely correlated to the increased number of introduced crop varieties.

In this light, the Sierra Leonean traditional farming systems of "sequential" or "relay" and mixed cropping in the uplands has proven to be a more robust system that greatly minimizes pests and diseases infestations. A mixture of several crops in one plot provides a buffer to attack by diseases and pests. For instance, the anthracnose (fungal) disease that affects rice does not affect cassava. Similarly, the spatial distribution of a mixture of crops in a mixed farm serves as barriers slowing down the spread of pests and diseases among the dif   ☐   ☐   ☐

Natural   are caused by lightning and sometimes by sun scorching. Prolonged drought periods often result in natural bush   because most of the biomass has become very dry. Slash and burn is an old farming practice in Sierra Leone. It was previously done on a small scale and used to kill big trees, shrubs and grasses in a shifting cultivation system whereby farming rotates from one piece of land to another. The problems today are the indiscriminate burning of large tracts of land during the Harmattan in preparation for the next season's farming. This result in wild bush   and large-scale slash and burn practices to clear grasslands for commercial farming.

Anthropogenic activities constitute the major threat to biodiversity. These activities include agriculture, livestock farming, forest exploitation,   A energy exploitation, mining, transportation, urbanization (infrastructure development) and waste disposal. The long civil   in Sierra Leone also exacerbated the threats (Koker and Kamara, 2002). The forest cover had been reduced from about 70% in 1990, to 5% with forest regrowth constituting 60% of the total area (Grubb et al., 1998). In addition to population displacement during the civil   the urban population has increased, putting tremendous pressure on land and habitat for food and fuel, thereby endangering biodiversity.

The degradation of habitats through urbanization has impacted strongly on the depletion of Agricultural biodiversity in Sierra Leone. Studies indicate that 85% of the species on the IUCN Red List is threatened by habitat loss, while clearing land for development and agricultural expansion have dramatically accelerated habitat loss. Agricultural biodiversity ecosystems have been degraded or altered by changes in land use and habitat destruction (development of tourism, deforestation, mining and aquaculture). Fragmentation of large areas of habitat (owing to   landslides, soil erosion, open-pit mining, road construction or other human activities) into smaller patches makes it   for isolated species to maintain large enough breeding populations to ensure their survival. It also diminishes the quality of the remaining habitats. Inland water ecosystems and wetlands (Inland Swamps) can also be altered and destroyed by the development of irrigation systems, dams, and reservoirs, as well as by introducing water drainages, canal and

8.8.3   Threats to Biodiversity in the Aquatic, Coastal, and Marine Ecosystems of Sierra Leone

Biodiversity in Sierra Leone has been subjected to serious threats. The most obvious threats include habitat loss and fragmentation of natural habitats due primarily to deforestation, wetland drainage and infrastructural development, overgrazing, poor mining practices, poor farming practices, inappropriate use of agrochemicals, pollution, bushfires, population pressure, civil conflict, poverty, illiteracy, lack of resources, limited trained human power, inappropriate policies, institutional weakness as well as socioeconomic factors.

According to Clark (1990), coastal zone refers to "all coastal areas that are subject to storm   ◻A
marsh, deltas, salt   ◻
water areas such as bays, lagoons, estuaries, deltaic waterway and near coast waters that include seagrass meadows, coral reefs,   beds submerged bars; the nearshore coastal waters and small coastal islands."

Along the coast activities such as   ◻A
(textile, chemical, and brewing), mining and mineral exploitation, tourism, marine transportation, marine and coastal infrastructure, waste dilution and domestic use water are bound to be on the increase. These activities require huge investment and appropriate infrastructure.

Increase in anthropogenic activities and introduced pollutants in coastal zones affect the complex food web and ecological relationships, which adversely   the biodiversity. In addition, human health and water quality may be adversely affected. The bulk of pollutants entering the sea are derived from the following sources; runoff and discharges from the land mainly through rivers (44%), atmosphere (33%), marine transportation; spills and operational discharges (12%), deliberate dumping of wastes (10%), and offshore development of mineral resources (1%) (GOSL, 2014).

In principle, every marine organism could be exploited on a sustainable basis. However, when more is taken than could be replaced overexploitation is the result. In Sierra Leone, there is evidence of overexploitation of certain categories of target species and a   reduction in others in response to growing demand and population growth. Out of seven major snapper species,   ◻*Dentex angolensis, D. congensis, D. canarensis, Pagellus belloti* and *Sparus caeruloesticus*) have been shown to be declining rapidly (Showers, 1996). There is evidence of overexploitation of the following species: *Pseudotolithus senegalensis, Drepane africana, Galeoides decadactylus, Dasyatis margarita* (Fombah, 1996). *Ilisha africana* is the only pelagic species known to have been overexploited. The coastal   *Arius latiscutatus* is slightly overexploited (Ndomahina and Mamie, 2002). The shrimps have reached the maximum sustainable yield levels of 3,000 mt (GOSL, 2014).

8.8.4 Impacts of Changes in Biodiversity and Its Socio-Economic and Cultural Implications

Biodiversity loss implication for Sierra Leone can be considered with wide-ranging effects. The majority of the population is rural and depends on biodiversity resources for their medicine, food, fiber income, and welfare. There is a challenge to the sustainable development of Sierra Leone due to its small land area and vast natural resources. Sierra Leone relies greatly on her natural resources derived from agriculture, forestry, ecotourism and mining. The trends and effects of unsustainability are visible. They include: farming in forested areas, exploitation of biodiversity resources, and destruction of natural habitat, degradation of land, high pollution, inability and improper waste management.

The contribution of both agriculture and   sectors to the GDP was estimated at 46% (SLARI, 2011). However, the increasing migration of young farmers from these sectors to greener pasture sectors such as mining has led to its proportionate decline. Except for coffee and cacao, most of the agricultural products are consumed locally with little spill over to neighboring countries.

The cultivation of predominantly economic crops, improved crop varieties and the use of monoculture system have led to reduced diversity within agricultural plant species. These factors also have the tendency of increased vulnerability of crop species, as old landraces that are resistant or tolerant to pest and diseases are lost. An example of this is the loss of *O. glabberima* and *O. barthii* rice species due to the predominant cultivation of the introduced genotypes of *O. sativa*.

8.8.5 Impact of Changes in Biodiversity on Forest and Related Ecosystem

The irregular supply of electricity and limited available gas contribute to increasing use of fuelwood and charcoal for cooking in both urban and rural areas. Additionally, people rely on the forest for medicinal plants, food, construction materials and other wood and non-woody materials. The habitat in the urban areas depends mainly on freshwater mini rivers and estuaries for their domestic activities. This poses a huge loss or degradation of the natural resources and results in reduced and irregular flows, dirty water and drying up of natural water bodies which affect the rural communities. However, fuelwood, poles, and charcoal are sources of income and livelihood for the rural poor. The balanced environmentally friendly approach is needed to exploit these natural resources. Another important effects of deforestation and land degradation on human well-being are soil erosion and consequent loss of soil fertility leading to reduce agricultural productivity. Some farmers mitigate these effects by adding fertilizer, which may have great effects on lives downstream or low lying areas such as lagoons and coastal areas. Fertilizer use has contributed to seaweed bloom in the lagoons and some beach areas in Freetown, Sierra Leone.

Despite the growing establishment of westernized health care system in Sierra Leone, conventional medicine is still popularly used. As such, many concerns are raised regarding the destruction of medicinal plants and agricultural lands. Most of the medicinal plants are now reported endangered or vulnerable and require replacement and propagation programs. The continuing decline in the native plant species may have negative impacts on the health of the rural population.

8.9 Concluding Remarks

A number of fauna and flora of Sierra Leone are data deficient and under serious threat of extinction. However, it is hoped that information on biodiversity in the current volume will serve as a valuable guide for future research. The biotic community has been impacted by the conversion of natural vegetation into mining, farming, housing, etc., leading to the alteration and loss of natural habitats and biodiversity. Several negative impacts occur in various ecosystems. The alarming climate change, increasing human population and development activities also necessitate robust ecological research, setting of adequate priorities targeting better understanding of evolutional trends and population dynamics of the biome's natural system for their conservation and sustainable utilization. There is also need of improved ecosystem services including maintenance of biodiversity, nutrient recycling and landscape, and aquatic ecologies among others.

Acknowledgment

The authors register their thanks and appreciation to all partners of research and development of biodiversity conservation of useful resources, management and prudent utilization.

Keywords

- animals
- ecosystem
- genetic diversity
- microbial

- plants
- Sierra Leone
- species

References

Aleem, A. A., (1979). Marine microplankton from Sierra Leone. *West. Ind. Mar. Sc.*, *8*, 291–295.

Aleem, A. A., (1980). Distribution and ecology of marine fungi in Sierra Leone (Tropical West Africa). *Bot. Mar.*, *23*, 679–688.

Alias, S. A., (1996). *Ecological and Taxonomic Studies of Lignicolous Marine Fungi in Malaysian Mangroves*. PhD Thesis. University of Portsmouth, U.K.

Allan, T., (1990). *Tropical Forestry Action Plans Inter-Agency Forestry Review*. Sierra Leone UNDP, Rome.

Bainbridge, V., (1972). The zooplankton of the Gulf of Guinea. *Bull. Mar. Ecol.*, *8*, 61–97.

Becker, P., (1994). *Assessment of Bird Species in Mangrove Forests*. WNM AFCOD/WP/No. 9.

BirdLife International (BLI), (2013a). *IUCN Red List for Birds*. www.birdlife.org.

BirdLife International (BLI), (2013b). *Species Fact sheet: Laniarius turatii*. www.birdlife.org.

Blinker, L., (2006). *Country Environment Profile (CEP) Sierra Leone*. Consortium Parsons Brinckerhoff, Cardiff, UK.

Bomah, A. K., (2002). *Analysis of the Threats/Pressures on Biodiversity and of the Sustainability of the use of Biological Resources in Sierra Leone*. BSAP. UNDP. Sierra Leone.

Bowen, M. D., Rollin, P. E., Ksiazek, T. G., Hustad, H. L., Bausch, D. G., Demby, A. H., et al., (2000). Genetic diversity among Lassa virus strains. *J. Virol.*, *74*(15), 6992–7004.

Burkill, H. M., (1985). *Useful Plants of West Tropical Africa: Revised Edition*, Families A to D, vol. 2.

Burkill, H. M, (1994). *Useful Plants of West Tropical Africa: Revised Edition*, Families E to I, vol. 2.

Clark, J., (1990). *Integrated Management of Coastal Sea Resources Draft Report*. FAO.

Cole, N. H. A., (1974). Climate, life forms and species distribution on the Loma Montane Grassland, Sierra Leone. *Biol. J. Linn. Soc.*, *69*(0), 197–210.

Cole, N. H. A., (1994). Habitat diversity of West African medicinal plants. In: *Proceedings of 14th AETFAT Congress on Biodiversity of African Plants*. Agricultural University, Wageningen, the Netherlands.

Earth Trends, (2003). *Biodiversity and Protected Areas – Sierra Leone*. http://earthtrends.wri.org. Retrieved 20 February 2011.

Edwards, A. J., Gill, A. C., & Obohweyere, P. O., (2000). *A Revision of Irvine's Fishes of Tropical West Africa Darwin Initiative Report*, 2 Ref. 162/7/145, pp. 155.

FAO (Food and Agricultural Organization), (2007). LPVDMAFFS (Livestock Production and Veterinary Division of Ministry of Agriculture, Forestry and Food Security), Country Report on The state of Sierra Leone's animal genetic resources: A contribution to the state of the world's animal genetic resource process, pp. 8–14.

FAO, (1990). *The "Guinea 90" Survey*. CECAF/ECAF Ser. 91/52.

FAO, (1996). *Country Report to the FAO International Technical Conference on Plant Genetic Resource*. Compiled by: Department of Agriculture and Forestry, Freetown, Sierra Leone, Leipzig.

FAO, (2015). *The Second Report on the State of the World's Animal Genetic Resources for Food and Agriculture*. Rome.

Fomba, S. N., (1994). Overview of mangrove rice production in West Africa 17–20. In: Wilson, R. T., & Wilson, M. P., (eds.). *Training in Mangrove rice Production. Instructors Manual*. Bartridge partners. Umberlergh. U.K.

Fombah, P. N., (1996). *The Biology of Drepana africana With Emphasis on its Growth, Feeding and Reproduction*. BSc. (Hons) Thesis USL.

Food and Agricultural Organisation (FAO), (1999), In: Zaid, A., Hughes, H. G., Porceddu, E., & Nicholas, F. W., (eds.). *Glossary of Biotechnology and Genetic Engineering*. FAO Research and Technology Paper 7.

Geisbert, T. W., & Feldmann, H., (2011). Recombinant vesicular stomatitis virus-based vaccines against Ebola and Marburg virus infections. *J. Infect. Disease, 204*(3), S1075–S1081.

Gleave, M. B., (1996). The length of the fallow period in tropical fallow farming systems: A discussion with evidence from Sierra Leone. *The Geograph. J., 162*(1), 14–24.

Government of Sierra Leone (GOSL), (2014). *Fifth National Report to the Convention on Biological Diversity*. UNEP. Freetown.

Granjon, L., DuPlantier, J. M., Catalan, J., & Britton-Davidian, J., (1997). Systematics of the genus *Mastomys* (Thomas, 1915) (Rodentia: Muridae). *Belg. J. Zool., 127*(1), 7–18.

Groom, M. J., Meffe, G. K., & Carroll, C. R., (2006). *Principles of Conservation Biology (3rd ed.)*. Sunderland, MA: Sinauer Associates. Website with additional information: http://www.sinauer.com/groom/.

Grubb, P., Jones, T. S., Davies, A. G., Edberg, E., Starin, E. D., & Hill, J. E., (1998). *Mammals of Ghana, Sierra Leone and the Gambia*. The Tenderine Press, Zennor, St. Ives, Cornwall, U.K.

Harcourt, C., Davies, G., Waugh, J., Oates, J., Coulthad, N., Burgess, N., Wood, P., & Palmer, P., (1992). Sierra Leone. In: Sayer, J., Harcourt, C., & Collins, N. M., (eds.). *The Conservation Atlas of Tropical Forests: Africa*, IUCN and Macmillan Publishers, U.K, pp. 244–250.

Homolka, S., Post, E., Oberhauser, B., George, A. B., Westman, L., Dafae, F., Rüsch-Gerdes, S., & Niemann, S., (2008). High genetic diversity among *Mycobacterium tuberculosis* complex strains from Sierra Leone. *BMC Microbiol., 8*, 103.

IMBO, (1996). Report on environmental baseline studies undertaken during Debeers marine diamond prospecting in Sierra Leone. *Bull. Mar. Biol. & Oceanogr. (Special Edition)*.

International Union for Conservation of Nature (IUCN), (2012). *IUCN Red List Categories and Criteria Version 3. 1 Second Edition*. Gland, Switzerland and Cambridge, UK, IUCN.

International Union for Conservation of Nature (IUCN), (2013). *Red List of Threatened Species [Online]*. Gland, Switzerland: International Union for Conservation of Nature, Available: http://www.iucnredlist.org/about (accessed 07/10/2013 2013).

Jones, S. M., Feldmann, H., Ströher, U., Geisbert, J. B., Grolla, A., Klent, H. D., et al., (2005). Live attenuated recombinant vaccine protects nonhuman primates against Ebola and Marburg viruses. *Nature Medicine, 11*(7), 786–790.

Klop, E., Lindsell, J., & Siaka, A., (2008). *Biodiversity of Gola, Sierra Leone*. Royal Society for the Protection of Birds, Conservation Society of Sierra Leone, Government of Sierra Leone.

Koker, U. S., & Kamara, B. M., (2002). *Assessment on the Impact of the Refugee and Displaced Communities as a Result of the 10-Year-Old Rebel War on Biodiversity of Sierra Leone*. BSAP. UNDP. Sierra Leone.

Lebbie, A. R., (2002). *Biodiversity Assessment and Identification of Priorities for Biodiversity Conservation in Sierra Leone*. BSAP. UNDP. Sierra Leone.

McCormick, J. B., (1987). Epidemiology and control of Lassa fever. *Curr. Top Microbiol. Immunol., 134*(0), 69–78.

McCormick, J. B., Webb, P. A., Krebs, J. W., Johnson, K. M., & Smith, E. S., (1987). A prospective study of the epidemiology and ecology of Lassa fever. *J. Infect. Disease, 155*(3), 437–444.

Ministry of Agriculture, (2005). Forestry and Food Security (MAFFS)/ National Agricultural Research Coordinating Council (NARCC). *Crop Production Guidelines for Sierra Leone*.

Mnzava, E. M., (1992). *Assistance for Forestry Planning Sierra Leone Forest Management*. UNDP. FAO.

National Biodiversity Strategy Action Plan (NBSAP), (2002). Biodiversity and of the sustainability of the use of biological resources in Sierra Leone. *Action Plan Report*. MCHE. Freetown.

National Biological Information Infrastructure (NBII), (2011). *Introduction to Genetic Diversity. U.S Geological Survey*. Archived from the original on February 25.

Ndomahina, E. T., & Mamie, J. C., (2002). Some studies on the food and feeding habits of the rough head Catfsh. *Arius latiscutatus* in the coastal waters of Sierra Leone. *J. Pure and Appl. Sci., 9*(0), 55–62.

Ndomahina, E. T., (2002). *An Assessment of the Coastal and Marine Biodiversity of Sierra Leone*. Consultancy of the Sierra Leone Maritime Administration.

Okoni-Williams, A. D., (2013). *Vegetation, Carbon and Nutrient Cycling in the Bush Fallow System: the Implementation to the Environment and Farming Communities in Sierra Leone*. PhD Thesis, University of Sierra Leone.

Okoni-Williams, A. D., Thompson, H. S., Koroma, A. P., & Wood, P., (2005). *Important Bird Areas in Sierra Leone: Priorities for Biodiversity Conservation*. Conservation Society and Government Forestry Division, MAFFS.

Oliver-Bever, B., (1983). *Medicinal Plants in Tropical West Africa*. Cambridge University Press.

Panagos, P., Jones, A., Bosco, C., & Senthil Kuma, P. S., (2011). European digital archive on soils maps (EuDASM) preserving important soil data for public free access. *Intern. J. Digital Earth, 4*(5), 434–443.

Showers, P. A. T., (1996). The abundance and distribution of the sparidae in Sierra Leone Waters - An Ecological Interpretation. *PhD Dissertation*, USL.

SLARI, (2011). *Sierra Leone Agricultural Research Institute*, Strategic Plan 2012– 2021.

Ter Meulen, J., Lukashevich, I., Sidibe, K., Inapogui, A., Marx, M., Dorlemann, A., et al., (1996). Hunting of peridomestic rodents and consumption of their meat as possible risk factors for rodent-to-human transmission of Lassa virus in the Republic of Guinea. *Amer. J. Trop. Med. Hyg.*, *55*(6), 661–666.

Tisdell, C., (2003). Socioeconomic causes of loss of animal genetic diversity: analysis and assessment. *Ecol. Econ., 45(3)*, 365–376.

United States Agency for International Development (USAID), (2007). Biodiversity and Tropical Forest Assessment for Sierra Leone. World Bank/Global Environment Facility (GEF). GEF Project Brief on a Proposed Grant from the Global Environment Facility Trust Fund in the Amount of USD 5. 0 Million to the Government of Sierra Leone for a Sierra Leone Protected Area Management Project, draft final version June 20. EPIQ IQC: EPP-I-00-03-00014-00, Task Order 02, pp. 70.

Wildlife of Sierra Leone – Wikipedia. Available at *https://en.wikipedia.org/wiki/Wildlife_of_ Sierra_Leone*.

Appendices

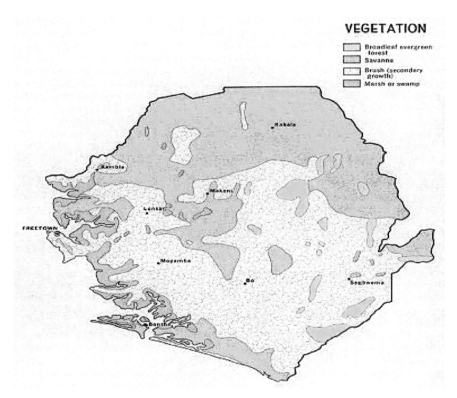

VEGETATION

Broadleaf evergreen forest
Savanna
Brush (secondary growth)
Marsh or swamp

Appendix 1   Vegetative cover in Sierra Leone, 1969.

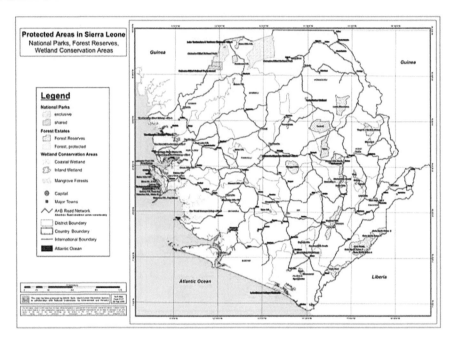

Appendix 2 Map of protected Areas in Sierra Leone [Adapted from United States Agency for International Development (USAID) (2007)].

CHAPTER 9

Biodiversity in South Africa

DANNI GUO and **JUDITH L. ARNOLDS**

South African National Biodiversity Institute, Kirstenbosch Research Centre, Private Bag X7, Claremont 7735, Cape Town, RSA, E-mail: d.guo@sanbi.org.za, j.arnolds@sanbi.org.za

9.1   Introduction and Background

South Africa is situated at the southern tip of Africa, and it has an amazing variety of wildlife, bird species, plant species, and minerals. In addition, its population comprises a unique diversity of people and cultures (Figure 9.1). South Africa covers an area of 122 million hectares, and this represents 2% of the land surface of the world. The 3,000 km² shoreline of South Africa stretching from the Mozambique in the east to the Namibia the west is surrounded by the Atlantic and Indian oceans, which meet up at the Cape Point at the southwestern corner. The contrast in temperature between the warm Agulhas Current in the east and south coast, and the cold Benguela Current

Figure 9.1   South Africa.

on the west coast, creates significant differences in climate, vegetation, and the marine life (South Africa Government Online, 2017).

South Africa is within the subtropical belt of high pressure, making it semiarid and dry with an abundance of sunshine (South Africa Government Online, 2017). Although most of South Africa is   [A it has however has considerable variation in climate as well as topography (Wikipedia, 2017). The country has an average annual rainfall of 450 mm, about 65% of the country receives less than 500 mm per year, and about 21% of the country receives less than 200 mm per year. Also South Africa has no   natural lakes, and the   A crop irrigation. Therefore, below average annual rainfall is quite common, and prolonged droughts and   occurs on a regular basis (South Africa Government Online, 2017).

Biodiversity or biological diversity is the total variety of living organisms in all ecosystems, and the genetic difference between them and the ecosystems in which they occur. Biodiversity is a natural wealth of the land we live on that provides our food, shelter and raw materials (DEAT, 2009). South Africa's exceptional biological diversity such as the variety of genes, species, ecosystems and ecological processes occurring in the country, is an asset of international, national and local value and   The rivers and wetlands, mountains and plains, estuaries and oceans, and coastline and landscapes contain an extremely rich and varied collection of life forms which are vital to the existence of all South Africans, and upon which the national economy is fundamentally dependent (DEAT, 1997). South Africa's biodiversity is increasingly threatened by human activities, which in turn threaten the very resource base upon which we depend (DEA, 2011). The impacts of human activities play an important role in the protection of biodiversity (Preston and Siegfried, 1995).

9.2 Vegetation of South Africa

The South African biomes (Figure 9.2, Table 9.1) are large, homogenous geographical areas defined by the major growth forms (trees, grasses, shrubs) and also by the climate (DEA, 2015). They are complex biotic communities, with its distinctive plant and animal species, and is maintained under the suitable climatic conditions of the region. Hence, the definition of a biome is complex as it extends beyond individual species to represent entire ecosystems under suitable climatic conditions and geological conditions (Mucina and Rutherford, 2006). The Fynbos Biome and the Succulent

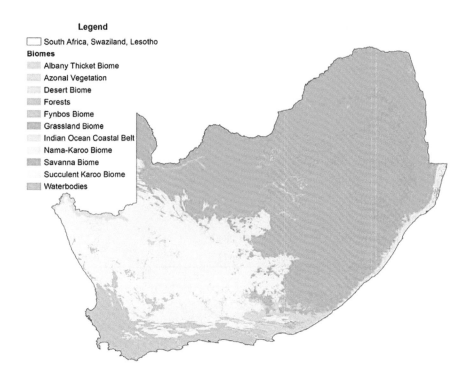

Figure 9.2 South African Biomes (*Source:* Mucina and Rutherford, 2006).

Karoo Biome, which together forms the smallest of the world's six Floristic Kingdoms, the Cape Floristic Region, and they are of major conservation concern. The other six biomes are Albany Thicket Biome, Desert Biome, Grassland Biome, Indian Ocean Coastal Belt Biome, Nama-Karoo Biome, Savanna Biome (Mucina and Rutherford, 2006). These biomes are not only threatened by agricultural expansion, overgrazing, and mining; but also by future climate changes and droughts. Three additional tiny biomes are the Azonal Vegetation Biome, Forests Biome, and Waterbodies (Guo et al., 2017a). Each biome consists of bioregions, which is defined as a spatial terrestrial structure based on similar biotic, physical features, and processes at a smaller regional scale (Mucina and Rutherford, 2006).

9.2.1 *The Fynbos Biome*

The Fynbos Biome has a high species density and diversity, and has over 9,000 vascular plant species, of which 70% are endemic (DEA, 2015).

Table 9.1 Areas Covered by Biomes

Biome	Area (Square km)	Area (%)
Albany Thicket biome	29,128	2.29
Azonal vegetation	28,982	2.28
Desert biome	7,166	0.56
Forests	4,715	0.37
Fynbos biome	83,944	6.62
Grassland biome	354,594	27.97
Indian Ocean Coastal Belt	14,282	1.12
Nama-Karoo biome	248,280	19.58
Savanna biome	412,545	32.54
Succulent Karoo biome	83,284	6.57
Waterbodies	673	0.05

Source: DEA (2007).

Scientists estimate that about 75% of South Africa's rare and endangered plants are found within the Fynbos Biome (DEA, 2015). The Fynbos Biome consists of mostly small leaved evergreen shrubs, reeds and grasses, and are recognized by the sclerophyllous and microphyllous nature of woody plants (DEA, 2015). This evergreen, fire-prone shrubland is frequently dominated by three plant families: shrubby Proteaceae and Ericaceae, and reed-like Restionaceae (Cowling et al., 1997; DEA, 2015).

The Fynbos Biome has three different fragmented vegetation types: Fynbos, Renosterveld and Strandveld.

The Fynbos consist of the bioregions (Mucina and Rutherford, 2006):

- Sandstone Fynbos
- Quartzite Fynbos
- Sand Fynbos
- Shale Fynbos
- Fynbos Shale Band Vegetation
- Silcrete, Ferricrete and Conglomerate Fynbos
- Alluvium Fynbos
- Granite Fynbos
- Limestone Fynbos
- The Renosterveld consist of the bioregions (Mucina and Rutherford, 2006):
- Shale Renosterveld

- Granite and Dolerite Renosterveld
- Alluvium Renosterveld
- Silcrete and Limestone Renosterveld

The Strandveld consist of the bioregion Western Strandveld (Mucina and Rutherford, 2006).

Fynbos is the dominant vegetation of the Cape Floristic Kingdom (CFR). The Cape Floristic Region is one of the world's six Floral Kingdoms and is only found within South Africa. The Cape Floristic Region (Figure 9.3) is home to 45% of the subcontinent's plant species and covers an area of 4% of southern Africa (DEA, 2011). It is situated in the Western Cape Province, and it is well known as a biodiversity hotspot. The Cape Floristic Region has the highest concentration of plant species in the world, and it contains an estimated about 9500 species, and of these species about 70% are endemic to CFR, and do not exist anywhere else in the world. This region also has a diverse animal species as well, including rare and endangered species (Cape Nature, 2017).

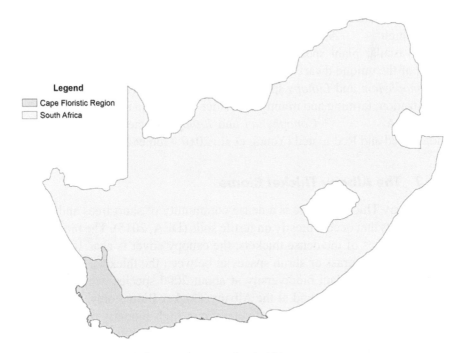

Figure 9.3 The Cape Floristic Region in South Africa.

9.2.2 The Succulent Karoo Biome

The Succulent Karoo Biome is one of only two biodiversity hotpots in the world located in a desert environment (Myers et al., 2000; DEA, 2015). The Succulent Karoo has the richest variety of succulent flora in the world, with 6356 plant species, and of which 40% are endemic, and 936 of the plant species are Red Listed (Driver and Maze 2002; DEA, 2015). This biome is the center of biodiversity for reptiles and other invertebrate groups (DEA, 2015). The Succulent Karoo Biome is mostly a shrubland covered by leaf-succulents or deciduous-leafed dwarf shrubs (DEA, 2015). The biome is dominated by the four plant families Aizoaceae, Asteraceae, Crassulaceae, and Euphorbiaceae (Desmet, 2007; DEA, 2015).

The Succulent Karoo Biome consists of the bioregions:

- Richtersveld;
- Namaqualand Hardeveld;
- Namaqualand Sandveld;
- Knersvlakte;
- Trans-Escarpment Succulent Karoo;
- Rainshadow Valley Karoo.

The Succulent Karoo biome is considered to be a biodiversity hotspot with a high   A⬚ 5,000 vascular plant species, including many rare and red listed species. Many of the unique dwarf succulent plants (Example: Figures 9.4 and 9.5) of *Conophytum* and *Lithops* species are under threat from climate change, urbanization, farming and mining, and since they are so tiny it makes survey   ⬚A   *Conophytum* and *Lithops* species now listed under endangered and Red Listed (Young et al., 2016; Guo et al., 2017b).

9.2.3 The Albany Thicket Biome

The Albany Thicket Biome is a dense community of short trees and herbaceous plants that occurs mostly on fertile soils (DEA, 2015). The biome consists of patches of the dense thickets, the canopy cover is near 100%, with patches of open grass or shrub spaces in between the thickets. The Albany Thicket has high plant biodiversity at about 2000 species, which includes 300 endemic species found at the Albany Thicket Biome only. In the past, the Albany Thicket has a wide diversity and high biomass of indigenous mammal herbivores, ranging from small antelope to large elephants (DEA, 2015). The thicket plants are adapted to browsing, and the seeds of many

Figure 9.4 *Conophytum roodiae* (Photo: A. J. Young).

Figure 9.5 *Lithopsdinterissp brevis* (Photo: A. J. Young).

species are dispersed by the mammals and birds. Even today, browsing her-bivores are still prominent. However, concentrated grazing by the domestic livestock especially goats, can easily eradicate some of the species and open up the dense thickets, leading to bare open grounds, which in turn result in reduction of the herbivores (DEA, 2015).

9.2.4   *The Desert Biome*

The Desert Biome vegetation is predominately characterized by annuals with drought resistant shrubs found in some regions, and the flora and fauna in this area are well adapted to arid conditions (DEA, 2015). The succulent plant species are well adopted to the dry conditions, and the local fauna species are adapted to living in the desert temperature cycles. In the Desert Biome, land uses include commercial and communal grazing, as well as wildlife ranching and management (DEA, 2015).

The Desert Biome consists of the bioregions (Mucina and Rutherford, 2006):

- Southern Namib Desert;
- Gariep Desert.

9.2.5   *The Forests Biome*

The Forests Biome are predominated by a tree layer or stratum, and range in height from 3 to 30 m (Mucina et al., 2006; DEA, 2015). The forests are the smallest biome in South Africa, and they are spread throughout the wetter areas of the country, embedded within other biomes. The Forests Biome has a high plant species diversity with a listing of 1438 plant species (Gelden-huys 1992; DEA, 2015). It has a unique fauna biodiversity, providing habitat for forest mammals and many bird species (DEA, 2015).

Forests Biome consists of the bioregions (Mucina and Rutherford, 2006):

- Zonal and Intrazonal Groups;
- Azonal Groups.

9.2.6   *The Grassland Biome*

The Grassland Biome is an ecosystem that is dominated by grasses (DEA, 2015), and it has a high plant biodiversity with 3,370 species (DEA, 2015). The Grassland Biome contains many endemic and threatened mammal, butterfly and bird species. In the past, grassland had a high biodiversity and large biomass of indigenous mammal grazers; but today they are being mostly replaced by the 6 million cattle and 13 million sheep. Wild herbivore grazers are returning to the game farms and conservation areas. The Grassland Biome due to its climate and landscape conditions are extremely suitable for agricultural cultivated crop expansion, plantations of exotic trees to provide timber and fuel. The Grassland Biome is also threatened due to mining and pollution, as it is also the location of the gold deposits and coal. Therefore, agriculture, urbanization, and mining activities are continuously transforming the biome (DEA, 2015).

The Grassland Biome consists of the bioregions (Mucina and Rutherford, 2006):

- Drakensberg Grassland;
- Dry Highveld Grassland;
- Mesic Highveld Grassland;
- Sub-Escarpment Grassland.

9.2.7   The Indian Ocean Coastal Belt Biome

The Indian Ocean Coastal Belt Biome has a mixture of plant species, with elements from both savanna and grasslands and coastal forests in this biome, and it includes areas of high endemism. The diversity of habitats of the Indian Ocean Coastal Belt Biome causes it to have a high animal diversity with a wide range of large mammal species, and it has 200 mammal species, as well as 540 species of birds. The Indian Ocean Coastal Belt is second only to the Fynbos for its species richness and endemism, and has 8200 species of plants with 23.5% endemic, 200 reptiles with 14.4% endemic, and 72 amphibians with 15.3% endemic (Conservation International 2014; DEA, 2015).

Indian Ocean Coastal Belt Biome consists of the bioregions (Mucina and Rutherford, 2006):

- Algae Beds;
- Seashore Vegetation;
- Estuarine Vegetation;
- Eastern Strandveld.

9.2.8   The Nama-Karoo

The Nama Karoo biome consists of grasses and bushes, and composes of mostly dwarf open shrubland, dominated by the six plant families: Aizoaceae,

Asteraceae, Liliaceae, Poaceae, Mesembryanthemaceae, and Scrophularia-ceae (Palmer and Hoffman 1997; DEA, 2015).In the Nama Karoo Biome, land use is dominated by extensive grazing of small and large livestock, and also an increasing wild life management. Invasion of alien species is a serious problem in the biome, with agricultural grazing worsening the alien invasion by overgrazing and land degradation, which causes an increase alien species expansion in some cases. Certain alien species once found are required to report to the Department of Agriculture, Forestry and Fisheries (DEA, 2015).

Nama-Karoo Biome consists of the bioregions (Mucina and Rutherford, 2006):

- Bushman land and West Griqualand;
- Upper Karoo;
- Lower Karoo.

Aloe dichotoma occurs in the arid regions of Namaqualand and Bushman land in South Africa, and it is limited by the biome regions, namely it exists only in Nama-Karoo and Succulent-Karoo regions (Figure 9.6). The Quiver tree is important to the regional ecosystem and environment because it pro-

Figure 9.6 *Aloe dichotoma* (Photo: W. B. Foden).

vides a source of moisture for a wide variety of mammals, birds and insects (Guo et al., 2016). Today, the *Aloe dichotoma* is threatened by agricultural expansion, overgrazing, and mining, as well as droughts and other climate changes (Foden, 2002).

9.2.9 The Savanna Biome

The Savanna Biome has a continuous layer of perennial grass and herbaceous species; and also a discontinuous layer of trees (Scholes and Archer, 1997; DEA, 2015). Savannas is essential because it provides grazing, fuelwood, timber, water resources, and can be commercially used for livestock production, and including ecotourism due to the number and variety of large mammals that are characteristic of these grassy ecosystems (DEA, 2015).

Savanna Biome consists of the bioregions (Mucina and Rutherford, 2006):

- Central Bushveld;
- Mopane;
- Lowveld;
- Sub-Escarpment Savanna;
- Eastern Kalahari Bushveld;
- Kalahari Duneveld.

9.2.10 The Azonal Vegetation Biome

The Azonal Vegetation Biome is a complex biome and refers to vegetation habitats existing in and around flowing or stagnant fresh and saline waterbodies. In Mucina and Rutherford (2006), Azonal Vegetation is represented by special soil substrate and hydrogeological conditions that causes an influence over the floristic composition and dynamics over the macroclimate. If the vegetation occurs exclusively within a climatic zone, then it is Intrazonal Vegetation; and if the vegetation occurs regardless of the climate or vegetation zones, then it is Azonal Vegetation.

Inland Azonal Vegetation Biome consists of the bioregions (Mucina and Rutherford, 2006):

- Freshwater Wetlands;
- Alluvial Vegetation;
- Inland Saline Vegetation.

Climate change including local climate variabilities, has been iden-
A□

Local climate variabilities are still tolerable, but extreme climatic events and prolonged climate change would prove to be serious in terms of the impact on natural biomes and ecosystems, and good political structures and policies are needed to deal with these issues. In Southern Africa, there are conservation and management   A □
the face of future climate change with prolonged droughts, but there are also other impacts such as overgrazing, land transformation and deforestation. Under semiarid climatic conditions, even a few degrees increase in temperature and a few millimeters decrease in rainfall could cause a decline in the biodiversity of plants and animals (Guo et al., 2016, 2017a, 2017b).

9.3 Wildlife Diversity in South Africa

South Africa is home to 7.5% of the world's plants, 5.8% of its mammals, 8% of its bird species, 4.6% of its reptiles and 5.5% of its insects (DEA, 2011). The biodiversity in terrestrial ecosystems is more comprehensively studied in South Africa than in the aquatic ecosystems. South Africa has an estimate of over 95000 known species (Figure 9.7, Table 9.2), which is a very large percentage of the world's species relative to the small surface area of the country (DEA, 2011). South Africa has about over 800 bird species, including the world's largest bird (the ostrich), and the heaviest flying bird (Kori bustard), and the world's smallest raptors (pygmy falcons) (Lonely Planet, 2017). There are more than 100 important bird areas occurring in South Africa, as well as five endemic bird areas, and this is a number only matched by one other country in the world, Madagascar (DEA, 2011).

South Africa not only has a vast varied plant life, a wealth of animal life exists in numbers and variety. It is home to the world's largest land mammal (the African elephant), the second largest (white rhino), and the third largest (hippopotamus); and it is also home to the tallest mammal in the world (giraffe), the fastest mammal in the world (cheetah), and the smallest mammal in the world (pygmy shrew) (Lonely Planet, 2017). South Africa's marine life is diverse as well, due to the result of the extreme contrast between the cold and warm ocean currents on the East and West Coast. South Africa has over 10,000 marine plant and animal species, which is almost 15% of the coastal species in the world, with about 12% of these species being endemic and occurring nowhere else (DEAT, 1998).

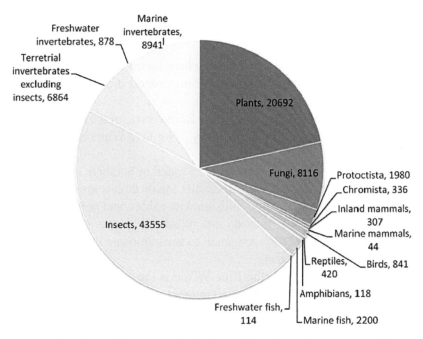

Figure 9.7 Numbers of known species in South Africa for major groupings of living organisms (*Source*: DEA, 2011; Nel and Driver, 2012).

Table 9.2 Species Richness of South African Taxa

Taxa	Number of described species in South Africa	Percentage of Earth's Total
Mammals	227	5.8%
Birds	718	8.0%
Amphibians	84	2.1%
Reptiles	286	4.6%
☐	112	1.3%
☐	2,150	16.0%
Invertebrates	77,500	5.5%
Vascular Plants	18,625	7.5%

Source: DEAT (1998).

The top 14 well-known endangered animals in South Africa are listed here and with the reasons why they are endangered, and the major threats (Drive South Africa, 2012; Getaway, 2017):

- Blue Crane: in danger of becoming extinct, due to direct poisoning, mining, agriculture, development and the further destruction of their grassland homes.
- Black Rhino: the Black Rhino, which seems to be the preference of poachers, is highly endangered with numbers dropping at a rapid rate year after year (Figure 9.8).
- White Rhino: the Northern White Rhino is one of the most endangered African animals and there are only 7 Northern White Rhinos left in the world.
- Vultures: 9 of vulture species are located in Southern Africa, and 7 of these are endangered or vulnerable. Major threats are loss of habitat, electrocution on pylons, collision with cables, and poisoning.
- African Wild Dog: they do not adjust to captivity easily, and many reserves and parks are too small to accommodate them. Major threats include human snaring.
- Pygmy Blue Whale: the Blue Whale is the largest known animal on earth, but the exact population numbers of the Blue Whale are hard to estimate and predict.

Figure 9.8   Black Rhino (Photo: Gmacfadyen, *Source:* Drive South Africa 2012).

- Cheetah: the Cheetah is smaller than the lion and leopard, which are both close on its heels heading towards the vulnerable list. It is vulnerable due to farmer   □
- African Penguin: the numbers of these awesome birds are decreasing rapidly, due to pollution and climate change.
- African Elephant: poachers after their ivory tusks are contributing to the endangerment of the African Elephant (Figure 9.9).
- Pickergill's Reedfrog: it is one of the most endangered amphibians in South Africa. It is under threat from coastal development, habitat fragmentation, and draining of water used for agricultural and urban development.
- Riverine rabbit: this nocturnal rabbit can only be found in the Karoo regions, and only lives in the deep silt   plains of Karoo rivers, and being endemic it can't be found anywhere else in the world, making it incredibly vulnerable to habitat loss. The major threats are loss of habitat due to cultivation, and livestock farming.
- Knysna seahorse: it occurs naturally in three estuaries around the country, namely Knysna, Swartvlei, and Keurbooms. However, the Knysna

Figure 9.9 African Elephant (Photo: Ralph Combs, *Source:* Drive South Africa 2012).

estuary is important to the   industry, and so habitat loss is the major threat.

- Golden Moles: it ranks high on the list of most endangered animals in South Africa, and it is threatened by the mining and agriculture development the grasslands.
- Yellow-breasted pipits: these birds occur mostly in the highland grasslands of the Drakensberg, and they are threatened from habitat loss due to commercial livestock farming.

9.4   Microbial Diversity in South Africa

Microorganisms exist in every niche on Earth, however, in South Africa there is a lack of research and a lack of legislation protecting the biodiversity of the microbial diversity, or of the microbial habitats. Microbial is classified as including the bacteria, and archaea (prokarya), fungi and yeasts (eukarya), and also viruses (Cowan et al., 2013).

South Africa has a wealth of   diversity, and many of these species hosts unique rhizospheric and endophytic microorganisms. Though plant virus diversity in South Africa has not been systematically investigated, but many have been found and characterized. Microbes are systematically investigated in South Africa, neither has nonpathogenic plant-associated microbes, nor the unique microbes living in the marine ecosystems and the hypersaline habitats been researched widely (Cowan et al., 2013). For example, the hypersaline pans in the Darling and Yzerfontein regions is well studied, but only about four reports have been published on the microbial diversity associated with South African salt pans (Cowan et al., 2013).

There are factors affecting discovery and systematic research on South African microbial diversity (Cowan et al., 2013):

- Microbial diversity and ecology are not priority research areas, in comparison with plants and animals.
- There is only small number of the academic community involved in environmental microbiology research.
- The very high cost of modern molecular technologies has prevented the growth of the microbial ecology research.

9.5   Genetic and Crop Diversity in South Africa

South Africa is a mega-diversity nation (Botha, 2004) with agricultural production constituting 50% of the source of threats to its biodiversity (Biggs et al., 2006). This highlights the complex link between understanding food security and biodiversity conservation within the country (Abdu-Raheem and Worth, 2013). In South Africa, agricultural activities range from intensive crop production and mixed farming in winter rainfall and high summer rainfall areas (WWF, 2012). Grains and cereals are South Africa's most important crops, occupying more than 60% of hectare under cultivation in the 1990s (Table 9.3). Maize, the country's most important crop, is most widely grown, followed by wheat, sugar cane and sunflowers. Citrus and deciduous fruits are exported, as are locally produced wines and flowers (DAFF, 2016).

Genetic diversity of 400 indigenous edible plant species and crop traditional varieties are conserved in gene banks. Agricultural biodiversity is declining at an accelerated rate in Southern Africa owing to increased demands from a rapidly growing population and increased competition for natural resources (Khumalo et al., 2012). Climate change, land-use and land-cover changes bring about habitat changes, losses and decline of biodiversity. The replacement of local crop varieties by improved or exotic varieties and species has been reported as the main cause of the genetic erosion of crops and other forms of agrobiodiversity (FAO, 2010; Khumalo et al., 2012). Biodiversity may be lowered as a result of invasive alien species, which are plants and microbes introduced to new regions mainly through human activities (Monier et al., 2004).

9.6   Threats to Biodiversity in South Africa

Threatened species constitute a widely used indicator of the status of biodiversity. There are several main issues that threatens terrestrial biodiversity in South Africa (DEAT, 1998; Driver et al., 2011):

Table 9.3 Production of Some Common Field Crops

Commodity	2015 (tons)
Maize	9,955,000
	663,000
Soybeans	1,070,000
Groundnuts	62,300
Dry beans	73,390
Wheat	1,501,190

Source: CEC (2016).

- Land transformation: land use, urbanization, agriculture, forestry, and mining;
- Environmental impacts: land degradation and ☐A cover and vegetation loss, soil erosion and soil transportation, alien plant invasion, contamination of land, climate change, crops, deforestation, and impacts of urban development;
- Other emerging issues: land restitution and land recapitalization, climate change and rural development, timber shortage, and acid mine drainage.
- There are a different set of issues that threatens the marine and coastal biodiversity in South Africa (DEAT, 1998):
- pollution;
- dredging and dumping;
- invasive alien species;
- climate change.

Human activities have changed South African ecosystems for many years, but the speed and extent of change increased rapidly with agricultural and industrial expansion. Current estimates suggest that mainly agriculture, developing urban areas and developments, deforestation, commercial   A☐
dams have led to the transformation and degradation of a substantial proportion of natural habitat (DEAT, 1998). Adding to habitat loss and degradation, the overexploitation of a number of species, the introduction of exotic species, and pollution of the soil, water and atmosphere have had major effects on South Africa's terrestrial, freshwater and marine biodiversity (DEAT, 1998).

Current South African assessments of the status of birds, mammals, and frogs have demonstrated that almost 10% of South Africa's birds and frogs are threatened, (Figure 9.10) and 20% of its mammals are threatened (DEA, 2011) (Figure 9.11). The country's plants are currently being assessed by the South African National Biodiversity Institute's threatened species program. Previous assessments show over 10% of plant species 22% of   species are threatened with extinction in South Africa (DEA, 2011). The South African ecosystems have been degraded and ecological processes reduced through fragmentation caused by several aspects of human activity (Table 9.4). These trends indicate that the situation is not improving and that the growing human population and unsustainable tempos of resource consumption will result in increased negative impacts on biodiversity (DEAT, 1998).

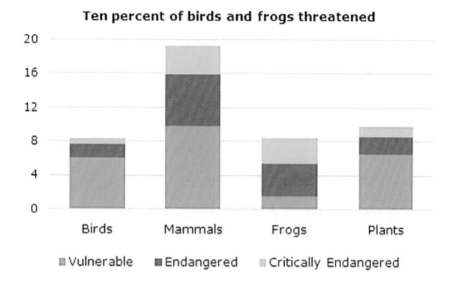

Figure 9.10 Threatened species per taxonomic group (*Source*: Endangered Wildlife Trust, 2002).

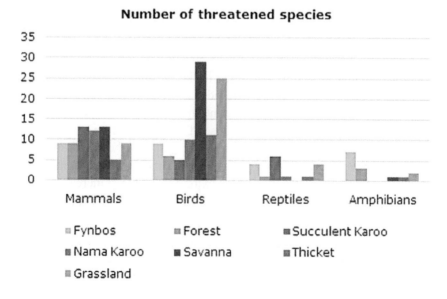

Figure 9.11 Threatened species per biome (*Source:* Endangered Wildlife Trust, 2002).

Table 9.4 Number of Threatened Species Per Taxonomic Group Per Biome

Group	Fynbos	Forest	Succulent Karoo	Nama Karoo	Savanna	Thicket	Grassland
Mammals	9	9	13	12	13	5	9
Birds	9	6	5	10	29	11	25
Reptiles	4	1	6	1	0	1	4
Amphibians	7	3	0	0	1	1	2

Source: Endangered Wildlife Trust (2002).

South Africa has the highest concentration of threatened plant taxa in the world in terms of area and total numbers. Of the 4,149 plant taxa whose conservation status has been evaluated, 3,435 are considered to be globally threatened with extinction. Most of these threatened species are found in the Cape Floristic Region, especially in the lowland Fynbos of the rapidly urbanizing of Cape Town (DEA, 2011). Nearly 1,900 of the 3,435 listed Red Data plant species in southern Africa are threatened totally or partially by alien invading plants. The South African Red Data Books show that 102 bird species (14%), 72 reptile species (24%), 17 amphibian species (18%), 52 species (17.6%) of mammals and 142 species (22%) of   are threatened (DEA, 2011).

The uncontrolled spread of invasive alien species is one of the key threats to indigenous biodiversity. This spread has negative impacts on the economy, in sectors as diverse as health, agriculture, water supply and tourism, and is likely to become much worse with climate change (DEA, 2011).

Climate change is widely attributed to the burning of fossil fuels, such as oil and coal, over the past few decades. This phenomenon has sharply increased the levels of carbon dioxide into the atmosphere. Carbon dioxide (CO_2) is called a 'greenhouse' gas because of its ability to trap heat. This and a number of other exacerbating factors are thought to be behind the increase in average global temperatures and changes in rainfall patterns. Warnings are that climate change will result in more extreme weather – increased   A and droughts, which could reduce agricultural production and threaten food security in South Africa (DEA, 2011).

9.7 Biodiversity Conservation Initiatives in South Africa

South African government is fully committed to biodiversity conservation, despite many difficulties. If one do not conserve biodiversity in South

Africa it will undermine the natural resource base upon which the population depend. Department of Environmental Affairs includes existing and future economic opportunities of using biodiversity in its policies as incentives, and realizes the risk to ecological processes, which are necessary to sustainability (DEAT, 1997).

9.7.1 Benefits Resulting From Species Harvested in the Wild

The advantages of conserving biodiversity are numerous. A big proportion of South Africa's population are directly dependent upon the biological resources for survival, including gathering, harvesting or hunting of animals and plants for food, medicine, shelter, fuel, building materials and trade (DEAT, 1997). The application of biological resources thus provides a significant buffer against poverty, as well as opportunities for self-employment in the informal sector. The South African fishing, hunting, wildflower, horticulture, natural product and wood-harvesting industries are all, to variable extents, dependent on species harvested from the wild (DEAT, 1997).

9.7.2 Benefits Resulting From the Direct Use of Ecosystems

Benefits arising from conservation of South Africa's biodiversity are not limited to the direct use of species. The ecosystems of South Africa are directly used for grazing, croplands, mining, recreation and tourism (DEAT, 1997). South Africa runs the risk of losing the economic advantages from resources and the option of excluding their use by future generations if cannot adequately conserve our resources (DEAT, 1997).

9.7.3 Advantages From the Direct Use of Ecosystems

Probably the most fundamental benefit of conserving biodiversity lies in the ecological service it provides. This service is important to fulfill human needs as well as those of all life on Earth. Some ecological services include:

- Maintenance of the hydrological cycle – the provision of clean water.
- Maintenance of atmospheric quality – provide clean air to breathe and to control the climate.
- Generation and conservation of soils – essential for agriculture and forestry.
- Nutrient cycling.

- Maintenance of a huge resource of genetic materials from which South Africa and other countries have developed crops, industrial products, etc.
- Most importantly, the protection and basis for adaptation which bio-diversity provides against   changes in climate and ecosys-tem processes – a big concern for South Africa where the climate is expected to become increasingly drier as the global climate changes (DEAT, 1997).

The South African government in collaboration with interested and affected parties has agreed to conserve the full spectrum of South Africa's biological diversity by applying different mechanisms such as legislation, planning controls, guidelines, and protected area designations (Figure 9.12). Legal measures and incentives have been introduced to conserve important ecosystems, habitats and landscapes outside of protected areas, including rangelands and their associated vegetation and indigenous wildlife resources (DEAT, 1998). Strengthening support for research on the improved under-standing of the structure and composition of South Africa's terrestrial,

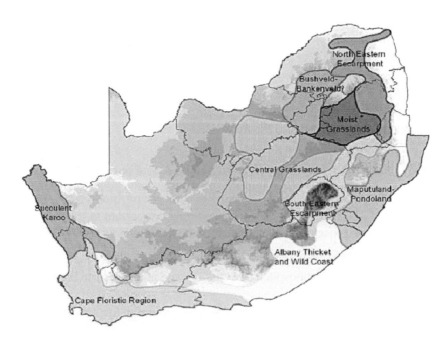

Figure 9.12 Nine broad priority areas under conservation action (*Source:* DEAT, 2009).

aquatic, marine and coastal ecosystems. To improve knowledge of and take appropriate action to conserve poorly known groups such as invertebrates, fungi and microorganisms. Also, to promote and support measures to manage   ⬜A
(DEAT, 1998).

9.8 Conclusion and Challenges to Biodiversity

South Africa needs to identify the processes or activities that have or are likely to have significant negative impacts on terrestrial, aquatic, marine and coastal biodiversity. Long-term monitoring of the effects of these processes and activities are vital. Department of Environmental Affairs needs to undertake necessary research to improve the understanding of the consequences of threatening processes or activities on ecological functions and processes, and other components of terrestrial, aquatic, and marine and coastal biodiversity. A mechanism to manage and collate this information needs to be developed to place it in the public domain, and to ensure that those decisions taken are based upon the best applicable knowledge available (DEAT, 1998).

Near the close of the twentieth century, the indigenous forests in South Africa have been reduced by 50%. Generally, an estimated 25% of South Africa's land has been transformed from its natural state (DEA, undated). Most riverine habitats have basically been changed, and very few naturally functioning freshwater systems have remained. Almost 50% of South Africa's wetlands have completely been lost through transformation to other land uses. Nearly all ecosystems in South Africa have been   ⬜ A changed by human activities. These changes include cultivation for commercial crops or subsistence agriculture; overstocking, overgrazing and poor land-use management; afforestation for commercial timber production; the spread of invasive alien trees, shrubs, herbs and fauna; urbanization and settlements; the damming of rivers; mining; transportation; industrialization; and subsistence and commercial harvesting of indigenous plant products (DEA, 2011). About 90% of South Africa falls within arid, semiarid or dry subhumid zones and is vulnerable to land degradation and   A Overgrazing is considered a threat to biodiversity in virtually all South African 'hotspots' of endemic species. This threat is particularly severe in the communally managed land of Maputaland, Pondoland, and the succulent Karoo (DEA, 2011).

Acknowledgments

Sincere thanks to SANBI (South African National Biodiversity Institute) for providing some of the data and reports used in this study. This project was in part financially supported by the National Research Foundation (NRF) Funding, Reference: IFR150206113775, Grant number: 96163.

Keywords

- challenges
- initiatives
- vegetation
- wildlife diversity

References

Abdu-Raheem, K. A., & Worth, S. H., (2013). Food security and biodiversity conservation in the context of sustainable agriculture: The role of agricultural extension. *South African J. Agricultural Extension, 41*, 1–15.

Biggs, R., Reyers, B., & Scholes, R. J., (2006). A biodiversity intactness score for South Africa. *South African J. Sci., 102*, 277–283.

Botha, M., (2004). *Implementing Laws for Conservation Action: Partnerships in the Biodiversity and Protected Areas Acts. [Online]* Available from: http://www.botanicalsociety.org.za/ccu/downloads/reports/Implementing%20laws%20for%20conservation%20action.doc (accessed 06/06/2017).

CapeNature, (2017). *Cape Floristic Region.* Accessed: 10–04–2017. Available: http://www.capenature.co.za/care-for-nature/biodiversity/cape-floristic-region/.

Conservation International, (2014). *Maputaland-Pondoland-Albany [Online].* Copy viewed on 17 March. Accessed: 1-04-2017. Available: http://www.conservation.org/where/priority_areas/hotspots/africa/Maputaland-Pondoland-Albany/Pages/default.aspx.

Cowan, D. A., Rybicki, E. P., Tuffin, M. I., Valverde, A., & Wingfield, M. J., (2013). Biodiversity: So much more than legs and leaves. *South Afr. J. Sci., 109*(11/12), Art. #a0037, pp. 9.

Cowling, R. M., Richardson, D. M., & Mustart, P. J., (1997). Fynbos. In Cowling, R. M., Richardson, D. M., & Pierce, S., (eds.). *Vegetation of Southern Africa.* Cambridge University Press, Cambridge UK, 99–130.

Crop Estimates Committee (CEC), (2016). *Crop Estimates [Online].* Available from: http://www.Sagis.org.za/CEC (accessed 13 June 2017).

Department of Agriculture, (2016). *Forestry and Fisheries (DAFF).* Economic reports [Online]. available from: http://www.daff.gov.za/statistics (accessed 10/06/2017).

Department of Environmental Affairs (DEA), (2007). *State of the Environment: Biomes [Online].* Accessed: 1–04–2017. Available: http://soer.deat.gov.za/268.html.

Department of Environmental Affairs (DEA), (2011). State of the environment. Biodiversity and Conservation. https://www.environment.gov.za/sites/default/files/docs/10ytearsreview_biodiversity_conservation.pdf (Accessed 15/06/2017).

Department of Environmental Affairs (DEA), (2011). State of the environment. *Threatened Species*. http://soer.deat.gov.za/434/_5623.html (accessed 17/06/2017).

Department of Environmental Affairs (DEA), (2015). In: Kharika, J. R. M., Mkhize, N. C. S., Munyai, T., Khavhagali, V. P., Davis, C., Dziba, D., et al., (eds.), *Climate Change Adaptation Plans for South African Biomes*. Pretoria.

Department of Environmental Affairs and Tourism (DEAT), (1997). *White paper on the conservation and sustainable use of South Africa's biological diversity*. Government Gazette no. 18163(3).

Department of Environmental Affairs and Tourism (DEAT), (1998). *South African National Report on the Convention on Biological Diversity*. South African National Report to the Fourth Conference of the Parties. https://www.cbd.int/doc/world/za/za-nr-01-en.pdf [Online] (accessed 19/06/2017).

Department of Environmental Affairs and Tourism (DEAT), (2009). *Biodiversity and Conservation*. A review of the Department of Environmental Affairs and Tourism: 1994–2009.

Desmet, P., (2007). Namaqualand-A brief overview of the physical and floristic environment", *J. Arid. Environments, 70*(4), 570–587.

Drive South Africa, (2012). *The Most Endangered African Animals*. Accessed: 10–04–2017. Available: https://www.drivesouthafrica.co.za/blog/the-most-endangered-african-animals/.

Driver, A., & Maze, K., (2002). The Succulent Karoo ecosystem plan [SKEP]. An introduction to SKEP, *Veld & Flora*, vol. 88.

Driver, A., Sink, K. J., Nel, J. N., Holness, S., Van Niekerk, L., Daniels, F., et al., (2012). *National Biodiversity Assessment 2011: An Assessment of South Africa's Biodiversity and Ecosystems*. Synthesis Report. South African National Biodiversity Institute and Department of Environmental Affairs, Pretoria.

Endangered Wildlife Trust (EWT), (2002). *The Biodiversity of South Africa 2002: Indicators, Trends and Human Impacts*. Struik, Cape Town.

Foden, W., (2002). *A Demographic Study of Aloe dichotoma in the Succulent Karoo: Are the Effects of Climate Change Already Apparent?* MSc Thesis, Percy FitzPatrick Institute of African Ornithology, Botany Department, University of Cape Town.

Geldenhuys, C. J., (1992). Richness, composition and relationships of the floras of selected forests in southern Africa. *Bothalia, 22*, 205–233.

Getaway, (2017). 10 endangered animals in South Africa. Accessed: 10–04–2017. Available: http://www.getaway.co.za/environment/conservation-environment/10-endangered-animals-south-africa-can-help/.

Guo, D., Arnolds, J. L., Midgley, G. F., & Foden, W. B., (2016). Conservation of Quiver Trees in Namibia and South Africa under a Changing Climate. *J. Geoscience and Environment Protection, 4*(7), 1–8.

Guo, D., Desmet, P. G., & Powrie, L. W., (2017a). Impact of the future changing climate on the Southern Africa biomes, and the importance of geology. *J. Geoscience and Environment Protection, 5*, 1–9.

Guo, D., Young, A. J., Desmet, P. G., & Midgley, G. F., (2017b). Climate change impacts on dwarf succulents in Namibia as a result of changes in fog and relative humidity. *J. Water Resource and Hydraulic Engineering, 6*(3), 57–63.

Khumalo, S., Chirwa, P. W., Moyo, B. H., & Syampungani, S., (2012). The status of agro-biodiversity management and conservation in major agroecosystems of Southern Africa. *Agriculture, Ecosystem and Environment, 157*, 17–23.

Lonely Planet, South Africa. [Online]. 2017. Accessed: 1–04–2017. Available: http://www.lonelyplanet.com/south-africa/wildlife/animals.

Monier, M., El-Ghani, A., & El-sawaf, N., (2004). Diversity and distribution of plant species in agro-ecosystems of Egypt. *Systematics and Geography of Plants, 74*, 319–336.

Mucina, L., & Rutherford, M. C., (2006). *The Vegetation of South Africa, Lesotho and Swaziland*. Strelitzia 19, South African National Biodiversity Institute, Pretoria.

Mucina, L., Geldenhuys, C. J., Rutherford, M. C., Powrie, L. W., Lotter, M. C., Von Maltitz, G. P., et al., (2006). Afrotemperate, subtropical and Azonal forests. In: Mucina, L., & Rutherford, M. C., (eds.). *The Vegetation of South Africa, Lesotho and Swaziland*. Strelitzia 19. SANBI. Pretoria.

Myers, N., Mittermeier, R. A., Mittermeier, C. G., Fonseca, G. A., & Kent, J., (2000). Biodiversity hotspots for conservation priorities, *Nature, 403*(6772), 853–858.

Nel, J. L., & Driver, A., (2012). *National Biodiversity Assessment 2011: Technical Report*. Freshwater component. CSIR Report No. CSIR/NRE/ECO/IR/2012/0022/A. Council for Scientific and Industrial Research, Stellenbosch, vol. 2.

Palmer, A. R., & Hoffman, M. T., (1997). Nama Karoo. In: Cowling, R. M., Richardson, D. M., & Pierce, S. M., (eds.). *Vegetation of Southern Africa*. Cambridge University Press, Cambridge, pp. 167–186.

Preston, G. R., & Siegfried, W. R., (1995). The protection of biological diversity in South Africa: Profiles and perceptions of professional practitioners in nature conservation agencies and natural history museums" (25), *South African Journal of Wildlife Research, 49*.

Scholes, R. J., & Archer, S. R., (1997). Tree-grass interactions in savannas. *Annual Review of Ecology and Systematics, 28*, 517–544.

South Africa Government Online, (2017). *About SA – Geography and Climate. [Online]*. Accessed: 10–04–2017. http://oldgov.gcis.gov.za/aboutsa/geography.htm.

South African National Biodiversity Institute (SANBI), (2017). *Red List of South African Plants*. Threatened Species Programme [Online]. Accessed: 17–05–2017. Available: http://redlist.sanbi.org/index.php.

Wikipedia, (2017). *Geography of South Africa [Online]*. Accessed: 1–04–2017. Available: http://en.wikipedia.org/.

World Wildlife Fund (WWF), (2012). *Agriculture: Facts & Trends: South Africa*. http://awsassets.wwf.org.za/downloads/facts_brochure_mockup_04_b.pdf (accessed 10/06/2017).

Young, A. J., Guo, D., Desmet, P. G., & Midgley, G. F., (2016). Biodiversity and climate change: Risks to dwarf succulents in Southern Africa. *J Arid Environments. 129*, 16–24.

Zietsman, L., (2011). *Observations on Environmental Change in South Africa*. Sun Press, Sun Media Stellenbosch.

CHAPTER 10

Biodiversity in Sudan

AHMED A. H. SIDDIG,[1,2] **TALAAT DAFALLA ABDEL MAGID,**[3]
HAGIR MAHAGOUB EL-NASRY,[1] **ABDELNASIR IBRAHIM HANO,**[1]
and **AMEER AWAD MOHAMMED**[4]

[1]University of Khartoum, Faculty of Forestry, Khartoum North, Sudan,
E-mail: asiddig@eco.umass.edu
[2]Harvard Forest, Harvard University – 324 North Main St., Petersham MA 1366, USA
[3]University of Bahri, College of Natural Resources, Khartoum North, Sudan
[4]Wildlife Research Centre, Khartoum North, Sudan

10.1 Introduction

Sudan is considered one of the largest countries in Africa with area approximately 1.8 million kilometers square and possesses 750 km coastline on the red sea, lies between the latitudes 10° and 22°N and longitude 22° and 38°E (Figure 10.1). The country is distinguished by gently sloping plains derived from iron rocks in the southern parts, clay soil in the middle and sandy soil in the west and north. There are many mountains like Jabel Mara in the west, and series of the Red Sea hills in the east.

The climate of Sudan is dominated by arid climate with some exceptional conditions in some parts of the country (Figure 10.2). Rainfall increases from less than 100 mm in the north to about 1,000 mm in the South (Adam and Abdalla, 2007). It rains usually in summer except some areas like the Red Sea hills have winter rains. Due to this wide variation in annual rainfall and temperature, Ayoub (1998), Badi (2004), and Mustafa (2007), stated that dry areas in Sudan comprise about 94% of the total area of the country including many subclimatic zones such as arid (30%), semiarid (20%), savannah (40%) and mountains and wetlands (10%).

The estimated Sudan's population in 2015 was about 40.0 million, 63% of them are rural people. The annual growth rate of the population is 2.6 % with the average population density of 43 people per sq. mile, which increases in the area between White Nile and Blue Nile (National Biodiversity Strategy and Action Plan – NBSAP, 2015).

The population had increased substantially during the previous 70 years, from 10.3 million in 1956 to 25.6 million in 1993 to 30.3 million in 2002 and to 40.0 million in 2015. Also the annual growth rate has increased from 1.9% to 2.7% as one of the highest in the world. Arab Organization for Agricultural Development (AOAD) (2006) reported that the majority of the popula-

Figure 10.1   Map of Sudan (*Source:* Wikipedia on August 14, 2017).

tions of Sudan are working in agricultural (58%) or pastoral activities. The remaining percentage is employed in manufacturing, construction, mining and civil services.

In 2003, Sudan's First National Communication under the United Nation Framework Convention on Climate Change (UNFCCC) reported that surface and groundwater resources are available in Sudan but are generally shared with neighboring countries. Ten countries, namely South Sudan, Burundi, Democratic Republic of Congo, Eritrea, Ethiopia, Egypt, Kenya, Rwanda, Sudan, Tanzania, and Uganda, share the Nile, which is the dominant surface water supplier. The three main non-Nilotic streams are Azum, Gash, and Baraka, which are also shared with Chad and Eritrea, respectively. Likewise, the Nubian sandstone aquifer is shared with Egypt, Libya, and Chad.

The wide variation of the climate, topography and soil factors are on different plants communities and ecosystems such as the desert,

Sudan map of Köppen climate classification

Figure 10.2   Map of climatic zones of Sudan based on Köppen climate Classification. Map was acquired from Wikipedia page on August 12, 2017.

semidesert, low or high rainfall savannah and special zones like the mountains, the Nile and   regions. These variations in vegetation communities encourage several patterns of natural habitats, which are maintaining the essential components of biological diversity particularly the wildlife.

Despite this great diversity in ecosystems and populations, current studies on the status of biodiversity are very limited. For instance, there is absence of information about plants composition, abundance and diversity as well as wildlife species. Also information about endangered species or focal populations it turns to be very limited although previous and current reports have warned from the risk of extinction. In addition, there is lack of knowledge about the anticipated impacts of climatic changes in Sudan on plants, animals and ecosystems. Therefore, in this chapter we provide brief overview about biodiversity of Sudan including description of current status of ecosystems and populations to the best to our knowledge, importance of some plants and animals in livelihoods, major threats and some conservation efforts, and crucial future directions needed for better biodiversity conservation.

10.2   Current Status of Biodiversity

10.2.1   Main Ecosystems

We can divide the ecosystems in Sudan into two major types; namely aquatic and terrestrial habitats. In the following paragraphs we brief on each of these two types of ecosystems and their occurrence in the country.

10.2.1.1   Aquatic (Marine and Freshwater) Ecosystems

Despite the apparently dry nature of Sudan but the country is characterized by wetland ecosystems those rich of biological diversity and fascinating forms of life. First, the *marine ecosystems* occur in the Red Sea in the eastern part of the country where the famous mangroves forests are dominated as well as three marines protected areas (e.g., Senganeeb and Donganab) are located. Second, there are many inland freshwater zones which include the River Nile and its tributaries and a large number of seasonal rivers and water courses such as the Gash and Baraka, originate within the Ethiopian highlands, and form two inland deltas in Sudan. Also, there are hundreds of lakes and ponds varying in size and biodiversity components distributed across the country.

10.2.1.2   Terrestrial Ecosystems

10.2.1.2.1   Forests, Grasslands, and Mountains

Sudan can be ecologically divided into five vegetation zones according to rainfall patterns from North to South (Abdel Magid and Badi, 2008). These are: (i) Desert: (0–75 millimeters of precipitation), (ii) Semi-desert: (75–300 mm), (iii) Low rainfall savannah on clay and sand: (300–800 mm), (iv) High rainfall savannah (800–1500 mm), and (v) Mountain Vegetation: (300–1000 mm). The ecological zones extend over a wide range from the desert in the extreme north to the savannah (Figure 10.2). According to the recently published, Land Cover Atlas of Sudan, Forests together with Rangeland represent 35.6% of the total country area (HCENR, 2014; NBSAP, 2015).

10.2.1.2.2 Agro-Ecological Zones

The agricultural sector has a vital role to play in achieving food Security by intensifying food production and providing employment opportunities in the rural area. Sudan has cultivable arable land estimated at 86 million hectare. However, less than 20% are used under three major farming subsectors:

The irrigated system is estimated at a million hectare. The subsector contributes an average of 21% of the total value of agricultural production, 100% of wheat and 25% of sorghum produced in the country. Although its contribution to sorghum production is low relative to the rain-fed subsector, it is more stable. In years of drought it plays an important role in meeting the consumption requirements.

The semimechanized rain fed with an area of about 6 million hectare the two main crops produce by this sector are sorghum and sesame, the crop yield depends on rainfall.

The agro-pastoral traditional rainfed sector is about 9 million hectares located in the western, Central and southern parts of Sudan. The main crops produced by this sector are sorghum, Millet, Groundnuts, Sesame, short Staple Cotton and Gum Arabic. The Sudan's forests follow the ecological clas-

A

the form of bush land and scattered trees and shrubs in the north and in dense forests of large trees in mixture of acacias and broad-leaved trees in the southern end of the savannah and mountain region. The Nile basin traverses the Sudan from south to north constituting the main drainage line. The Nile and the mountain region incorporate distinct sites that are characterized by availability of water resources supporting agriculture, livestock and forests developments. Rich forests grow along the Nile and its tributaries. Forests and woodlands cover about 64.36 million ha, while rangelands are estimated to cover 24 million ha (Elsiddig et al., 2011). Forage from rangelands is estimated to provide, depending upon the region, from 55–80% of the national herd feed requirements. Livestock husbandry in its various forms is practiced by an estimated 40 percent of the population. It plays a great role as a food system, store of value, wealth, and means of access to power and authority in places where the banking systems and market economy do not reach the targeted areas or function properly (HCENR, 2014).

10.2.2 *Populations and Taxonomic Groups*

10.2.2.1 Forests Diversity

The natural cover of Sudan is generally poor, sparse and scanty vegetation due to its location in such arid environment of the country; except along the Nile banks and water courses where ephemeral herbs and grasses occur after the short rainy seasons. The forests resources in Sudan were estimated by FAO in 1990 to be about 19% of the total area of the country. The national forest inventory which was done by Forest National Corporation (FNC) with the cooperation of FAO in 1995 documented that the forest cover was 24.9% of the total area of Sudan (Mukhtar and El Wakeel, 2002). Recent estimates based on the Afri-cover assessment and displayed in the NBSAP of 2015 showed that the forest cover accounts for only 11.9% of the area of the country with annual deforestation rate that exceeds 2.4%. The reserved forests are about 837 forests distributed all over the country (Badi, 2004; Elsiddig et al., 2007, 2011). They constitute a great potential for biodiversity conservation and play vital role as an important component of natural resources and land use. Within these forests is thought to be there are over 500 unique species of trees and 185 species of shrubs based on experts opinions but these information haven't been investigated yet through field surveys (Mukhtar and ElWakeel, 2002; NBSAP, 2015). Al-Amin (1990) in one of the prominent trees dendrology studies in Sudan has broadly classified the forest composition to three regions/ecosystems: (i) *Arid &semiarid region* (includes Northern, River Nile, Khartoum, White Nile, North Kordofan, North and West Darfur and Red Sea states) where the vegetation is typically includes some scattered perennial vegetation, bushes and woody succulent thorny or leafless shrubs and dominated by *Acacia tortilis, Capparis decidua* and *Maerua crassifollia* association. (ii) *Dry Savanna ecosystem* (includes Kassala, Gadaref, West Kordofan, and East and Central Darfur states) in which species richness and density increase with noticeable abundance and dominancy for species like *Acacia seyal, Balanites aegyptiaca, Faidherbia albida, Ziziphus spina-chirsti,* and *Calotropis procera.* (iii) Wet Savanna ecosystem (Blue Nile, South Kordofan, and South Darfur States) is essentially different and considered the most richest and dense region in Sudan in terms of trees and shrubs composition. Among the famous trees species occur at this zone are *Adansonia digitata, Balanites aegyptiaca, Borassus aethiopium, Cordia africana, Diospyros mespiliformis, Tamarindus indica, Prosopis africana, Sclerocarya birrea, Terminalia brownie, Crateva adansonii,* and *Comberetum* sp.

In the basin and along the banks of the Nile and tributaries species like *Phoenix dactylifera*, *Acacia nilotica*, and *Tamarix aphylla* are found.

10.2.2.2   Wildlife Diversity

Sudan is very rich in wildlife resources, which are considered an essential resource due to its economic, environmental, social, cultural and recreational values. Abdelhameed and Nimir (2007) stated that out of 13 orders of mammals endemic in Africa 12 orders are in Sudan. Nimir (1984) indicated that about 224 species and subspecies are classified under 91 genera of animals were documented in Sudan.

Abdelhameed and Nimir (2007) mentioned that there are about 83 main wildlife species and their distribution in 19 nature reserves in Sudan (Appendices 1 and 2). 931 kinds of birds were recorded by Nikolaus (1987). About 24 families of Nile   A
snakes were cited by Nimir (1995). Notwithstanding there is no recent wildlife surveys, on the other hand the report by UNEP (2007) is considered the most recent report about the status of Sudan's wildlife diversity. Here is a summary of occurrence of wildlife populations in different habitats as stated by UNEP (2007).

10.2.2.2.1   Arid Regions

The Red Sea Mountains, as well as those on the Ethiopian border and in Northern Darfur, are home to isolated low-density populations of Nubian ibex (*Capra ibex nubiana*), wild sheep (*Ammotragus tragelaphus*) and several species of gazelle. Larger predators are limited to jackal (*Canis mesomelas*) and leopard (*Panthera pardus*). Due to the lack of water, wildlife in the desert plains is extremely limited, consisting principally of Dorcas gazelle (*Gazella dorcus*) and smaller animals.

10.2.2.2.2   The Nile Riverine Strip

The Nile riverine strip is heavily populated and supports birdlife, reptiles and smaller animals (including bats).

10.2.2.2.3   The Sahel Belt (Central Dry Land Agricultural Belt)

In the Sahel belt, expansion of mechanized agriculture and moving pastoralists has eliminated much of the wild habitat. Also the region hosts migratory birds, particularly in the seasonal wetlands and irrigated areas.

10.2.2.2.4   The Marra Plateau

The forests of Jebel Marra historically hosted significant populations of wildlife, including lion and greater kudu (*Tragelaphus strepsiceros*). Limited surveys in 1998 reported high levels of poaching at that time. Due to the current conflict in Darfur, there is very limited information on the current status of wildlife in this region.

10.2.2.2.5   The Nuba Mountains

The wooded highlands of the Nuba Mountains historically host large populations of wildlife. Nimir (1984) reported the presence of greater kudu, dik-dik (*Madagna saltiana*), porcupine (*Hystrix cristata*), python (*Python sebae*), and guinea fowl (*Numida meleagris*). The greater kudu has also been reported at Jebel Karan east of Abu Gibeiha. The animals found in Nuba Mountains include Dorcas gazelle (*Gazella dorcus*), Dama gazelle (*Gazella damaruficollis*), Sommering gazelle (*Gazella soemmeringii*), Oryx (*Aegoryxal gazel*), Ostrich (*Struthio camelus),* Striped hyaena *(Hyaena hyaena*), Spotted hyaena *(Crcuta crocuta),* Jackals (*Canisme somelas*), and rats.

10.2.2.3   Genetic Diversity

Forest management has always been a tool for managing forest recourses. Any environmental activities at war and conflict affected sites should consider sound management methodologies such as inventory, planning, organization and control (Abdel Magid and Badi, 2008; Elsiddig et al., 2011). These entire components are obligatory for forest management, and it will act as a means of identifying what can be done to enhance and protect the values and aspects of resources that are most important.

Ex situ conservation and management activities or programs (forestry). The herbarium of FRC is the largest in Sudan incorporating over 7,000 forest specimens with a computerized appendix. The Agricultural Plant Genetic Resources and Research Centre (APGRC) are entrusted with conservation of plant genetic resources for food and agriculture. More than 11,000 accessions representing more than 60 crops have been acquired.

In situ conservation for Animal Genetic Resources: Some of the livestock indigenous breeds whose numbers are at risk cannot be genetically improved fast enough to adapt to climate change. Consequently, efforts to conserve them both in situ and ex situ need to be   ⌐A livestock research stations were established in various parts of the country in order to study, improve and preserve the local breeds. However, the Animal Resources Research Corporation (ARRC) runs several research stations for breeding management,   A

breeds (Butana, Kenana and Baggara) are studied in three research stations. Livestock keepers play a great role in ex situ conservation in a number of ways. In the extensive production systems adaptive traits are prioritized to production traits as a hazard reduction strategy. Another strategy to conserve local breeds is to diversify livestock types. Marriage between the various multiethnic groups and political alliances helps in the distribution and dissemination of animal herds on the largest possible area of land and under wide environmental conditions. Agro-pastoralists used to close some grazing areas and preserve it for the time when there is drought as a coping strategy against wet season failure (Anon, 2015).

Ex situ conservation for Animal Genetic Resources: An insemination center was established to disseminate this technology and now other few centers are providing this service, e.g., Climax Company and Animal Production Research Centre at Kuku.

10.3 Importance and Uses of Biodiversity

Perhaps these diverse ecosystems with great number of animals, plant, and microorganisms have created plethora of benefits to millions of people in Sudan and contributed significantly to their livelihoods and welfare (Elsiddig, 2011; Siddig, 2014). Abdel Magid (2001) reported that all sorts of biological diversity provide millions of dollars to the national economy through trades in agricultural crops, wildlife, and forests products. Specific and direct biodiversity values to the communities include shelters, bush-meat, fuel, food, medicines, fibers, gums, fodder, and livestock welfare (Table 10.1). Moreover, biodiversity proved to have substantial contribution towards local communities' development creating permanent or seasonal work opportunities, income generation and eventually reduces poverty stress. In the following table (i.e., Table 10.1) we provide a summary of some of the benefits that forest trees provide as described in NBSAP (2015).

10.4 Efforts of Biodiversity Conservation

10.4.1 Establishment of Biodiversity Institutions

After the Battle of Omdurman and the start of Condominium rule in 1898, forestry activities started in Sudan in 1901. The Woods & Forests Ordinance was promulgated in 1901 and the Department of Woodlands & Forests was

Table 10.1 Common Sudanese Trees Species and Their Uses as Described in the National Biodiversity Strategy Action Plan for 2015

Use	Trees species
Timber and building materials	*Acacia nilotica, Albizia aylemeri, Ailanthus excelsa, Balanites aegyptiaca, Borassus aethiopium, Burkea africana, Cassia siamea, Celtis integrofolia, Cordia africana, Crateva adansonii, Dalbergia sissso, Dalbergia melanoxylon, Daniellia oliveri, Detarium microcarpum, Kigelia africana*
Antiques	*Dalbergia melanoxylon*
Fruits, oil and honey bee production	*Acacia nioltica, Adansonia digitata, Balanites aegyptiaca, Borassus aethiopium, Cordia africana, Diospyros mespiliformis, Grewia tenax, Ficus sycamorus, Hyphaene thebaica, Gardenia lutea, Tamarindus indica, Prosopis africana, Sclerocarya birrea, Terminalia brownie, Moringa oleifera, Ximenia americana*
Medicinal uses	*Salvadora persica, Acacia polyacantha, Adansonia digitata, Cissus quadrangularis, Cordia sinensis, Crateva adansonii, Cymbogogon citratus, Tamarindus indica, Stereospermum kunthianum, Albizia anthelmintica, Faedharbia albida, Fagonia cretica, Grewia villosa, Hydnora abyssica, Psidum guajava, Sterculia setigera*
Gums	*Acacia senegal, Acacia seyal, Acacia polyacantha, Sterculia setigera*
Fodder	*Balanites aegyptiaca, Ficus sycamorus, Ziziphus spina-christi, Faedherbia albida, Acacia senegal, Acacia mellifera, Stereospermum kunthianum, Combretum aculeatum, Maerua crassifolia, Cadaba grandulosa, Bauhinia rufescens*

established the same year. The Ordinance was replaced in 1908 by the First Forest Act. Adoption and implementation of administrative and legislative measures continued ever since. The national forest service of Sudan is one of the oldest statutory institutions in the country (Abdel Nour, 2017). In 1902, the unit of Wildlife protection and hunting was established under direct supervision of General British Governor in Sudan. Currently, there are many institutions deal with biodiversity resources these including Ministry of the Environment and Physical Development, Ministry of Interior's

Wildlife Protection Administration, Forest National Corporation, and Ministry of Agriculture.

10.4.2 Establishment of Protected Areas

Sudan started to establish protected areas according to London convention for protection of African plants and animals (1933), which is driven from. According to the Wildlife Protection Act of 1935 which was based on London convention, protected areas were established and classified into three categories, which are national parks (*with highest degree of protection*) and it has been classified in category two according to the IUCN categories, game reserves (*limited use of resources could be permitted*) and game /birds sanctuaries (*protection for specific species*) (UNEP, 2007; Abdelhameed and Nimir, 2007). The total protected area is approximately 14.2 million ha representing 5.7% of the Sudan (248,800,000 sq. ha). The proposed protected areas are estimated about 3.4 million ha (1.4%) of the total country area (Abdelhameed and Nimir, 2007). The protected areas in Sudan are distributed among many ecological zones. But the distribution of protected areas of dry lands habitat occurred approximately in the northern latitude 12° N of the Sudan. According to UNEP (2007), Sudan has seven existing or proposed marine protected sites (Appendices 1 and 2), with a total area of approximately 576,000 km^2, and 18 existing or proposed terrestrial and freshwater protected sites, with a total area of approximately 3,946,303 km^2.

10.4.3 National Biodiversity Strategy and Action Plan (NBSAP)

Along the line of conservation efforts and the country's strategy to cope with biodiversity crises, Sudan has initiated many national action plans on biodiversity. The most recent strategy was put together for the period between 2015 and 2020 and called National Biodiversity Strategy and Action Plan (NBSAP). Increasing conservation efforts and establishment of new protect areas to satisfy community needs is central goal for this NBSAP of 2015. Other objectives were to: (i) enforcing laws and legislations; (ii) increasing governmental commitments regarding funding and institutional reforms; (iii) developing an effective system for raising awareness and actively involving communities in conservation practices; and 4) fulfilling international commitments and conventions.

10.4.4   Regional and International Agreements

Sudan has acknowledged many international and regional environmental conservation efforts such as ratification on many Multilateral Environmental Agreements (MEA) – (Table 10.2). However, generally developing countries are facing many challenges in implementing these MEA. Sudan as a developing country is no exception. These challenges can be summarized as follows: The huge gap between developed and developing countries in terms of technology and scientific information, and establishment of solid regional organizations. Implementation of MEA in Sudan requires restructuring of the existing national institutions, formulation of national programs between the different MEA's national Focal points especially in similar activities. Table 10.2 summarizes the regional and international biodiversity related conventions that Sudan has been involved in which including the agreement, year of ratification, country status, and important objectives of the agreement.

10.5   Threats and Conservation Challenges

In spite of the ongoing conservation efforts and declaration of new protected areas in Sudan, however, habitat destruction and decline of plants and animals is thought to be a major biodiversity dilemma in the country due to several human and natural threats as discussed by Takona (1999), Goda (2007), Abdel Magid (2001), UNEP (2007), Siddig (2009), and HCENR (2014). Dominant human disturbances in Sudan are including deforestation and habitat destructions, urbanization, overexploitation of plants and animals, agricultural expansion and land clearance, and pollution. On the other hand, the most important natural threats face biodiversity and protected areas in Sudan are desertification and erosion, climatic changes including drought and flooding, fire, landslides.

In addition, Siddig (2014) has provided comprehensive analysis to the challenges of biodiversity conservation in Sudan as related to governance and political instability in the country which we can reiterate them in the following:

- Limited management programs and budgets to implement the national strategies of biodiversity related sectors.
- Absence of rigorous national plans for integrated use of lands and natural resources.

Table 10.2 Regional and International Environmental Conventions That Sudan Has Signed and Considered Active Member

Convention	Convention Date	Status/date of Sudan ratification	Main remark/objectives
London Convention on the ☐ their Natural state	1933	1933	The convention deals with the preservation of animals and plants naturally by establishing protected areas and regulating the collection and hunting of wild animals.
Paris International Convention for the protection of Birds	1950	1953	The main objective of this convention is to protect birds in their natural habitats.
Rome International Convention for the Preservation of Flora	1951	1971	The convention prohibited the Trans-boundary movement of plant pathogens.
African Convention on the Conservation of Natural Resources. Algeria	1968	1973	The convention encourages the African states for resources protection and development.
Convention on wetlands of International Importance specially as water fowl Habitat Ramsar	1971	2005	The convention deals with the conservation of wetlands for the protection of waterfowls and establishment of protected areas to serve the cultural, economic and tourism purposes.
Paris Convention concerning the protection of the world culture and natural Heritage	1962	1973	The convention is calling for an effective system to protect cultural environment and natural heritage of out sanding universal value.
Washington convention on International trade in endangered ☐ (C.I.T.E.S)	1973	1982	The convention deals with the protection of endangered species through export and import licenses in endangered, and threatened species and species, which its jurisdiction.

Table 10.2 (Continued)

Convention	Convention Date	Status/date of Sudan ratification	Main remark/objectives
Bonn convention on the conservation of Migratory species of wild animals	1979	2002	council for technological advice to conserve migratory species.
Jeddah regional convention for the conservation of the Red Sea and Gulf of Eden Environment	1982	1985	The Sudan is one of six countries which deals with the sustainable use of Marine and coastal resources of the Red Sea.
Vienna convention for the protection of the Ozone layer	1985	1993	The instrument provide for exchange of information, transfer of technology, promotion of public awareness present substances which deplete the ozone layer.
United Nation Framework Convention on Climate Change (UNFCCC)	1992	1994	The convention regulates levels of the green house gases (GHG) and its concentration in the atmosphere to combat climatic changes. Such a level should ecosystems to adapt naturally to climatic change, to ensure that food production is not threatened and to enable economic development to proceed in a sustainable manner.
Convention on Biological Diversity (CBD)	1992	1995	The objectives of this convention are to conserve biological diversity, to promote the sustainable use of its components and to encourage equitable sharing of the genetic resources.

Convention	Convention Date	Status/date of Sudan ratification	Main remark/objectives
International convention to combat experiencing serious Drought and or Africa (UNCCD)	1994	1995	The objectives of this convention are to combat fects of drought, improved land productivity, rehabilitation, conservation and sustainable management of land and water resources.
Koyoto protocol	1997	2005	Stabilization of atmospheric concentrations of greenhouse gases at a level that would prevent dangerous anthropogenic interference with climate system. The protocol commits developed countries and countries making the transition to market economy to reduce their overall emissions of six greenhouse gases by an average of 5.2% below 1990 levels between 2008–2012.
Nagoya Protocol on Access and	2010	2014	The aim of this protocol is the implementation of one of the three objectives of the CBD: the fair and equitable genetic resources.

- Absence of effective awareness raising programs related to the importance of biodiversity conservation and public engagement.
- Enforcement of the laws and legislations is not up-to-speed due to weak institutions.
- Wars and   have led thousands of people to   A   -lages to seek shelters inside the biodiversity-rich forests and protected areas. These refugee camps have dramatically contributed to ecosystems degradation and decline in numbers of plants and wildlife species.
- Presence of funding issues, shortage in equipment, and limited training opportunities.

10.6   Biodiversity Information Gap

Despite the fact that most of Sudan is characterized by harsh and dry environment, but the country is thought to be rich by its diverse plants, animals and microorganisms those play significant role in people's livelihoods. However, this claimed richness is only based on experts knowledge or regional extrapolations (which are much appreciated) but not supported by policy-driven and systematic field-based surveys. Consequently, information about current status and trends of these populations as well as their habitats is extremely limited (UNEP, 2008; Siddig, 2014). Furthermore, there is consensus among Sudanese conservationists and environmental scientists that biodiversity of Sudan is under continuous decline due to decades of massive pressure on habitats and populations from people, wars and unrest, abandonment of the government, and after all drought and desertification.

On the other hand, documentation and dissemination of   A mation including biodiversity data is common issue in Sudan. It is very surprising that there is an ongoing and severe gap of electronically available information about composition, abundance status, and distribution of plants, microbes, insects, amphibians,   birds, etc., of the country. This   A ciency not only at the populations level, but also there are much unknowns about ecosystems and genetic diversity.

During our literature search for this chapter writing efforts, we have encountered great   any recent information. Even the little information we reported here are either from reports of former and recent environmental projects those supported by international agencies (e.g., FAO, UNDP, UNEP, USAID, etc.) or personal communications with colleagues.

While there is some knowledge about forest trees, range plants, crops, mammals, birds and   A pods, amphibians, reptiles. Similarly, majority of the protected areas including reserved forests as crucial ecosystems are understudied and information about their current status and threats is scarce. For instance, more than 80% of the wildlife research in Sudan has been and is being conducted in Dinder National Park with absence of any information from other 18 protected areas.

Given this lack of information, current efforts for establishing new protected areas,   ⌐A
national biodiversity strategies must consider running biodiversity monitoring and assessment programs with the hope to better discover the diversity of the country, improve documentation and dissemination, and involve policymakers in such information-based conservation initiatives.

Of course initiation of successful biodiversity and ecological assessments, monitoring, documentation and reporting will not be complete and realistic unless some preliminary efforts and commitments have been readily made. These efforts may include strengthening institutions and facilities, capacity building and training, conducting baseline and long-term monitoring initiatives, adoption of technology, and importantly raising awareness and enforcing laws.

10.7 Urgent Management Interventions and Future Directions

According to NBSAP (2015), the global 20 Aichi Biodiversity Targets were selected and used as Sudan national strategic goals. The main components of NBSAP include plant, forestry, range, farm animal, wildlife, marine and in-land waters biodiversity and biotechnology and biosafety. Considering each component circumstances, specific and realistic component targets corresponding to global targets were set. Proposed actions are based upon these component targets. Sudan has political commitment to mainstream biodiversity components and ecosystem as high development priorities. There are many opportunities for mainstreaming and endorsement such as the National Constitution, policies, strategies, and legislations. In the Proposed National Constitution, government should consider efforts to address biodiversity components and ecosystem as an integrated element of sustainable development policies (macroeconomic, agricultural policy and climate change). Management and conservation of biodiversity components and ecosystem frameworks have been integrated into sustainable development planning. The following are identifiable recommendations:

- Surveying and documenting current status of plants, wildlife, and microbes populations in Sudan. Decisions about their conservation depend on this information.
- Devoting special attention to the endangered species and species of need as well as understudied groups such as microbes and amphibians, for example.
- Surveying different ecosystems and agro-ecological zones including those related to cultivated plant species, natural range plants, wild food plants and weeds within the different production systems.
- Establishment of a national information system with information sharing mechanisms on the state of in-situ and ex-situ conditions, with due consideration to establishing an early warning system as part of the national information system.
- Capacity building of the existing institutions, in terms of coordination, human resources and physical infrastructure.
- Strengthening breeding capacities in research institutes and universities in terms of human resources and physical capacities to ensure effective and wider coverage of germplasm enhancement.
- Establish effective linkages between conservation and use of the conserved genetic resources. Initiate and support on-farm conservation activities for cultivated plant species.
- Revision the current national strategies, policies and legislation for animal, forestry and agricultural sectors.
- Efforts should be exerted to give power to rural communities and to build their capacities for the mutual management of natural resources. It is recognized that such development will lead to better management and utilization of forests and tree resources.
- Programs to preserve forest resources must commence with the local people who are both sufferers and instruments of destruction, and who will tolerate the burden of any new management system.
- Interventions for the sustainable use of Non Wood Forest Products (NWFP) (forestry) include improvement of management practices including; in-situ conservation of NWFP, Agroforestry systems, conservation and introduction of forest laws, conservation measures and reforestation activities, controlled grazing and browsing, introduction of energy saving systems at household levels and introduction of community-based forest management systems. This will be most effective with an integrated land management practice, which also aims at increasing agricultural security and productivity.

- Initiate a national research programs for the development of the wild food including domestication activities.
- Establishment of in-situ conservation sites for wild plant species including wild relatives of crops and natural range plants.
- Establishment of more protected areas to cover more than 17% of the country area;
- Establishment of at least one protected area in each State.
- Sustainable access of using the natural resources by the indigenous and the local community.
- Develop effective systems and tools for monitoring, evaluate and management of Invasive Alien Species.
- Also future research directions should focus on knowing abundance and distribution of some indicators species as well as assessing the impacts of human disturbances and climatic changes on ecosystems and these focal species.

Keywords

- **biodiversity information gap**
- **conservation challenges**
- **conservation efforts**
- **threats**

References

Abdel Magid, T D., (2001). *Forest Biodiversity in Sudan With Particular Reference to NWFPs*. Forest National Corporation, Sudan.

Abdel Magid, T. D., & Badi, K. H., (2008). *Ecological Zones of the Sudan*. Paper presented at the workshop on the importance of Wetlands, in Sudan. Nile Trans Boundary Environmental Action Project.

Abdel Nour, H. O., (2017). *Sudan Forests National Corporation Institutional Analysis*. UN Environment.

Abdelhameed, S. M., & Nimir, M. B., (2007). *Protected areas in Sudan – Wildlife Research Center*. Sudan Money Print Press.

Abdelhameed, S. M., Nimir, M. B., & EL Jack, A. O., (2008). *The Status of Protected Areas in Sudan*. Scientific Conference of Animal Resources Research Corporation.

Adam, H. S., & Abdalla, H. A., (2007). *Metrology and Climatology*. Desertification and desert cultivation studies institute (DADCSI) and UNESCO chair on desertification studies – University of Khartoum – University of Khartoum Press.

Al Amin. H. M., (1990). *Trees and Shrubs of the Sudan*. Ithaca Press.

Anon, Sudan's country Report Contributing to the State of the World's Biodiversity for Food and Agriculture, 2015. Republic of Sudan. Ministry of Agriculture and Forestry. Quality Control and Export Development Unit.

AOAD, (2006). *Arab Agricultural Statistics Yearbook*. Khartoum, vol. 26.

Ayoub, A. T., (1998). Extent, severity and causative factors of land degradation in the Sudan. *J. Arid Environments, 38*, 397–409.

Badi, K. H., (2004). *Changing Forest Cover and Rainfall in Central Sudan During (1930–2000)*. MSc Thesis University of Khartoum – Faculty of Forestry.

Elsiddig, E. A., Mohamed, A, G., & Abdel Magid, T. D., (2011). Forest Plantations/Woodlots in the eastern and north-eastern African countries of Kenya, Tanzania, Uganda, Burundi, Rwanda, Ethiopia and Sudan. *Sudan Report, African Forest Forum*.

Elsiddig, E. A., Mohamed, A. G., & Abdel Magid, T. D., (2007). *Sudan Forestry Sector Review*. Forests National Corporation and National Forest Programme Facility.

Goda, S., (2007). *A Forestation in Arid Lands With Particular Reference to the Sudan*. Desertification and desert cultivation studies institute (DADCSI) and UNESCO chair on desertification studies, University of Khartoum.

HCENR, (2014). *Sudan's Fifth National Report to the Convention on Biological Diversity*. The Higher Council for Environment and Natural Resources, Ministry of Environment and Physical Development. Khartoum, Sudan.

IUCN, UNDP, & HCENR, (2000). *National Biodiversity Strategy of Sudan*.

Ministry of Agriculture, (2014). *Animal Wealth and Irrigation*, Khartoum State (MAAI). Annual Report.

Mukhtar, M. E., & El Wakeel, A. S., (2002). *Biodiversity in Forest Plant of Sudan*. National biodiversity strategy and Action Plan (NBSAP) - (SUD/97/G31/A/IG). Higher council for environment and natural resources (HCENR).

Mustafa. M. A., (2007). *Desertification Processes*. Desertification and desert cultivation studies institute (DADCSI) and UNESCO chair on desertification studies - University of Khartoum -University of Khartoum press.

NBSAP, (2015). *National Biodiversity Strategy and Action Plan 2015–2020*. Higher Council for Environment and Natural Resources (HCENR), Ministry of Environment, Natural Resources and Physical Development. Sudan.

Nikolaus, G., (1987). Distribution Atlas of Sudan's Birds with notes on habitat and status. *Bonn. Zool. Manographien, Nr. 25*, 1–322.

Nimir, M. B., (1984). *Land Use Conflicts in the Dinder Region Presented at the Seminar on Environmental Change and Desertification in Sudan*, p. 10.

Nimir, M. B., (1995). *Management of Protected Areas in the Sudan*. Sub-regional Symposium on protected areas in Arab region. UNESCO.

Siddig, A. A. H., (2009). *Assessment of Climate Change Impacts on Wildlife Habitats of Dry Lands Sudan - Case of Al-Sabaloka Game Reserve*. Master thesis submitted to University of Khartoum, Desertification and Desert Cultivation Studies Institute, Sudan.

Siddig, A. A. H., (2014). Biodiversity of Sudan: Between the harsh conditions, political instability and civil wars. *Biodiversity Journal, 5*(4), 545–555.

Takona, N. Y., (1999). *Survey of the Habitats and Wildlife in Selected Sites in Jebel Marra*. MSc Thesis University of Khartoum - Faculty of Forestry.

UNEP, (2007). *Report of Wildlife and Protected Areas Management in Sudan*.

Wikipedia, (2017). Sudan https://en. wikipedia. org/wiki/Sudan.

WRC, (2008). *Ecological Survey of Wildlife Resources in North and South Kordofan States*. Report by wildlife research centre and funded by IFAD project Development of natural resources in Kordofan region - Sudan.

Appendices

Appendix 10.1   The Protected Areas of Sudan

No	Protected area	Types	Date of declaration	Area (km²)	Habitat(s)
1	Dinder National Park	National Park	1935	10,290	Low rainfall Savenna
		Biosphere Reserve			
		Ramsar Site			
2	Radom National Park	National Park	1980		
		Biosphere Reserve			
3	Jabal Dair	National Park	2010	330	Mountain
		Biosphere Reserve			
4	Gazali National Park	National Park	2017	—	Desert
5	Dongonab Bay	Marine National park	2006	300,000	Semi-desert Marine/ tidal
			2004	2,808	
		Ramsar Site			
		World Natural Heritage Site			
6	Sanganeb	Marine National park	1990	26,000	Semi-desert Marine
		World Natural Heritage site			
7	Jebel Hassania	National Park	2003	850,000	Semi-desert
8	Wadi Howar	National Park	2002	1,455,300	Desert
9	Sabaloka	Game Reserve	1946	116,000	Semi-desert
10	Tokor	Game Reserve	1939	630,003	Semi-desert
11	Taia Basunda El Galabat	National Park	2002	567	Semi-desert
12	ErkawitSinkat	Game Sanctuary	1939	12,000	Semi-desert
13	Erkawit	Game Sanctuary	1939	82'000	Semi-desert
14	Khartoum (Sunut) Forest	Bird Sanctuary	1939	1,500	Semi-desert Riparian site

Appendix 10.2 The Proposed Protected Areas in Sudan

No	Protected area	Types	Area (km²)	Habitat(s)
1	Suakin Archipelago*	National Park Ramsar Site	150,000	Marine
2	KhorKilab	National Park	2,000	Marine
3	Abu Hashish	National Park	2,000	Marine
4	Shuab Rumi	National Park	4,000	Marine
5	Port Sudan	National Park	100,000	Marine
6	Jebel Gurgei Massif	Game Reserve	10,000	Semi-desert
7	Red Sea Hills	Game Reserve	15,000	Semi-desert
8	Lake Nubba	Bird Sanctuary	10,000	Freshwater lake
9	Jebel Aulia Dam	Bird Sanctuary	100,000	Freshwater lake
10	Khashm el Girba Dam	Bird Sanctuary	10,000	Freshwater lake
11	Sennar Dam	Bird Sanctuary	8,000	Freshwater lake
12	Jebel Elba	Nature Conservation Area	480,000	Desert
13	Jebel Marra massif	Nature Conservation Area Important Bird Area	150,000	Desert
14	Gebel al Barkal	World Culture Heritage Site	—	Desert

* Degazetted on 1992.

Source: From UNEP (2007) and Abdelhameed et al. (2008).

CHAPTER 11

An Overview of Biodiversity in Tanzania and Conservation Efforts

ALLY K. NKWABI, JOHN BUKOMBE, HONORI MALITI, STEPHEN LISEKI, NICEPHOR LESIO, and HAMZA KIJA

Tanzania Wildlife Research Institution (TAWIRI), Box 661 Arusha, Tanzania,
E-mail: nkwabikiy@yahoo.com, ally.nkwabi@tawiri.or.tz

11.1 Introduction

Tanzania is a major source of globally significant biodiversity, ranking among the top countries in tropical Africa in terms of the number of distinct ecoregions represented, and in species richness/species endemism (Salehe, 2011). The country has extensive national parks, conservation areas, the Eastern Arc' mountains, wetlands, coastal forests, marine and fresh water systems as outstanding reservoirs of plant and animal species making Tanzania one of the world's greatest reservoirs of biodiversity. The natural resources that the country harbors include forests, water, minerals, fish, wildlife and soils. Such resources are of paramount importance to the existence of ecosystems that range from wetlands, marine to highlands or mountain ecosystems. These ecosystems support the livelihoods of a majority of Tanzanians and the country's economy in general. They provide goods and services, which include food, water, fuel-wood, building materials, medicine, biological diversity as well as raw materials to industries. Indeed, they are very vital for socioeconomic development.

Tanzania hosts six out of the 34 globally known biodiversity hotspots. The country has extensive diversity of species with at least 55,266 known and   species and is among 15 countries globally with the highest number of endemic as well as threatened species. It accounts for more than one-third of total plant species in Africa and ranks twelfth globally in terms of bird species. The country has designated about 40% of its total surface area to forest, wildlife and marine protected area and it is a home to about 20% of Africa's large mammals. Among the protected areas of Tanzania, the status of Selous Game Reserve, Ngorongoro Crater, and Serengeti National Park were upgraded to "World Heritage Sites." Furthermore, Lake Manyara National Park, the Ngorongoro Crater and Serengeti National Park have been designated as biosphere reserves.

This chapter explains and presents an overview of biodiversity of Tanzania with special emphasis to species numbers and status in the country; it also explains the trends and threats facing biodiversity. In addition, physical features as well as major ecosystems of the country are described.

11.2   Geographical Location of Tanzania

The country is the largest in East Africa and the second largest in the SADC Region, it lies within the African Great Lakes Region. It is located on the eastern coast of Africa and has an Indian Ocean coastline approximately 800 kilometers long. The country lies between latitudes 1°S and 12°S and longitudes 30°E and 40°E with a total area of 945,234 km^2 of which 886,037 km^2 is surface land (URT, 2001). It possesses several offshore islands, including Unguja and Pemba (2,400 km^2) as well as Mafia islands (URT, 2009). Tanzania is the 13th largest country in Africa and the 31st largest in the world, it borders Kenya and Uganda to the north; Rwanda, Burundi, and the Democratic Republic of the Congo to the west; and Zambia, Malawi, as well as Mozambique to the south (Figure 11.1).

11.3   Physical Features

Tanzania harbors different physical features including: High Mountains and highlands where northern and southern highlands are found, lowland such as Kilombero valley and all along the coastline and moderate plain land mostly occurring in the central part of the country. These different types of physical features have resulted for Tanzania to have different vegetation zones. The main type of physical features each with distinct associated elevations and features are explained in the following subsections.

11.3.1   The Lowland Coastal Zone

This zone comprises the coastal plain and lowlands with altitude ranging from 0 to 550 meters above mean sea level. Coastal Plains that extend along the coastline of Tanzania Mainland for about 800 km long from the border with Kenya in the north, to the border with Mozambique in the South.

11.3.2   The Highland Zone

This zone encompasses the dissected highlands, up to 2,100 m, which flank the deep trough of Lake Tanganyika to the west and extend, with isolated

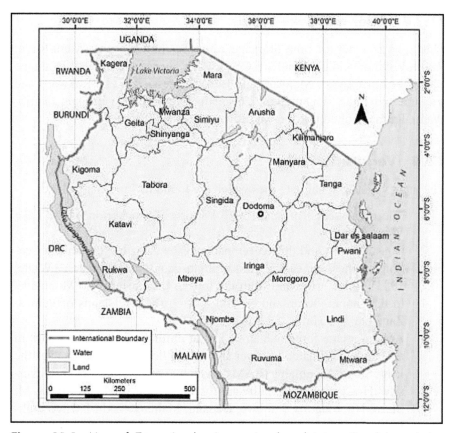

Figure 11.1 Map of Tanzania showing regional and international boundaries (*Source:* URT, 2014a).

blocks of the Uluguru, Nguru, Usambara and Pare Mountains, to the north-east border. In the northeast, tectonic and volcanic activity has produced spectacular mountain peaks and highlands in the Eastern Riff Zone, including the snow-capped Mt. Kilimanjaro (5,895 m), Mt. Meru (4,966 m), the Ngorongoro crater and Ol Donyo Lengai. Lakes Natron, Eyasi and Manyara lie on the riff valley floor. In the south are the Southern Highlands, including the Livingstone ranges, as well as Rungwe, Poroto, and Mbeya Mountains.

11.3.3 The Plateaus Zone

The plateau zone comprises of gently undulating country with an elevation ranging from 800–1,500 m. The zone covers much of the western, northern and southern of Tanzania. The plateaus zone covers Tabora, Rukwa (North and Center) Mbeya, Kigoma and Part of Mara Regions.

11.3.4 River and Lake Basins

This zone is made of nine drainage water basins (Pangani, Wami/Ruvu, Rufiji, Ruvuma and the Southern Coast, Lake Nyasa, the Internal Drainage Basins of Lake Eyasi, Manyara and Bubu depression, Lake Rukwa, Lake Victoria and Lake Tanganyika). The main rivers are the Pangani, Rufiji, Wami, Ruaha, Ruhudji, Ruvu, and Ruvuma.

11.4 Vegetation

The main vegetation habitat types which are found in Tanzania include:

1. The moist forest mosaic (Lake Victoria Phytochorion of the Lake Victoria basin).
2. Coastal forest and thicket remnants of the Zanzibar-Inhambane section of the Guinea-Congolian phyto-geographical region (White, 1983) which are found from the foot of the Eastern Arc Mountains to the Indian Ocean shores, and in the offshore Islands of Pemba, Zanzibar and Mafia.
3. Afro-montane forest which occurs at altitudes from about 2,000 m a.s.l and is estimated to cover about 2 million ha of land, such as the Eastern Arc Mountains (EAMs), which is one of 34 global biodiversity 'Hotspots.' Other montane forests are on Mt Kilimanjaro, Mt Meru, Ngorongoro, Rungwe, Hanang, Mahenge and Matengo highlands, Mahale Mountains and Ufipa Plateau. Mt Kilimanjaro's large altitudinal range (700–5,895 m) supports rich biodiversity ranging from savannah bushland, grassland, pastureland and cropland in the low-lying areas, indigenous forest at mid-altitudes, and alpine vegetation on the higher slopes.
4. *Acacia-Commiphora* thorn bush savannah grassland (mostly in the dry and semiarid northern parts of Tanzania, Figure 11.2).
.   *Brachystegia–Julbernardia* savannah (Zambezian and the Guinea-Congolean Zone).

11.5 Climate

The climate of Tanzania varies from place to place in accordance with geographical location, altitude, relief and vegetation cover. Tanzania's climate ranges from tropical in the low altitudes to temperate in the highlands. The country has two major rainfall regimes: first is unimodal (October–April)

Figure 11.2 *Acacia-Commiphora* thorn bush savannah grassland in Serengeti National Park, May 2016.

and the second is bimodal occurring from October to December and March to May (Mwandosya et al., 1998; Chang'a et al., 2010). The former is experienced in southern, central, and western parts of the country, and the latter in the north from Lake Victoria extending east to the coast (Chang'a et al., 2010). The bimodal regime is caused by seasonal migration of the

Intertropical Convergence Zone (ITCZ), which is relatively narrow belt of very low pressure and heavy precipitation that forms near the earth's equator (Mwandosya et al., 1998; McSweeney et al., 2010). In general, the mean annual rainfall varies from 500 to 2,500 millimeters, with average of annual precipitation over the entire country being 1,042 mm. The average length of dry season is 5 to 6 months with average temperature ranging between 24°C and 34°C, depending on location. Within the plateau, mean daily temperatures range between 21°C–24°C.

11.5.1 Agro-Climatic Zones

Agriculture is one of very important economic activity to Tanzania because more than 80% of the poor populations live in rural areas and almost all of them are involved in the farming sector. However, the agro-climatic zones are nowadays affected by the impacts of Climate Change, which are already having their toll in the livelihoods of people and in the agricultural sectors of the economy in the country. Frequent and severe droughts in many parts of the country are being felt with their associated consequences on food production and water scarcity, among others. The severe drought that occurred in the year 2003 hit most parts of the country, leading to food shortages and insecurity, water scarcity, hunger and acute shortage of power signify the vulnerability of the country to impacts of climate change. The grouping of agro-climatic zones was based on altitude, precipitation pattern, dependable growing seasons and average water holding capacity of the soils and physiographic features as described in the following subsections.

11.5.1.1 Coastal Zone

This zone has two subzones and areas, the North and South, with altitude under 0–550 m. The North subzone is comprised of Tanga (except Lushoto), Coast and Dar es Salaam. Soils and Topography in this subzone is infertile sands on gently rolling uplands, alluvial soils in Rufiji, sand, and infertile soils. Rainfall is bimodal, 750–1,200 mm falling in October through December and March through June. The south subzone covers the eastern Lindi and Mtwara (except Makonde Plateau), it has fertile clays on uplands and river flood plains. Rainfall in the south subzone is unimodal, ranges from 800–1,200 mm, falling in December through April.

11.5.1.2 Arid Lands

The arid lands zone has two subzones and areas, the north (altitude ranging from 1,300–1,800 m) and South (altitude 500–1,500 m). The North subzone is encompassed of Serengeti National Park, Ngorongoro Conservation Area and Part of Masai land. Soils and Topography in this subzone are of volcanic ash and sediments. The texture of the soils is variable and very susceptible to water erosion. Rainfall is unimodal, unreliable, ranging from 500–600 mm falling in March–May. The South subzone comprised of Masai Steppe, Tarangire and Mkomazi National Parks, Pangani and Eastern Dodoma. The soil type of this subzone occurs in rolling plains of low fertility which is susceptible to water erosion. The soil type of flood plain in Pangani river area is of saline, alkaline soil. Rainfall in South subzone is unimodal and also unreliable, ranging from 500–600 mm falling in March–May.

11.5.1.3 Semi-Arid Lands

This zone has two subzones and areas, the central and southern. The central subzone, which has altitude ranging from 1000 to 1500 m is comprised of Dodoma, Singida, Northern Iringa and some part of Arusha as well as Shinyanga Regions. This subzone comprises of well-drained soils with low fertility on undulating plains with rocky hills and low scarps. Alluvial hardpan and saline soils occur in Eastern Rift Valley and Lake Eyasi, while black cracking soils occurring in Shinyanga region. Rainfall is unimodal, unreliable, 500–800 mm falling in December through March. The South subzone with altitude ranging from 200–600 m is comprised of Morogoro (except Kilombero and Wami Basins and Uluguru Mountains), Lindi and Southwest Mtwara. Soils and Topography in this subzone is moderately fertile loams and clays in flat or undulating plains with rocky hills, occurring in South and central Morogoro have infertile sand soils. Rainfall in Southeastern subzone is unimodal ranging from 600–800 mm falling in December through March.

11.5.1.4 Southern and Western Highlands

The zone has three subzones and areas, Southern, Southwestern and Western. The Southern subzone (with altitude ranging from 1,200–1,500 m) covers a broad ridge of North Morogoro to North of Lake Nyasa and covering part of Iringa and Mbeya Regions. Soil and Topography in this subzone is moderately fertile clay with volcanic soils on undulating plains to dissected hills and mountains in Mbeya region. Rainfall is reliable with local rain shadows, it ranges from 800–1,400 mm falling in December through April. The South-western subzone (altitude ranging from 1,400–2,300 m) is in Ufipa plateau in Sumbawanga. Soils and Topography in this subzone is undulating plateau above Rift Valleys and sand soils of low fertility. Rainfall is unimodal and reliable, it ranges from 800–1,000 mm falling in November and April. The Western subzone (altitude 100–1,800 m) is along the shore of Lake Tang-anyika in Kigoma and Kagera. Soil and Topography in this subzone is loam and clay of low fertility in hills, with alluvium and pounded clays in the valleys. This subzone is situated in north–south ridges separated by swampy valleys in the regions mentioned above. Rainfall is bimodal, ranging from 1,000–2,000 mm falling in October–December and February–May.

11.5.1.5   Northern Highlands

The zone has two subzones and areas, Northern and Granite. The Northern subzone (altitude ranging from 1,000–2,500 m) encompasses foot of Moun-tain Kilimanjaro and Mountain Meru as well as Eastern Rift Valley to Eyasi. Soil and topography in this subzone is volcanic soils deep and fertile loams from lavas and ash on volcanic uplands. Generally, soils in dry areas are prone to water erosion. Rainfall is bimodal, varies widely from 1,000–2,000 mm falling in November–January and March–June. The Granite subzone (altitude ranging from 1,000–2,000 m) covers Mountains Uluguru in Moro-goro, Pare Mountains in Kilimanjaro and Usambara Mountains in Tanga and Tarime highlands in Mara. Soils and topography in this subzone are steep Mountain side to highland plateaus. Soils are deep, arable and moder-ately fertile on upper slopes, shallow and stony on steep slopes. Rainfall is bimodal and very reliable from 1000–2000 mm falling in October–Decem-ber and March–June.

11.5.1.6   Alluvial Plains

This zone has four subzones and areas located in different regions of Tan-zania including Kilombero (Morogoro), Rufiji (Coast), Usangu (Mbeya), Wami (Morogoro): Soils and Topography in Kilombero subzone (altitude from 750–1,200 m) are clay plains with alluvial fans on east and west. Rainfall is unimodal, very reliable; it ranges from 900–1,300 mm falling in November through April. Soils and topography in Rufiji (altitude <500 m), which is located in the coast are alluvial, sandy upstream, loamy down steam in floodplain within wide mangrove swamp delta. Rainfall is uni-modal, often-inadequate 800–1,200 mm falling in December through April.

Soils and topography in Usangu subzone (altitude 2,400–5,000 m), which is located in Mbeya is seasonally flooded clay soils in North and alluvial fans in South. Rainfall is unimodal, ranging from 500–800 mm falling in December through March. Soils and topography in Wami subzone (altitude ranging from 400–1,000 m), which is located in Morogoro are moderately alkaline black in East, alluvial fans with well-drained black loam in West. Rainfall is unimodal, ranging from 600–1,800 mm falling in December–March.

11.6 Socio-Economics

Tanzania is one of the 49 Least Developed Countries (LDCs) in the world, with about 35% of the population living below the poverty line (Estrin and Mickiewicz, 2012; Sumner, 2012). However, Tanzania recorded improved economic performance at macrolevel in the past six years, whereby GDP grew at 6.2% in 2012 (URT, 2014a). Agriculture (including livestock) is the dominant sector in Tanzanian economy, providing livelihood, income and employment to over 80% of the population and it accounted for 56% of GDP and about 60% of export earnings, the sector has registered an average annual growth rates of 4.8% compared to the average growth of 3.1% during 2010 to 2012. Other sectors that contribute to the economy are fisheries, mining, tourism and forestry, water, marine and coastal resources, energy, industry and wildlife.

11.7 Types of Ecosystems in Tanzania

The ecosystems in Tanzania are broadly divided into two main categories, namely: terrestrial and aquatic (marine, freshwater and wetlands) ecosystems. Terrestrial ecosystems include forests, mountains, dry lands (arid and semiarid), savannah and agricultural lands. Tanzania's forest covers about 48 million hectares (about 55% of the total land area) with woodlands being the most common, occupying about 51% of the total land area equivalent to 93% of the forest area. The remaining 7% is composed of lowland forests, humid montane forests, mangrove forests and plantations (URT, 2014a). The estimated total volume of trees is 3,100 million m^3, of which 97% comes from trees of natural origin, and only 3% from planted trees (Mauya et al., 2014; Mugasha et al., 2016). Almost half of the total volume is found in protected areas, and therefore not legally accessible for extraction. Most of the un-gazetted forest and woodland resources are found in coastal forests, miombo woodlands and in the village (agricultural) land, (URT, 2014b).

Tanzania has about 88.6 million hectares of land suitable for agricultural production, including 60 million hectares of rangelands suitable for livestock grazing (Rwehumbiza, 2014). The dry lands refer to the arid and semi-arid areas.

Aquatic ecosystem is comprised of two categories including marine and freshwater ecosystems. The marine ecosystems of Tanzania occupy an area of about 241,500 km² and support a mixture of cover types with patches of closed-canopy forest embedded within woodland savanna, grassland and wetland areas (Burgess et al., 2000). Marine ecosystems include coastal forests, mangroves, coral reefs, sea grass beds, sandy beaches, rocky shores and numerous islets. All mangroves areas are gazetted as protected areas, covering 115,500 hectares on Tanzania mainland and 18,000 hectares in Zanzibar. Coral reefs present one of the most productive and biologically diverse marine ecosystems hosting over 700 species of   A⌐
tebrates, making them an important   resource supporting about 90% of artisanal marine   ⌐A   ² (Tumbo et al., 2015). The extent of sea grass beds and the relative species densities are yet to be established. Notably, coastal and marine ecosystems of Tanzania are characterized by numerous rocky shores that offer stable substrata for luxurious growth of algae. Proliferation of some green algae, e.g., *Ulva* is sometimes considered as indication of nutrient pollution in the area (Ocean Road beach in Dar es Salaam and Malindi in Zanzibar provide good examples of this phenomenon). More than half of the original extent of Tanzania's coastal forests has been degraded or converted to agricultural land (Tobey and Torell, 2006; Burgess et al., 2010). The remaining coastal forests are highly fragmented; the majority of the 66 listed forests cover less than 15 km² each with some patches as small as 1 km² (Semesi, 1992; Burgess et al., 2000; Tobey and Torell, 2006).

Tanzania is endowed with considerable freshwater resources that include lakes, rivers, springs, natural ponds, underground sources, man-made reservoirs and wetlands. Lakes cover about 6% of the land area, including the great lakes (Lake Victoria, Lake Tanganyika and Lake Nyasa) that are transboundary. Other lakes include Lake Rukwa and a chain of Rift Valley lakes (Lakes Natron, Eyasi, and Manyara). These lakes harbor high endemism in
   ⌐   ⌐   ⌐ ⌐ ⌐   ⌐
There is a diverse network of permanent and seasonal rivers including Kilombero, Ruaha, Wami, Ruvuma, Mara, Kagera, Malagarasi, and Pangani, their tributaries and associated small streams. With exception of a few rivers found within protected areas (e.g., those under protected montane

forests and Ramsar sites), many rivers are not protected, and thus exposed to decreased ecological integrity as well disruption of ecosystem goods and services they provide. Dams cover over 850 km² including Mtera (610 km²), Nyumba ya Mungu (180 km²), Hombolo (15.4 km²) and Kidatu (10 km²). In addition to their importance in terms of hydropower production, they are source of water and ▢mpoundments serve as important wildlife habitat. Only "Nyumba ya Mungu" has the of partial
▢ ▢ ▢ ▢

Tanzanian wetlands estimated to occupy about 88,300 km² roughly 10% of the total land area of Tanzania Mainland, 58% of which are lakes and swamps (Kamukala and Crafter, 1993). Major wetlands are found along River systems such as the ▢ A
vosi system, Kilombero and Ihefu. Other important wetlands are the alkaline lakes and endorheic swamps, including the soda Lakes of Natron, Manyara, Burigi, Tarangire, Bahi, and Yaida swamps. Four of these wetlands namely Malagarasi-Muyovozi, Lake Natron Basin, Kilombero valley ▢A
▢ ▢ ▢

11.8 Conservation of Wildlife in Tanzania

Tanzania's mega-biodiversity is not evenly distributed over ecosystems. Species and genetic resources inhabit both in protected and nonprotected areas. There are six biological hot spots that have value as centers of high species diversity, high level of endemism and contain high proportion of world's total population. These areas are the 'Eastern Arc' Mountain Forests, the coastal forests, the great lakes, the ecosystems of the alkaline Rift valley lakes and the grassland savannas. These areas are subject to anthropogenic influence and environmental impacts that require regular monitoring for devising appropriate strategies to ensure their conservation and sustainable use.

The wildlife conservation in Tanzania was started since 1891 when laws controlling hunting were enacted by the German rule (Songorwa, 1999; Nelson et al., 2007). The main objective of setting the rules was to regulate off-take, hunting methods and trade in wildlife. The German rule emphasized more fully on protecting endangered species not to be hunted. The Game Reserve was established in 1905 by the German's rule, which latter formed what is now known as the Selous Game Reserve in 1921. Currently the networks of protected areas in Tanzania are categorized into terrestrial and marine: extending from sea habitats over grasslands to the top of Kilimanjaro, the tallest mountain in Africa.

11.8.1   Terrestrial Protected Areas

In order to enhance conservation of wildlife, Tanzania has gazetted about 33.56% of her total land area as wildlife protected areas (Table 11.1), with a further 15% set aside as forest protected area. For sustainability, Tanzania has 17% of her land area devoted to wildlife conservation in protected areas where no human settlement is allowed and 18% of its surface area to protected areas where wildlife coexist with humans. The protected areas of varying status include National Parks, Game Reserves, Marine Parks, Forest Reserves and Wildlife Management Areas. In addition, Tanzania has 109 Forest reserves (Figure 11.3).

11.8.1.1   National Parks

There are 16 National Parks, which cover 6.07% of the country's total land surface. These are areas of high biodiversity values and their main purpose is conservation of habitats and wild animals, which constitute unique naturally occurring biodiversity of Tanzania. These National Parks include Serengeti, Gombe, Mahale Mountains, Saanane Island, Tarangire, Manyara, Arusha, Kilimanjaro and Mkomazi. Others are Saadani, Mikumi, Ruaha, Kitulo, Katavi and Udzungwa Mountains. In the National Parks only nonconsumptive utilization such as photographic tourism, education and research are permitted. Serengeti National Park, for example, is one of the world heritage sites, which are famous for their endless plains to signify the wonders of nature, extensive savannah woodlands and migrating wildebeest.

Table 11.1 Categories of Protected Areas and Their Coverage in Tanzania

Category	Number	Area (km²)	Percentage of Tanzania's total area
National Parks	16	57,365.05	6.07
Ngorongoro Conservation Area	1	8,292.00	0.89
Game Reserves	28	114,782.47	12.14
Game Controlled Areas	42	58,565.02	6.20
Wildlife Management Areas	38	29,518.40	3.12
Ramsar Sites	4	48,684.00	5.13
Total	129	317,207.00	33.56

Source: URT (2014b).

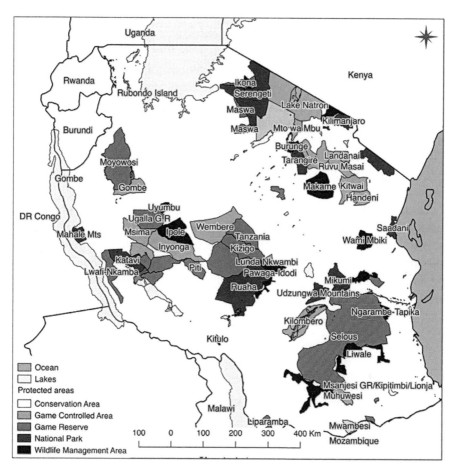

Figure 11.3 Map of Tanzania showing terrestrial protected areas (Map produced by Hamza Kija: TAWIRI GIS Office, 2017).

In Tanzania, wildebeest (*Connochaetes taurinus*) is the most abundant and remarkable big game species. Some wildebeest populations migrate annually to new grazing grounds. The most spectacular is the Serengeti-Mara ecosystem migration, where thousands of wildebeest migrate from Tanzania to Kenya and vice versa throughout the year seeking for fresh grazing and better quality water.

In addition to savannah woodlands, some National Parks such as Katavi, Gombe and Mahale Mountains are found in the famous wet patches of miombo woodlands in the Western part of Tanzania. On the other hand, Udzungwa National Park, which is part of the Eastern Arc chain of Mountain blocks, is an important montane forest park possessing very critical endemic species of   and fauna. It possesses essential habitats to some of the rare and endemic species of primates such as the Sanje River Mangabey (*Cercocebus sanjei*), and Iringa Red Colobus monkey (*Procolobus gordonorum*) (Burgess et al., 2007). The Highland Mangabey (*Rungwecebus kipunji*), which is critically endangered, is found within the forests of the Livingstone Mountains just outside the Udzungwa Mountains. The forests of Udzungwa Mountains are among the top ten important birds' conservation areas in Africa, whereby endemic species like the Udzungwa forest partridge (*Xenoperdix udzungwensis*) and Rufous-winged Sunbird (*Cinnyris rufipennis*   □   □   gess et al., 2007).

Generally, National Parks in Tanzania form the core of a much larger protected ecosystem whereby they preserve the country's rich natural heritage by providing secure breeding grounds where its fauna and   can thrive, safe from the   interests of a growing human population. The existing park system in Tanzania protects a number of internationally recognized strongholds of biodiversity and World Heritage sites, thereby redressing the balance for those areas of the country affected by deforestation, agriculture and urbanization.

11.8.1.2   Game Reserves

There are 28 Game Reserves, all covering 9.8% of the total land surface area. Activities related to consumptive such as hunting and nonconsumptive tourism such as photographic tourism, research and education are permitted.

11.8.1.3   Ngorongoro Conservation Area

Ngorongoro Conservation Area (NCA) covers 1 % of the country's total land surface area. This is a unique area in terms of conservation of archaeology, culture, wildlife and water catchments. In NCA, Maasai tribe settlements and pastoralism practice (presenting a rare scenario of coexistence among human beings, livestock and wildlife), photographic tourism, education and research are permitted.

11.8.1.4   Game Controlled Areas

The country has gazetted 42 Game Controlled Areas, which cover 9.6% of the total land surface area where licensed hunting, photographic tourism, human settlements and other human activities, research and education are permitted.

11.8.1.5 Partial Game Reserves

This is a category of wildlife conservation area meant for protection and conservation of endemic, rare and endangered species, and conservation of wildlife species of a national or international conservation importance. Kihansi Falls of the Udzungwa Mountains, for example, is in this category because Kihansi gorge is a home for the critically endangered species of amphibian of the Kihansi Spray Toad (*Nectophronoides asperginis*) (Poynton et al., 1998), endemic coffee plant *Coffea kihansiensis* (Davis and Mvungi, 2004), and a recent new described species of butterfly, the *Charaxs mtuiae* (Mtui's Kihansi Black Charaxs) (Collins et al., 2017). In 1994, the Government of Tanzania established Lower Kihansi Hydropower Project (LKHP) to meet the country's demand for power supply and went into operation in December 1999. Implementation of the LKHP facility required diversion of 98% of Kihansi River water that used to pass through the Kihansi River Gorge. Reduction of the river flow from Kihansi River had effects on climatic condition and ecological systems of the gorge, with one major catastrophe being the extinction of the *Nectophrynoides asperginis* in the wild (IUCN, 2016).

11.8.1.6 Wildlife Management Areas

Currently there are 38 Wildlife Management Areas (WMAs) country-wide at different stages of development of which 17 WMAs have attained Authorized Association (AAs) status. Wildlife management areas were established to strengthen involvement of community in wildlife conservation. They were first addressed by Wildlife Policy 1999, aimed to be village land use option, which competes with other land uses within the village land. Villagers were expected to get direct benefits from the wildlife they conserve in their village land through maintenance of WMAs. However, the WMAs face many challenges like inadequate manpower, low funds for running conservation activities, villagers' livestock predation by wild animals, and many others, which the government and many donors funded organizations try to address. But the most acute problem and least addressed is strong political intervention against the management of these WMA's.

11.8.1.7 Ramsar Sites

The Convention on Wetlands came into practice for the United Republic of Tanzania on 13 August 2000. At present, Tanzania has four sites designated as Wetlands of International Importance, with a surface area of about 4,868,424 hectares (Table 11.2).

11.8.2 Marine Protected Areas

Tanzania has 21 Marine Protected Areas, which include four Marine Parks and 17 Marine Reserves. Out of the 32,000 km^2 of the territorial sea of Tanzania Mainland only 2,173 km^2 (about 6.5%) has been gazetted as Marine Protected Areas (MPAs). The Protected Areas in Tanzania Mainland are: Mafia Island Marine Park (MIMP), Mnazi Bay and Ruvuma Estuary Marine Park (MBREMP), and Tanga Coelacanth Marine Park (TaCMP), and 15 Marine Reserves. In Zanzibar there is one Marine Park and two Marine Reserves (Figure 11.4).

11.8.2.1 Marine Parks

Tanzania established Marine Parks as a special area of the reserves where various community users and habitation are encouraged. However, it is highly regulated and continuous education of the community is provided to ensure longevity of the fauna in the areas where they live.

11.8.2.2 Marine Reserves

Tanzania gazetted marine reserve as an area allocated to help towards the preservation of the coastal and marine biodiversity. In the Marine Reserves fishing and inhabitation is prohibited and only limited entry such as for tourism and research is permitted.

11.9 Species Diversity

Tanzania is a tropical country with a high level of ecosystem diversity, the total number of species found vastly exceeds that of most countries. The

Table 11.2 Names and Size of Ramsar Sites in Tanzania

Name	Area (ha)
Kilombero Valley Floodplain	796,735
Lake Natron Basin	224,781
Malagarasi-Muyovozi Wetlands	3,250,000
□ □	A

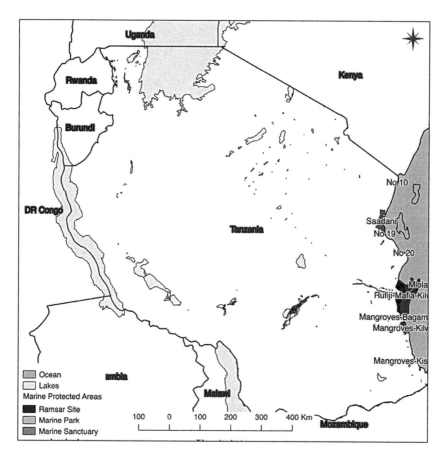

Figure 11.4 Map of Tanzania showing location of marine protected areas.

current review from different sources annotated checklist of Tanzanian flora and fauna species lists a total of 55,266 species of plants, invertebrates and vertebrate animals. Species of livestock and crops are not included in a total number presented. However, Gideon et al. (2012) and COSTECH (2012) presented a checklist of 14,336 species of flora and fauna for Tanzania. The differences in the reported numbers of species could be attributed by the fact that the search effort devoted in the current survey used a combination of various digital resources including the FishBase for all fishes in Tanzania (Froese and Pauly, 2017), International Ornithological Committee (IOC) *World Bird List* (Gill and Donsker, 2016), the reptile database (Uetz and Hošek, 2017), as well as lists of mammals of Tanzania (www.https://

en.wikipedia.org/wiki/List_of_mammals_of_Tanzania). In addition to the checklist, the Tanzania Biodiversity Information Facility (TanBIF) hosted at the Tanzania Commission for Science and Technology (COSTECH) that avails species occurrence data over the internet, the International Union for Conservation of Nature (IUCN, 2016) that gave information on species lists, endemism and threats facing biodiversity in Tanzania and search engine like Google were also used for information concerning biodiversity of Tanzania.

11.9.1 Fauna Diversity

Tanzania has a rich and diverse fauna, including a wide variety of endemic species and subspecies, and unique habitats. The total number of fauna species is at least 44,610 with over 372 species of mammals, 1,152 species of birds, 39,644 species of insects, over 2,240 species of fish, 197 amphibians, about 366 species of reptiles, 144 echinoderms, 120 species of porifera, and about 365 species of cnidarians distributed in all over the country. Tables 11.3 and 11.4 show the number of orders, Families, Genera and number of species including the relative proportion to the total number in the world based on current survey.

Table 11.3 The Known Number of Fauna Species in Tanzania Including Endemic and Threatened Species

Group of animal	Order	Families	Genera	Number of Species in Tanzania	Worldwide species	% Worldwide
Actinopterygii	41	90	520	2,250	32,156	7.0
Amphibia	2	15	31	197	6,802	2.9
Birds	30	107	317	1,152	9,026	12.7
Mammalia	17	50	148	372	5,416	6.9
Reptilia	3	26	114	366	9,232	3.96
Insecta	8	347	4,996	39,644	1,020,169	3.9
Echinoderm				144	6,600	2.18
Porifera				120	5,000	2.4
Cnidaria				365	10,107	3.61
Total				44,610	1,104,508	4.038

Source: Modified from URT (2014b); Gill and Donsker (2016).

Table 11.4 Number of Insect Species Presented Depending on Their Major Orders and Family in Tanzania and Their Comparison With the World Species

Order	Families	Genera	Number of Species in Tanzania	Worldwide species	% Worldwide
Coleoptera	77	1,494	5,416	400,000	1.3
Lepidoptera	59	1,325	5,753	150,000	3.8
Diptera	78	725	2,052	120,000	1.7
Hymenoptera	51	614	15,419	130,000	11.9
Hemiptera	51	376	694	82,000	0.8
Orthoptera	16	345	843	20,500	4
Odonata	11	73	235	5,900	3.9
Isoptera**	4**	44**	112**	2,600**	4.3**

Source: www. Modified from Insectoid.info/checklist/insect/Tanzania.

** Total number of species and families presented based on checklist of Isoptera recorded in East Africa (Kenya, Uganda and Tanzania) by Wanyonyi et al. (1984).

11.9.2 Threatened, Endangered, and Vulnerable Species

The country is currently facing tremendously increase in threatened species. The 2013 IUCN Red List indicate that there are 914 threatened species recorded in Tanzania, accounting for about 4% of threatened species globally and the country was ranked 15th globally with regard to the number of threatened species (URT, 2014b). However, the current review discovered an increase in number of threatened species in the country. According to the 2016 IUCN Red List, there are at least 1087 threatened species; of which 490 are fauna and 597 flora. Example of fauna listed by IUCN Red list which are critically endangered, endangered and/or vulnerable include terrestrial animal species such as Black rhinoceros (*Diceros bicornis*) and African elephant (*Loxodonta africana*, Figure 11.5), Wild dog (*Lycaon pictus*), Chimpanzee (*Pan troglodytes*), Cheetah (*Acinonyx jubatus*), Shoebill (*Balaeniceps rex*), Wattled Crane (*Bugeranus carunculatus*), Grey Crowned-crane (*Balearica regulorum*, Figure 11.6). Listed flora includes species such as *Pterocarpus angolensis* and *Dalbergia melanoxylon*. Aquatic species, which are threatened include coelacanth (*Latimeria chalumnae*), dugongs (*Dugong dugon*) and sea turtles.

Despite of the biodiversity endemism, like many other countries, Tanzania is experiencing increasing pressures on biodiversity due to a number of threats caused by natural and human drivers. The threats include over-

Figure 11.5   Some of the threatened mammal species grazing in the grassland of Serengeti National Park, Tanzania (a = black rhinoceros, b = African elephant).

Figure 11.6   Grey Crowned Crane in *Acacia* woodland of Serengeti National Park, August 2016.

exploitation of natural resources, habitat fragmentation and degradation, as well as climate change (Foley et al., 2014). Tanzanian habitats accommodate animal species which are   ⌐A

Tanzania have about 38 species of mammals, 52 species of birds, 34 species of reptiles, 61 species of amphibians, 180 species of marine and freshwater
A
are listed as globally threatened due to anthropogenic activities and natural drivers (Table 11.5).

11.9.3 Flora Diversity

More than 50% (equivalent to 44.4 million ha) of Tanzania total land area is covered by indigenous vegetation. Currently there is no published species list of plants in Tanzania. However, a number reports have indicated that Tanzania has at least 11,000 plant species but the sources are highly scattered. In this chapter, an assessment plant groups used an unpublished List of East African Plants (LEAP) (EAH, 1996; Knox and Berghe, 1996). In this list a total of 9,320 species were encountered in 7 classes, 1,917 genera and 244 families and 99 orders. The classes included in the LEAP master are Cycadopsida, Equisetopsida, Liliopsida, Lycopodiopsida, Magnoliopsida, Pinopsida, Polypodiopsida and Psilotopsida. However, a total of 20 classes are published in the IUCN red list (Table 11.6). The classes excluded from the LEAP master list are Anthocerotopsida, Bryopsida, Jungermanniopsida, Isoetopsida, Gnetopsida, Ginkgoopsida, Florideophyceae, Equisetopsida, Marattiopsida, Marchantiopsida, Takakiopsida and Ulvophyceae. This indicates that the LEAP master list is an incomplete list, which needs to be updated. The class Magnoliopsida appeared the leading group with the highest diversity of plant species Table 11.6, and is the one with the highest diversity of plants in the IUCN red list.

11.9.4 Threatened and Endangered Plant Species

A total of 641 threatened and endangered species in 516 Genera in 136 families, 65 orders and 11 classes of plants were recorded from the IUCN red listed plants for the country (Table 11.6). Examples of selected plants which are overexploited in the country include *Pterocarpus angolenesis* (lower Risk/Near threatened), *Dalbergia melanoxylon* (lower risk/near threatened), *Uvariodendron gorgonis* (endangered) and *Erythrina schliebenii* (critically endangered) and *Holmskioldia gigas* (critically endangered) (URT, 2014b). Of the aforementioned plant species, *D. melanoxylon* is listed as CITES II species in Tanzania.

Table 11.5 Current Status of Fauna Species in Tanzania

Group	Total species in Tanzania	Critically Endangered	Endangered	Extinct	Extinct in the Wild	Least Concern	Near Threatened	Vulnerable
				Number of species				
Large Mammals	372	4	9	0	0	175	18	14
Small mammals	1,152@	1	7	0	0	104	1	3
Birds	366	6	18	–	–	968	36	28
Reptiles	197	5	17	0	0	78	6	12
Amphibians	993	21	29	0	1	111	3	11
Marine Fish	1,257	0	3	0	0	574	23	33
Freshwater Fish	0	54	20	1	0	460	6	70
Both Marine and ☐	30,589	0	0	0	0	24	0	0
Terrestrial invertebrates	8,270	18	14	0	0	323	5	19
Marine invertebrates	785	1	4	0	0	216	89	51
Freshwater invertebrates		0	8	0	0	168	13	10

Source: Modified from URT (2014b); Gill and Donsker (2016).

Table 11.6 Status of Plant Species in Tanzania as Reported in IUCN RedList 2016

Class	Total number of Species in Tanzania	Plants Listed in the IUCN	CR	EN	VU	Subtotal***	% of Total Species in Tanzania	% of Species in the IUCN list	NT	LC
Anthocerotopsida	**	2	*	*	*	*	*	*	*	*
Bryopsida	**	41	*	1	*	1	0.01	0.004	*	*
Cycadopsida	5	307	2	*	*	2	0.02	0.008	2	*
Equisetopsida	75	7	*	*	*	*	*	*	*	*
Liliopsida (Monocotyledon)	2088	4,492	17	46	31	92	0.1	0.4	11	273
Jungermanniopsida	**	45	*	1	*	1	*	*	*	*
Isoetopsida	**	51	*	*	*	*	*	*	*	*
Gnetopsida	**	97	*	*	*	*	*	*	*	*
Ginkgoopsida	**	1	*	*	*	*	*	*	*	*
Florideophyceae	**	58	*	*	*	*	*	*	*	*
Equisetopsida	**	7	*	*	*	*	*	*	*	*
Lycopodiopsida	37	21	*	*	*	*	*	*	*	*
Magnoliopsida	6751	16,146	33	182	267	482	5.2	2.16	67	161
Marattiopsida	**	1	*	*	*	*	*	*	*	*
Marchantiopsida	**	11	*	*	*	*	*	*	*	*

Table 11.6 (Continued)

Class	Total number of Species in Tanzania	Plants Listed in the IUCN	CR	EN	VU	Subtotal***	% of Total Species in Tanzania	% of Species in the IUCN list	NT	LC
Pinopsida	7	606		2		2	0.02	0.008	1	*
Polypodiopsida	356	332	3	12	3	18	0.2	0.004	*	*
Psilotopsida	1	4	*	*	*	*	*	*	*	*
Takakiopsida	**	1	*	*	*	*	*	*	*	*
Ulvophyceae	**	1	*	*	*	*	*	*	*	*
TOTAL OVERALL	9,320	22,238	55	244	301	598	6.4	0.5	*	*

Note: 1. *Source:* LEAP of Tanzania, IUCN RedList 2016 and Tanzania plants list.

2. ** = data are not available in the LEAP of Tanzania and the LEAP master for East African plants.

3. * = data were not available.

11.10 Endemism: Notes on Selected List of Species in Tanzania

Documenting species of limited occurrence will warrant additional attention for conservation. Tanzania is home to several endemics across flora and fauna taxa, some of which are threatened by extinction if effective measures are not taken to ensure their existence. Some endemics from the different groups of organisms are presented in Section 11.10.1–11.10.2.

11.10.1 Fauna Endemism

Tanzania reveals a high degree of fauna species endemism, which can be attributed by the complex topographical conditions and biological isolations in some areas resulting in unique microclimate and distinct ecological conditions that supports many endemic species (Myers et al., 2000; Foley et al., 2014).

11.10.1.1 Birds

A total of 34 bird species are endemic to Tanzania (Burgess et al., 2007; Plumptre et al., 2007; Gill and Donsker, 2016). Some of the species included in the list are; the Usambara Eagle-owl (*Bubo vosseleri*), the Pemba Scops-owl (*Otus pembaensis*), the Grey-breasted Francolin (*Pternistis rfopictus*), the Nyasa (Lilian's) Lovebird (*Agapornis lilianae*) which has now been introduced into Kenya and Burundi, the Pemba Green-pigeon (*Treron pembaensis*), the Uluguru Bush-shrike (*Malaconotus alius*), the Banded Green Sunbird (*Anthreptes rubritorques*) and the Pemba Sunbird (*Cinnyris pembae*). Others are the Iringa Akalat (*Sheppardia lowei*), the Kilombero Weaver (*Ploceus burnieri*), Beesley's Lark (*Chersomanes beesleyi*), the Pemba White-eye (*Zosterops vaughani*), the Usambara Hyliota (*Hyliota usambara*), Reichenow's Batis (*Batis reichenowi*), the Kipengere Seedeater (*Crithagra melanochroa*), and the recently described Rubeho Warbler (*Scepomycter rubehoensis*, threatened) and Ruaha Red-billed Hornbill (*Tockus ruahae*). The Udzungwa Forest-partridge (*Xenoperdix udzungwensis*) is an endemic genus and species.

11.10.1.2 Arthropods

There is a total of 134 butterfly species that are endemic to Tanzania. Among them are *Papilio ufipa, Charaxs usambarae, Acraea punctimarginea,*

Euphaedra confina, Anthene leptala, Anthene madibirensis, Anthene uke-rewensis, and the Tanzanian Diadem *Hypolimnas antevorta* (Davenport, 2002). Some species such as *Euxanthe wakefieldi, Papilio polystratus* and *Salamis parhasus* are endemic and forest dependent hence are vulnerable to extinction due to deforestation. Other endemic insects include many beetles from the Uluguru Mountains including, e.g., *Euripogena leleupi, Euripogena rotundicollis* and *Euripogena uluguruana*. Nummelin and Nshubemuki (1998) and Grebennikov (2008) list more than 20 species of caddisflies (Trichoptera), mostly from the Usambara and Uluruguru forests.

11.10.1.3 Reptiles

Tanzania has a total of 87 reptile species that are endemic, including, e.g., Uluguru One-horned Chameleon (*Kinyongia oxyrhina*), Giant Fischers (*Kinyongia matschiei*) and the Three Horned Chameleon (*Trioceros deremensis*), Turquoise Dwarf Gecko (*Lygodactylus williamsi*), Pemba Day Gecko (*Phelsuma parkeri*) and Ukinga Girdled Lizard (*Cordylus ukingensis*). Others are Ornate Shovelsnout Snake (*Prosymna ornatissima*), Werner's Green Tree Snake (*Dipsadoboa werneri*), Usambara Garter Snake (*Elapsoidea nigra*), Matilda's horned viper (*Atheris matildae*), and Horned Bush Viper (*Atheris ceratophora*). Additionally, three reptile genera (*Loveridgea, Xyelodontophis* and *Adenorhinos*) are endemic to the country (Plumptre et al., 2007; Gideon et al., 2012).

11.10.1.4 Amphibians

There is a total of 91 amphibian species that are endemic to Tanzania, including the Mazumbai warty frog (*Callulina kisiwamsitu*), Barbor's forest tree frog (*Leptopelis barbori*), Uluguru banana frog (*Afrixalus uluguruensis*), Usambara big-fingered (*Probreviceps macrodactylus*), Keith's striped frog (*Phlyctimantis keithae*), Mette's Reed Frog (*Hyperolius pseudargus*), the running frog (*Kassina jozani*), The Usambara torrent frog or Tanzania rocky river frog (*Arthroleptides martiensseni*), Nike's Squeaker (*Arthroleptis nikeae*), the Kihansi spray toad (*Nectophronoides asperginis*), Tree toad (*Churamiti maridadi*), Usambara Blue-bellied Frog (*Hoplophryne rogersi*), Amani Forest Frog (*Parhoplophryne usambarica*), Scarlet-snouted Frog (*Spelaeophryne methneri*) (Plumptre et al., 2007; Poynton et al., 2007).

11.10.1.5 Mammals

The country is home to about 20% of Africa's large mammals (Foley et al., 2014), with a total of 31 species endemic to Tanzania. The mammals species include the Primates Sanje mangabey monkey (*Lophocebus kipunji*) in Udzungwa Mountains, the Sanje River Mangabey (*Cercocebus sanjei*), Uhehe red Colobus (*Piliocolobus gordonorum)*, Zanzibar Red Colobus (*Procolobus kirkii*), shrews (Peter's musk Shrew (*Crocidura gracilipes*), Amani Musk shrew (*Crocidura tansaniana*), Uluguru Musk Shrew (*Crocidura telefordi*), Usambara Musk Shrew (*Crocidura usambarae*) and Tanzania Mouse Shrew (*Myosorex geata*); Fruit-eating bats (Pemba flying fox, *Pteropus voeltzkowi*), Insect-eating bats (Tanzania Woolly bat, *Kerivoula africana*); and Rodents (Swynnerton's Bush Squirrel, *Paraxrus vexillarius*).

11.10.2 Flora Endemism

Tanzania is endowed with more than 1,500 plants that are unique (Dauby et al., 2016; Sosef et al., 2017) among over 11,000 vascular plant species. The unique species in the country inhabit parts of biodiversity hotspots such as the Coastal Forests of Eastern Africa and the Eastern Afromontane. The Eastern Afromontane Hotspot is one of the biological wonders of the world, with globally significant levels of diversity and endemism. Some groups of endemic plants to Tanzania have been listed at the Living National Treasures website (LNT, 2017) which are also listed in Mabberley (2008) include the African violets *Saintpaulia goetzeana* and *Saintpaulia shumensis*, *Aloe dorotheae*, the orchids *Polystachya longiscapa* and *Ancistrorhynchus refractus*, a palm *Dypsis pembana*, a cycad *Encephalartos sclavoi*, *Ecbolium tanzaniense*, *Cola usambarensis*, the Lake Latumba Coral Tree, *Allanblackia stuhlmannii*, *Impatiens kilimanjari*, *Isoglossa variegata*, and *Uvariopsis bisexualis*. Endemic genera include *Stephanostema*, the orchids *Neobenthamia* and *Sphyrarhynchus*, *Mwasumbia*, *Sanrafaelia*, *Streptosiphon*, *Farrago*, *Urogentia*, and *Neohemsleya*. In total, 21 species in 9 families were listed in Mabberley's Plant-book. The listed families with respective number of species in bracket are: Compositae (2), Gentianaceae (1), Gramineae (5), Leguminosae (1), Melastomataceae (2), Orchidaceae (3), Rubiaceae (5), Sapotaceae (1) and Triuridaceae (1).

11.11   Challenges and Threats to Biodiversity in Tanzania

Tanzania's biological diversity is generally threatened. Resembling many other countries in the world, Tanzania is affected by rapid loss of biodiversity. Socioeconomic development for most of the people in Tanzania is limited to dependency on biological resources, there by continuously putting pressure on these resources available. Efforts provided by biologists to find a way on how natural resources can be conserved, recognize five main categories of direct threats to biodiversity in Tanzania:

- conversion, loss, degradation, and fragmentation of natural habitats;
- overharvesting or overexploitation of particular species;
- infrastructures and new economic developments;
- invasive nonnative species that harm ecosystems or species;
- pollution or contamination that harms natural habitats or species;
- climate change effects that harm natural habitats including species.

11.11.1   *Conversion, Loss, Degradation, and Fragmentation of Natural Habitats*

The direct threat to biodiversity in Tanzania comes in the form of the conversion, loss, degradation, and fragmentation of natural ecosystems (Nkwabi et al., 2011; Nkwabi, 2016; Nkwabi et al., 2017). Habitat conversion, degradation and loss due to expansion of agricultural activities as well as need for settlements (Estes et al., 2012), overgrazing by livestock, fuelwood needs and bush-meat snaring and hunting in Tanzania are in increase. Inadequate land use planning and agreements, low capacity to monitor and enforce laws and regulations have attributed to the threat to the biodiversity. In addition low technology in agriculture and livestock development leads to need for large space of land for practices with relatively low productivity in return.

11.11.2   *Overharvesting or Overexploitation of Natural Resources*

Poaching for ivory, trophies and meat has traditionally been the major cause of the species decline in Tanzania. Although illegal hunting tourism remains a significant wildlife utilization factor in the country, currently the most important perceived threat is the loss and fragmentation of habitat caused by ongoing human population expansion and the associated activities and requirements that lead to rapid land conversion. Loss of freshwater and marine biodiversity is more seriously affecting the provision of ecosystem services. In Tanzania, there is a great force to transform ecosystems; for example, forests, wetlands, and dry lands are heavily converted to agricultural and grazing land. From the undergoing anthropogenic activities the country possesses important species with declining populations that are globally endangered and threatened. Therefore, the general trends of biodiversity in the country depict a situation of concern. Most of the ecosystems, be it terrestrial or aquatic, are deteriorating with decreasing capacity to provide essential services while a significant number of species are on the decline and some of them are even on the brink of extinction.

Human activities, such as mining, agriculture, burning and felling of trees for   A is on the increase, resulting in the decline of forest and woodland (Figure 11.7). The Miombo woodland and forest in WMAs of Ruvuma landscape in southern Tanzania, for example, has been reduced by 50%. In the Riverine forest mosaic, the Magazine and Kisungule forests, which has undergone extensive wood cutting, has been reduced to a stunted forest devoid of birds' life, and containing only a few species of   A   □ try species (Nkwabi et al., 2017).

The Kagera River forests though a part of the Ibanda/Rumanyika Game Reserve, are also under threat from wood fuel gathering and logging activities. Coastal forests have also disappeared at an alarming rate, mainly as a result of agriculture, pit sawing, charcoal production and mining of salt, limestone, beach sand and hydrocarbons. Large areas of mangrove forests have been cleared in preparation of salt pans (Figure 11.8), destroying   □ breeding areas as well as other organisms, such as birds. Other activities, such as lime making and sand extraction, are not only destroying the breeding areas for marine species but are also causing beach erosion. The loss of coastal forests and their fragmentation has in turn led to the concentration of a variety of endemic species in a very restricted area. For example, the blue dwarf gecko (*Lygodactylus williamsi*) is found only on *Pandanus* stems in Kimboza Forest Reserve. The expanding settlements have been encroaching on wildlife habitats, reducing their size considerably, and threatening the existence of certain animals. For example, the African elephant and black rhinoceros, though still found in some coastal forests, are believed to be in danger of extinction.

Poaching of animals for meat and trophy, particularly in the protected areas have resulted into fragmentation of wildlife habitats, and this poses

Figure 11.7   Setting fire in the miombo woodland (c), clearing of forest for
agriculture (d), An illegal logging (e & f) conducted at Kisungule and
Kimbanda Wildlife Management Area, November 2014.

another major threat to biological diversity (Kideghesho et al., 2006; Kideghesho et al., 2013). African elephant and black rhinoceros have, particularly, been very vulnerable to poaching. Consequently, the black rhinoceros
has been reduced to less than 300 animals, mainly in Selous and Mikumi
National Parks (Fyumagwa and Nyahongo, 2010; TAWIRI, 2010; Foley et
al., 2014). Tanzania is also in danger of losing the elephant as a species
of   commercial value. Other threatened or endangered animal
species include Ader's duiker (*Cephalophus adersi*) which is restricted to
eastern Zanzibar, the Abbott's duiker (*Cephalophus spadix*) which occurs
in the forests of Kilimanjaro and the Eastern Arc mountain chain, and some
bird species, such as Uluguru bush-shrike (*Malaconotus alius*), Sokoke pipit
(*Anthusso kokensis*), Sokoke scops-owl (*Otus ireneae*) and Long-billed For-

Figure 11.8 Mangrove forest cleared for salt pan preparation in Lindi region.

est-warbler (*Artisornis moreaui*). Therefore, poaching of animals for meat and trophy as well as birds trade might pose a major   in the population of animals.

Human activities are also threatening the aquatic resources. Dynamite   is not only destroying coral reefs, which are important breeding grounds for many   A
targeted   A
have also contributed to the loss of biodiversity, particularly in both major and small lakes of Tanzania. While the introduction of Nile perch, a *Piscivorous* spp. in the lake has led to the disappearance of several indigenous species, the habit of some   to catch   A
considerably affected the population of some species, such as *Labeo* sp.

11.11.3 Infrastructures and New Economic Developments

Road networks are economically important to both people and protected areas because they facilitate communication among societies as well as transportation of goods and services (Timothy, 1999; Ferraro, 2001). Trombulak and Frissell (2000) reported the existence of roads in wildlife ecosystems significantly influences ecological and environmental condition through habitat fragmentation and loss due to increased anthropogenic activities. Such activities therefore, may influence wildlife population (Kociolek et al., 2011). The Proposed construction of Mto wa Mbu-Mugumu road passing through Serengeti National Park predicted to acting as a big population pull factor to Serengeti area, it may directly affect biodiversity through clearing of vegetation, roadkill and blockage of the migratory corridor for wildebeest, Thomson's Gazelle (*Eudorcas thomsonii*), Grant's gazelle (*Nanger granti*) and zebra (*Equus burchellii*) moving between Serengeti and Masai-Mara National Reserve in Kenya (Dobson et al., 2010). Roads passing through protected areas have been found to have adverse impacts on wild animals. Road kills caused by vehicles are considered as the cause of biodiversity loss in wildlife sanctuaries (Kioko et al., 2015). The lessons from Mikumi National Park and other protected areas elsewhere where public roads pass across indicate negative effects and ecological impacts associated with roads. Drews (1995) estimated a minimum of 3 kills per day during road accidents at Mikumi national park indicating that infrastructure development had major impact in dispersal of wildlife.

A proposal by Tata Chemical Industries Ltd (Lake Natron Soda Ash Plant) in collaboration with the Tanzanian Government to construct a factory that would produce 500,000 tons of soda ash per year and employ 150 permanent staff sounds economically promising, but with negative ecological impacts. Lake Natron is the only regular breeding site for Lesser Flamingos (*Phoenicopterus minor*) in the Eastern part of Africa (Mlingwa and Baker, 2006). The 1.5–2.5 million lesser   represent three quarters of the world population of the species. Therefore, Lake Natron is an Important Bird Area (IBA), which, if left undisturbed, has adequate food and nesting sites for   (Mlingwa and Baker, 2006). Also, apart from the   A mingo population, will also cause a negative impact on mammal populations and vegetation around the area and in the northern areas of Gelai to Longido. In addition, the opening of the area to hundreds of workers may give rise to poaching and encroachments for bush meat and charcoal trade as result affecting biodiversity population and dispersal.

There is a proposed Uranium mining at the area between the Selous Game Reserve and Selous-Niassa Wildlife Corridor. The area is exceptionally rich in wildlife species with several packs of wild dogs, giraffe (*Giraffa camelopardalis*), common waterbuck *(Kobus ellipsiprymnus)*, sable antelope (*Hippotragus niger*) herds of buffalo (*Syncerus caffer*) and elephants are observed in all parts of the corridor. Despite its biodiversity and wildlife potential, the Selous-Niassa Wildlife Corridor is threatened by an increasing human population and activities, which are incompatible with conservation interests. The environmental implications that are likely to arise from this economic opportunity include: blockage of the wildlife corridor and interference with migratory routes of animals. The uranium mining will have a negative consequence by increasing pressure on the natural resources and potentially more illegal logging, cultivation poaching and pollution. Also, it will accelerate loss/disturbance of biodiversity due to vegetation clearance, disturbance to biodiversity through carpeting of vegetation cover, increased potential for accidents to wildlife and people as well as health impacts to fauna from the drinking of contaminated water and from heavy metals taken up with forage.

11.11.4 Invasive Non-Native Species That Harm Native Ecosystems or Species

Invasive Alien Species (IAS) are a threat to ecosystem integrity and native species in some ecosystems of Tanzania. In fact, the country has the undesirable distinction of providing an example of one of the most ecologically damaging deliberate introductions of a nonnative species in the world. The Nile Perch (*Lates niloticus*), native to Africa but not to Lake Victoria, was introduced into the lake in 1950s to increase a fishery, but as a top predator it fed on Lake Victoria's native cichlids, many of which were endemic to the lake. The introduction has attributed to the extinction or depletion of several hundred native cichlid species. The IUCN's Invasive Species Specialist Group considers *Lates niloticus* as one of the world's 100 worst invasive species.

The water hyacinth (*Eichhornia crassipes*), native to South America, was   reported on Lake Victoria in 1989 and quickly spread. Until about 10 years, later tens of thousands of hectares of the water surface were covered by the plants (Kateregga and Sterner, 2007). This ecological invasion disrupted   ⬜A

gered cichlid   ⬜A

Neochetina bruchi and *Neochetina eichhorniae*, which feed on water hyacinth, were introduced into the lake as biological control agents. By 2005, the water hyacinth had been dramatically reduced (Kateregga and Sterner, 2007). Heavy rains in late 2006 raised water levels and swept nutrients into the lake, and the area covered by water hyacinth again expanded dramatically, and the problem continues to resist an easy solution.

In Zanzibar, the introduced Indian House Crow (*Corvus splendens*) provides an example of the consequences of an invasive species on native biodiversity (West, 2010; Gichua et al., 2014). One of the island's most striking features is the extremely high density of crows and the notable lack of native songbirds. Indian House Crow is detrimental to indigenous bird species via predation or competitive displacement by attacking, mobbing birds and destroying eggs as well as nests (Chongomwa, 2011). They have also been known to attack people (Fraser et al., 2015), and are a   human and animal health risk as a fecal contaminator of human environments and water sources (Ryall, 1992; Ryall and Meier, 2008). These crows have also been   A □ as *Salmonella* spp., *Shigella* serotypes, *Proteus* spp., *Vibrio* spp., *Pseudomonas* spp., *Escherichia coli*, *Campylobacter* spp. (Al-Sallami, 1990; Cooper, 1996). Crow predation on reptiles and amphibians, and damage to crops and poultry is substantial (West, 2010; Fraser et al., 2015). This bird, Indian House Crow (Figure 11.9), which is native to India, was introduced to the island in the 1890s and was recognized as a potential pest as early as 1917 (Ryall, 2010).

The increasing threats from invasive alien plants have triggered adverse effects worldwide. Alien invasive species have been described as being a major cause of species extinction (Gurevitch and Padilla, 2004) and may lead change local ecosystem through competition with or predation on local species as well as alteration of ecosystem functioning (D'Antonio and Vitousek, 1992). The impacts of alien plant species on natural and managed ecosystems are a worldwide environmental and socioeconomic problem (Lowe et al., 2000; Brooks et al., 2004; Pyšek and Richardson, 2010), because they out compete native species. Plant invasions in Tanzania have been reported in some protected areas (Wakibara and Mnaya, 2002; Foxcroft et al., 2006; IUCN, 2015). For example, many invasive weedy plant

Figure 11.9   Indian House Crow at Mheza district in Tanga region.

species are invading in the rangelands of western in Mahale Mountains National Park and northern Tanzania, including in Rubondo Island National Park, the Serengeti and Ngorongoro ecosystem (Lotter, 2004; Clark et al., 2011; Bukombe et al., 2013; Kija et al., 2013). *Azadirachta indica* in Saadani National Park, *Cedrela odorata*, a potential invasive woody species in Kimboza Forest Reserve, Morogoro, *Castilla elastica* and *Cordia alliodora* in Amani Nature Reserve (Dawson et al., 2009). The Mexican prickly poppy (*Argemone mexicana*), and *Datura stramonium*, called Jimson weed or datura, are two aggressive and toxic invasive plants from North America, have been documented in Tanzania (Foxcroft et al., 2006; Gallagher, 2016). *Parthenium hysterophorus*, also called whitetop (Figure 11.10), is another alien invasive plant that is raising concern in Tanzania (Kilewa and Rashid, 2014). A "crowed-sourced" monitoring and mapping project is tracking the spread of this species (https://partheniumafrica.crowdmap.com). *Lantana camara*, considered to be one of the world's most aggressive invasive plants, is said to be widespread in Tanzania (Eustace and Lekule, 2016).

Invasive alien plant species clearly have profound impacts on ecological health and well being. An invasive alien species is a species that is not native to the environment/ecosystem it inhabits, and has negative impacts on ecosystem health, human health, and/or economics (Pyšek et al., 2009). Invasive Alien Species are characteristically adaptable, aggressive and have a high reproductive capacity and hence can thrive well in areas beyond their range. The spread of invasive species is now recognized as a second major

Figure 11.10 Distribution of *Parthenium hysterophorus* weed in farm and residential area in Arusha region.

threat to biodiversity in the world as they can outgrow natural species over an area. This work has compiled a total of 78 plant species, which are invasive in Tanzania (Table 11.7) (TanBIF, 2013; URT, 2014b). These species mostly fall into four major classes: Magnoliopsida (78.2%), Liliopsida (17.9%), Pinopsida (1.3%) and Pteridopsida (2.6%). However, there are other alien plant species (61 species), which are potentially invasive in the country.

11.11.5 Pollution or Contamination that Harms Natural Habitats or Species

Pollution can be a major threat to ecosystems and species. Pollution from untreated sewage discharged from coastal cities and beach tourism facilities can cause significant damage to nearby coral reefs (Muthiga et al., 2008). Sediment from coastal agriculture and construction can also damage reefs. Fertilizer and pesticides used on fields can be washed into nearby streams, rivers, and wetlands, threatening fish, amphibians, insects, crustaceans, mollusks, and other aquatic species. Plastic pollution involves accumulation of plastic products in the environment leading to adverse effect on lands, waterways and oceans. They are used mostly because of the cheapness in their production and manipulation for use in many aspects. In Tanzania plastics are involved in many uses including, among others, electronic equipment, some vehicle parts, large to small package containers, building materials, agricultural and fishing gears, and carriages. However, plastic bags being the

Table 11.7 Status of Invasive Alien Plant Species in Tanzania

Class	Order	Family	Genus	Invasive Alien Species	Potentially invasive Alien species
Filicopsida	0	0	0	0	0
Florideophyceae	0	0	0	0	0
Isoetopsida	0	0	0	0	0
Liliopsida	7	7	13	14 (17.9%)	14
Magnoliopsida	21	27	55	61 (78.2%)	47
Pinopsida	1	1	1	1 (1.3%)	0
Pteridopsida	1	1	2	2 (2.6%)	0
Rhodophyceae	0	0	0	0	0
Total numbers	30	36	71	78	61

Source: TanBIF (2013).

cheapest carriage are mostly used only once and disposed unmanaged and therefore contributing most to the environmental pollution. The plastic bags are a frightening problem because they are long-lived products that are difficult and costly to recycle and most end up in the environment where they take around 300 years to degrade.

11.11.6 Climate Change Effects That Harm Natural Habitats or Species

Climate change is a potential threat of unknown magnitude, which may accentuate other direct threats already discussed above, especially habitat loss, degradation, and fragmentation, and the threat from invasive species. Climate change influences the abundance and distribution of living organisms by affecting habitats where they live over much of the continent (Bellard et al., 2012; Dickinson et al., 2015). Large parts of Tanzania currently experience a tropical, semiarid climate, in which rainfall is extremely variable from year to year. Rainfall variation is strongly influenced by sea surface temperature anomalies associated with the El Nino-Southern Oscillation (ENSO) (Trenberth et al., 2002; Collins et al., 2010). Drought and extreme rainfall events are the norm, not an exception. African biodiversity is, in many respects, the product of long-term natural cycles of climate change over tens of millions of years. The long-distance seasonal migrations of African ungulates are adaptations to track this natural climate variability (Marshall and Weissbrod, 2011).

11.12 Biodiversity Conservation Initiatives in Tanzania

Tanzania has, for a long time, been committed to conserving her biological resources and the environment. Since colonial times protected areas were established for the protection of both plants and animals. Various other measures were also implemented to conserve the environment, including soil and water conservation programs such as "Hifadhi Ardhi Dodoma" (HADO), "Hifadhi Ardhi Shinyanga" (HASHI) and promotion of use of terrace agriculture in hilly areas of Kilimanjaro Region. Important initiative that Tanzania has taken to address the issue of biodiversity conservation is the establishment of the Tanzania National Parks and Wildlife Division. Tanzania established National Environment Management Council in 1983 as an advisory body to the government in environment matters. This was followed by the establishment of the Division of Environment in 1990 under the Ministry of Tourism, Natural Resources and the Environment, and later elevated to the Vice President's Office in 1995. In 1989, Tanzania prepared the Tanzania Forestry Action Plan (TFAP) as an important step in strengthening the country's efforts towards sustainable management of her natural resources. The TFAP addressed such issues as sustainable land husbandry, community and farm forestry, forest management, bioenergy development, forest industries, bee-keeping, wildlife management and conservation of ecosystems and biodiversity. In June 1992 Tanzania signed a Convention on Biological Diversity focusing to implement the conservation and sustainable use of biological diversity, and for integration of conservation and sustainable use of biological diversity into relevant programs and policies.

11.12.1   Wildlife Sector in Tanzania

Administratively the wildlife sector in Tanzania has divided its mandates into Central and Local Governments for efficiency and effectiveness in managing wildlife resources. The Central Government includes ministries, executive agencies, Ngorongoro Conservation Area Authority, Tanzania National Parks Authority, Parastatal organization and independent departments, while the Local Government includes District Councils, Wards and Village Councils. The role of Central Governments is to provide clear national policy and regulatory framework stimulate and promote participation of various stakeholders in the implementation of policy, manage core wildlife protected areas and providing professional standards and technical assistance in conservation and utilization of resources. District Councils are responsible for conservation of wildlife outside national parks, game reserves and Ngorongoro Conservation Area, and protect people's lives and properties from dangerous and destructive wild animals.

11.12.2   Agencies Responsible in Managing the Wildlife and Wetlands in Tanzania

11.12.2.1   Wildlife Division (WD)

WD is responsible for the management of Game Reserves (GRs), Game Controlled Areas (GCAs) and all wildlife outside protected area boundaries and Wetlands. Also the WD facilitated the establishment of Wildlife Management Areas (WMAs), creates awareness and disseminates information about wildlife management to the village communities in their village lands.

Currently all responsibilities under Wildlife Division have been addressed by Tanzania Wildlife Management Authority. Wildlife Division has left with wildlife policy making and implementation of the policy.

11.12.2.2 Tanzania National Parks Authority (TANAPA)

TANAPA is a parastatal organization responsible for the management of all National Parks in the country, the authority manages and regulate the use of areas designated as National Parks by such means and measures to preserve the country's heritage, encompassing natural and cultural resources, both tangible and intangible resource values, including the fauna and flora, wildlife habitat, natural processes and wilderness quality.

11.12.2.3 Ngorongoro Conservation Area Authority (NCAA)

NCAA is a parastatal organization responsible for the management of Ngorongoro Conservation Area and regulating the utilization of resources under the NCA. Upon its establishment, a multiple use philosophy was adopted giving the NCAA a legal mandate to conserve the environment and wildlife and develop Maasai pastoralists who formerly inhabited the whole of the Serengeti ecology.

11.12.2.4 Tanzania Wildlife Research Institute (TAWIRI)

The government initiated TAWIRI in order to administer wildlife research in Tanzania with an overall objective of providing scientific information and advice to the government and wildlife management authorities on the sustainable conservation of wildlife.

11.12.2.5 College of African Wildlife Management (CAWM)

The government established CAWM to provide need-based training to protected area and wildlife managers by offering a variety of awards for long courses in Wildlife Management.

11.12.2.6 Tanzania Wildlife Management Authority (TAWA)

Recently, in October 2015, the government had taken an important initiative to address the issue of biodiversity conservation by establishing 'Tanzania Wildlife Management Authority' also known by its acronym as "TAWA."

The authority has taken all responsibility of protecting and conservation of Wildlife outside the jurisdiction of Tanzania National Parks and Ngorongoro Conservation Area Authority. The authority administer areas that are designated as Game Reserves, Game controlled areas, Wetlands Reserves, and Ramsar Sites. TAWA administer protection and utilization of wildlife in corridors, dispersal areas, open areas, Wildlife Management Areas, village land, public and private land.

11.13 Concluding Remarks

Conservation and sustainable utilization of biological diversity is a major challenge to all countries of the world striving to achieve sustainable development. Biodiversity is central to sustainable development in every country worldwide. The earth's biological resources are vital to humanity's economic and social development. Biodiversity is an asset of tremendous value to present and future generations. Tanzania has taken efforts to protect these resources against destruction and loss by, among other things, setting aside land as protected areas in the form of national parks, forest reserves, nature reserves, game reserves, game controlled and wildlife management areas. However, these areas and adjacent lands have long been subjected to a number of emerging issues and challenges, which complicate their management, thus putting the resources at risk of over exploitation and extinction. These issues and challenges include, government policies, failure of conservation (as a form of land use) to compete effectively with alternative land uses, habitat degradation and blockage of wildlife corridors, over exploitation of biodiversity resources and illegal resource extraction. The government should immediately seek proper adaptation measure to cope with these upcoming events. Besides, focusing only on ecological aspects will not provide a long-term security of biodiversity conservation in the country since people still substantially depend on these resources for their existence. Therefore, it is recommended that the government should actively involve local community participation in natural resource management to secure the future of country's remaining biological diversity. A persistent program on monitoring and management of country's biodiversity is essential. Government laws concerning biodiversity issue needs to be modified by considering country's current sociopolitical context.

Climate change problems and their potential impacts on the biodiversity should be addressed by the adoption of a variety of mitigation and adaptation measures. The measures should include controlling anthropogenic activities such as deforestation, adoption of proper land management practices (including agroforestry), changing energy technologies (e.g., the use of -cient wood stoves and biogas), in case of agriculture breeding crops that will be adaptable to changes of climate, i.e., draft resistance crops, short-term harvested crops, etc. Additionally, the government should involve adopting the integrated land and water management practices, and augmenting collaborations between the conservation and sustainable use of biodiversity in climate change scenarios. The government is urged to assess and identify invasive species and develop effective strategies for their control. This can be done by educating the public about the types of invasive species found in Tanzania and raise awareness of their relevance so that their control can start at the grass root level. More research is required to understand vulnerability of different ecosystems to new invasions by the alien invasive species in Tanzania so that mitigation measures will be proposed and practiced.

Keywords

- climate
- socio-economics
- species diversity
- vegetation

References

Al-Sallami, S., (1990). A possible role of crows in the spread of diarrhoeal diseases in Aden. *J. Egypt. Public Health Assoc., 66*(3–4), 441–449.

Bellard, C., Bertelsmeir, C., Leadley, P., & Thuiller, W., (2012). Impacts of climate change on the future of biodiversity. *Ecol. Lett., 15*(4), 365–377.

Brooks, M. L., D'Antonio, C., Richardson, D. M., et al., (2004). Effects of invasive alien plants on fire regimes. *BioScience, 54*(7), 677–688.

Bukombe, J., Kitte, A. M., Mney, P., & Mwita, M., (2013). *A Preliminary Survey of Alien Invasive Plant Species in Serengeti: A Call for Prompt Preventive Action.* Paper presented at the Proceedings of the 8th TAWIRI Scientific Conference 6th to 8th December 2011, Arusha, Tanzania.

Burgess, N. D., Bahane, B., Clairs, T., & Danielsen, F., (2010). Getting ready for REDD+ in Tanzania: A case study of progress and challenges. *Oryx, 44*(3), 339–351.

Burgess, N. D., Butynski, T. M., Cordeirom, N. J., & Stuart, S. N., (2007). The biological importance of the Eastern Arc Mountains of Tanzania and Kenya. *Biol. Conserv., 134*(2), 209–231.

Burgess, N., Klerk, H., Fjeldsa, J., & Rahbek, C., (2000). A preliminary assessment of congruence between biodiversity patterns in Afrotropical forest birds and forest mammals. *Ostrich, 71*(1–2), 286–290.

Chang'a, L. B., Yanda, P. Z., & Ngana, J., (2010). Spatial and temporal analysis of recent climatological data in Tanzania. *J. Geogr. Reg. Plann., 3*(3), 44–65.

Chongomwa, M. M., (2011). Mapping locations of nesting sites of the Indian house crow in Mombasa. *J. Geogr. Reg. Plann., 4*(2), 87–97.

Clark, K., Lotter, W. D., & Runyoro, V., (2011). *Invasive Alien Plants: Strategic Management Plan in the Ngorongoro Conservation Area.* Arusha, Tanzania: United Republic of Tanzania, Ministry of Natural Resources and Tourism.

Collins, M., An, S. I., Cai, W., et al., (2010). The impact of global warming on the tropical Pacific Ocean and El Niño. *Nat. Geosci., 3*(6), 391–397.

Collins, S., Congdon, C., & Bampton, I., (2017). Review of the *Charaxes gallagheri* complex resulting in two new species and undescribed morphs of *Charaxes gallagheri* van Son 1962 (Lepidoptera, Nymphalidae, Charaxinae). *Entomol. Afr., 22*, 19–30.

Cooper, J. E., (1996). Health studies on the Indian house crow (*Corvus splendens*). *Avian Pathol., 25*(2), 381–386.

COSTECH, (2012). *Checklist of Tanzanian Species Version 1.* Dar es Salaam: Tanzania Biodiversity Information Facility (TanBIF).

D'Antonio, C. M., & Vitousek, P. M., (1992). Biological invasions by exotic grasses, the grass/fire cycle, and global change. *Annu. Rev. Ecol. Syst., 23*(1), 63–87.

Dauby, G., Zais, R., Blach-Overgaard, A., et al., (2016). RAINBIO: A mega-database of tropical African vascular plants distributions. *PhytoKeys, 74*, 1–18.

Davenport, T. R. B., (2002). *Endemic Butterflies of the Albertine Rift valley-An Annotated Checklist.* Mbeya, Tanzania: The Wildlife Conservation Society.

Davis, A. P., & Mvungi, E. F., (2004). Two new and endangered species of *Coffea* (Rubiaceae) from the Eastern Arc Mountains (Tanzania) and notes on associated conservation issues. *Bot. J. Linn. Soc., 146*(2), 237–245.

Dawson, W., Burslem, D. F., & Hulme, P. E., (2009). Factors explaining alien plant invasion success in a tropical ecosystem differ at each stage of invasion. *J. Ecol., 97*(4), 657–665.

Dickinson, M., Prentice, I. C., & Mace, G. M., (2015). *Climate Change and Challenges for Conservation.* London: Grantham Institute Briefing paper No 13, in partnership with the Centre for Biodiversity and Environment Research at University College London.

Dobson, A. P., Borner, M., Sinclair, A. R. E., et al., (2010). Road will ruin Serengeti. *Nature, 467*(7313), 272–273.

Drews, C., (1995). Road kills of animals by public traffic in Mikumi National Park, Tanzania, with notes on baboon mortality. *Afr. J. Ecol., 33*(2), 89–100.

EAH., (1996), *LEAP of Tanzania* East African Herbarium (EAH), Nairobi Kenya (accessed March 2017). (Unpublished data).

Estes, A. B., Kuemmerle, T., Kushnir, H., Radeloff, V. C., & Shugart, H. H., (2012). Landcover change and human population trends in the greater Serengeti ecosystem from 1984–2003. *Biol. Conserv., 147*(1), 255–263.

Estrin, S., & Mickiewicz, T., (2012). Shadow economy and entrepreneurial entry. *Rev. Dev. Econ., 16*(4), 559–578.

Eustace, A., & Lekule, N. M., (2016). Jan-Mar-2016 Coden: IJPAJX-CAS-USA, Copyrights@ 2016 ISSN-2231–4490 Received: 1 st Oct-2015 Revised: 27 th Oct-2015 Accepted: 1 st Nov-2015 Research article Spatial distribution pattern and abundance of *Ageratum conyzoides* (Asteraceae family) in riverine and non-riverine habitats of MWEKA, vol/issue: 6(1).

Ferraro, P. J., (2001). Global habitat protection: limitations of development interventions and a role for conservation performance payments. *Conserv. Biol., 15*(4), 990–1000.

Foley, C., Foley, L., Lobora, A., et al., (2014). *A Field Guide to the Larger Mammals of Tanzania*. Princeton: Princeton University Press.

Foxcroft, L. C., Lotter, W. D., Runyoro, V. A., & Mattay, P. M. C., (2006). A review of the importance of invasive alien plants in the Ngorongoro Conservation Area and Serengeti National Park. *Afr. J. Ecol., 44*(3), 404.

Fraser, D. L., Aguilar, G., Nagle, W., Galbraith, M., & Ryall, C., (2015). The house crow (*Corvus splendens*): A threat to New Zealand? *ISPRS Int. J. Geo-Inf., 4*(2), 725–740.

Froese, R., & Pauly, D. E., (2017). FishBase (version Oct 2016). Digital resource at www.catalogueoflife.org/col. In: Roskov, Y., Abucay, L., Orrell, T., Nicolson, D., Bailly, N., Kirk, P., et al., (eds.). *Species 2000 & ITIS Catalogue of Lifeed*, Leiden, the Netherlands: Naturalis.

Fyumagwa, R. D., & Nyahongo, J. W., (2010). Black rhino conservation in Tanzania: Translocation efforts and further challenges. *Pachyderm, 47*, 59–65.

Gallagher, D., (2016). American plants in Sub-Saharan Africa: A review of the archaeological evidence. *Archaeol. Res. Afr., 51*(1), 24–61.

Gichua, M., Njoroge, G., Shitanda, D., & Ward, D., (2014). Invasive species in East Africa: Current status for informed policy decisions and management. *J. Agric. Sci. Technol., 15*(1), 45–55.

Gideon, H., Nyinondi, P., & Oyema, G., (2012). *Checklist of Tanzanian Species*. Dar es Salaam, Tanzania: Tanzania Commission for Science and Technology (COSTECH).

Gill, F., & Donsker, D., (2016). *IOC World Bird List (v 6. 4)*. http://www.worldbirdnames.org/ http://avibase.bsc-eoc.org/checklist.jsp?region=tz&list=howardmoore/IOC, Retrieved 12 January 2017.

Grebennikov, V. V., (2008). A featherwing beetle without wings: re-discovery and second species of Rioneta (Coleoptera: Ptiliidae) from the Uluguru Mountains, Tanzania. *Zootaxa, 1732*, 45–53.

Gurevitch, J., & Padilla, D. K., (2004). Are invasive species a major cause of extinctions? *Trends Ecol. Evol., 19*(9), 470–474.

IUCN, (2015). *Invasive species in United Republic of Tanzania* Species Survival Commission (SSC).

IUCN, (2017). *The IUCN Red List of Threatened Species*. Version 2016–3 2016, Retrieved 17 January 2017.

Kamukala, G. L., & Crafter, S. A., (1993). *Wetlands of Tanzania*. Paper presented at the Proceedings of a Seminar on the Wetlands of Tanzania, 1991, Morogoro, Tanzania.

Kateregga, E., & Sterner, T., (2007). Indicators for an invasive species: Water hyacinths in Lake Victoria. *Ecol. Indic., 7*(2), 362–370.

Kideghesho, J. R., Nyahongo, J. W., Hassan, S. N., Tarimo, T. C., & Mbije, N. E., (2006). Factors and ecological impacts of wildlife habitat destruction in the Serengeti ecosystem in northern Tanzania. *Afr. J. Environ. Assess. Manag., 11*, 17–32.

Kideghesho, J., Rija, A. A., Mwamende, K. A., & Selemani, I. S., (2013). Emerging issues and challenges in conservation of biodiversity in the rangelands of Tanzania. *Nat. Consrv., 6*, 1–29.

Kija, B., Mweya, C. N., Mwita, M., Kiza, H., & Fyumagwa, R., (2013). Prediction of suitable habitat for potential invasive plant species *Parthenium hysterophorus* in Tanzania: A short communication. *Int. J. Ecosyst., 3*(4), 82–89.

Kilewa, R., & Rashid, A., (2014). Distribution of Invasive Weed *Parthenium hysterophorus* in Natural and Agro-Ecosystems in Arusha Tanzania. *Int. J. Sci. Res., 3*(12), ISSN (Online): 2319–7064.

Kioko, J., Kiffner, C., Jenkins, N., & Collinson, W. J., (2015). Wildlife roadkill patterns on a major highway in northern Tanzania. *Afr. Zool., 50*(1), 17–22.

Knox, E. B., & Berghe, E. V., (1996). The use of LEAP in herbarium management and plant biodiversity research. *J. East Afr. Nat. Hist., 85*(1), 65–79.

Kociolek, A. V., Cleveenger, A. P., St Clair C. C., & Proppe, D. S., (2011). Effects of road networks on bird populations. *Conserv. Biol., 25*(2), 241–249.

LNT, (2017). *Checklist of Endemics.* http://lntreasures. com/tanzaniap. html, Retrieved 1 July, 2017.

Lotter, W. D., (2004). *Progress Report and Updated Recommendations for the Management of Invasive Alien Plants in the Ngorongoro Conservation Area.* Arusha, Tanzania.

Lowe, S., Browne, S., Bodjelas, S., & De Poorter, M., (2000). *100 of the World's Worst Invasive Alien Species: A Selection From the Global Invasive Species Database.* New Zealand: Invasive Species Specialist Group Auckland.

Mabberley, D. J., (2008). *Mabberley's Plant-book (Third ed.).* Cambridge: Cambridge University Press.

Marshall, F., & Weissbrod, L., (2011). Domestication processes and morphological change: Through the lens of the donkey and African pastoralism. *Curr. Anthropol., 52*(S4), S397–S413.

Mauya, E. W., Mugasha, W. A., Zahabu, E., Bollandshas, O. M., & Eid, T., (2014). Models for estimation of tree volume in the miombo woodlands of Tanzania. *Southern Forests: J. For. Sci., 76*(4), 209–219.

McSweeney, C., Lizcano, G., New, M., & Lu, X., (2010). The UNDP climate change country profiles: Improving the accessibility of observed and projected climate information for studies of climate change in developing countries. *Bull. Am. Meteorol. Soc., 91*(2), 157–166.

Mlingwa, C., & Baker, N., (2006). Lesser Flamingo *Phoenicopterus minor* counts in Tanzanian soda lakes: Implications for conservation. In: Boere, G. C., Galbraith, C. A., & Stroud, D. A., (eds.). *Waterbirds Around the World.* The Stationery Office, Edinburgh. The Stationery Office: Edinburgh, UK., pp. 230–233.

Mugasha, W. A., Mwakalukwa, E., Luoga, E., et al., (2016). Allometric models for estimating tree volume and aboveground biomass in lowland forests of Tanzania. *Int. J. For. Res.,* Article ID 8076271, 8076213 pages http://dx. doi: org/8076210.8071155/807201 6/8076271.

Muthiga, N. A., Costa, A., Motta, H., Muhando, C., Mwaipopo, R., & Schleyer, M., (2008). *Status of Coral Reefs in East Africa: Kenya, Tanzania, Mozambique and South Africa.* Status of Coral Reefs of the World.

Mwandosya, M. J., Nyenzi, B. S., & Lubanga, M. L., (1998). *The Assessment of Vulnerability and Adaptation to Climate Change Impacts in Tanzania.* Dar es Salaam: Centre for Energy, Environment, Science and Technology.

Myers, N., Mittermeier, R. A., Mittermeier, C. G., Fonesca, G. A. B., & Kent, T., (2000). Biodiversity hotspots for conservation priorities. *Nature, 403*(6772), 853–858.

Nelson, F., Nshala, R., & Rodgers, W. A., (2007). The evolution and reform of Tanzanian wildlife management. *Conserv. Soc., 5*(2), 232–261.

Nkwabi, A. K., (2016). *Influence of Habitat Structure and Seasonal Variation on Abundance, Diversity and Breeding of Bird Communities in Selected Parts of the Serengeti National Park, Tanzania.* PhD thesis, University of Dar es Salaam, Dar es Salaam.

Nkwabi, A. K., Liseki, S., Bukombe, J. K., et al., (2017). Species richness and composition of butterfly with reference to anthropogenic activities in the wildlife management areas, Southern Tanzania. *Int. J. Fauna. Biol. Stud., 4*(1), 34–40.

Nkwabi, A. K., Sinclair, A. R. E., Metzger, K. L., & Mduma, S. A. R., (2011). Disturbance, species loss and compensation: Wildfire and grazing effects on the avian community and its food supply in the Serengeti Ecosystem, Tanzania. *Austral Ecol., 36*(4), 403–412.

Nummelin, M., & Nshubemuki, L., (1998). Seasonality and structure of the arthropod community in a forested valley in the Uluguru Mountains, eastern Tanzania. *J. East Afr. Nat. Hist., 87*(1), 205–212.

Plumptre, A. J., Behengana, M., Devenport, T. R. B., Kahindo, C., Kitoyo, R., et al., (2007). The biodiversity of the Albertine Rift. *Biol. Conserv., 134*(2), 178–194.

Poynton, J. C., Howell, K. M., Clarke, B. T., & Lovett, J. C., (1998). A critically endangered new species of Nectophrynoides (Anura: Bufonidae) from the Kihansi Gorge, Udzungwa Mountains, Tanzania. *Afr. J. Herpetol, 47*(2), 59–67.

Poynton, J. C., Loader, S. P., Sherratt, E., & Clarke, B. T., (2007). Amphibian diversity in East African biodiversity hotspots: Altitudinal and latitudinal patterns. *Vertebr. Conserv. Biodivers., 16*, 1103–1118.

Pyšek, P., & Richardson, D. M., (2010). Invasive species, environmental change and management, and health. *Annu. Rev. Environ. Resour., 35*, 25–55.

Pyšek, P., Křivánek, M., & Jarošík, V., (2009). Planting intensity, residence time, and species traits determine invasion success of alien woody species. *Ecology, 90*(10), 2734–2744.

Rwehumbiza, F. B., (2014). *A Comprehensive Scoping and Assessment Study of Climate Smart Agriculture Policies in Tanzania.* Pretoria, South Africa: Food, Agriculture and Natural Resources Policy Analysis Network (FANRPAN).

Ryall, C., & Meier, G., (2008). House crow in the Middle East. *Wildlife Middle East News, 3*(3), 7.

Ryall, C., (1992). Predation and harassment of native bird species by the Indian house crow, *Corvus splendens*, in Mombasa, Kenya. *Scopus, 16*(1), 1–8.

Ryall, C., (2010). Further records and updates of range expansion in house crow *Corvus splendens. Bull. Br. Ornithol. Club., 130*, 246–254.

Salehe, J., (2011). *The Forests and Woodlands of the Coastal East Africa Region.* Dar es Salaam: World Wide Fund for Nature (WWF).

Semesi, A. K., (1992). Developing management plans for the mangrove forest reserves of mainland Tanzania. *Hydrobiologia, 247*(1), 1–10.

Songorwa, A. N., (1999). Community-based wildlife management (CWM) in Tanzania: Are the communities interested? *World Development, 27*(12), 2061–2079.

Sosef, M. S., Dauby, G., Blach-Overgaard, A., et al., (2017). Exploring the floristic diversity of tropical Africa. *BMC Biol., 15*(1), 15. https://doi.org/10.1186/s12915-12017-10356-12918.

Sumner, A., (2012). Where do the world's poor live? A new update. *IDS Working Papers 2012. 393*, 1–27.

TanBIF, (2013). Tanzania Biodiversity Information Facility: http://www.tanbif.or.tz/ lin2/tanbif_linnaeus. php? menuentry=zoeken TanBIF-COSTECH-Tanzania.

TAWIRI, (2010). *Tanzania Elephant Management Plan 2010–2015*. Arusha: Tanzania Wildlife Research Institute.

Timothy, D. J., (1999). Cross-border partnership in tourism resource management: International parks along the US-Canada border. *J. Sustainable Tourism, 7*(3–4), 182–205.

Tobey, J., & Torell, E., (2006). Coastal poverty and MPA management in mainland Tanzania and Zanzibar. *Ocean Coast. Manag., 49*(11), 834–854.

Trenberth, K. E., Caron, J. M., Stepaniak, D. P., & Worley, S., (2002). Evolution of El Niño–Southern Oscillation and global atmospheric surface temperatures. *J. Geophysical Research: Atmospheres, 107*(D8), doi: 10.1029/2000JD000298.

Trombulak, S. C., & Frissell, C. A., (2000). Review of ecological effects of roads on terrestrial and aquatic communities. *Conserv. Biol., 14*(1), 18–30.

Tumbo, M., Mangora, M. M., Paulinem, N. M., & Kuguru, B., (2015). *Review of Literature for a Climate Vulnerability Assessment in the RUMAKI Seascape, Tanzania*. Dar es Salaam: WWF Tanzania Country Office.

Uetz, P., & Hošek, J., (2017). The Reptile Database (version Dec 2015). In: Roskov, Y., Abucay, L., Orrell, T., Nicolson, D., Bailly, N., Kirk, P., et al., (eds.). *Species 2000 & ITIS Catalogue of Life*. Digital resource at www.catalogueoflife.org/, College edition, Leiden, the Netherlands: Species 2000, Naturalis.

URT, (2001). *National Report on the Implementation of the Convention on Biological Diversity of Tanzania*. Dar es Salaam: Division of Environment, Vice President's Office.

URT, (2009). *Fourth National Report on the Implementation of the Convention on Biological Diversity of Tanzania*. Dar es Salaam: Division of Environment, Vice President's Office.

URT, (2014a). *Fifth National Report on the Implementation of the Convention on Biological Diversity*. Dar es Salaam, Tanzania: Division of Environment, Vice President's Office.

URT, (2014b). *State of the Environment Report II*. Dar es Salaam: Division of Environment, Vice President's Office.

Wakibara, J. V., & Mnaya, B. J., (2002). Possible control of *Senna spectabilis* (Caesalpiniaceae), an invasive tree in Mahale mountains National Park, Tanzania. *Oryx, 36*(4), 357–363.

Wanyonyi, K., Darlington, J., & Bagine, R., (1984). Checklist of the species of termites (Isoptera) recorded from East Africa. *J. East Afr. Nat. Hist. Soc. Natl. Mus., 181*, 1–10.

West, L., (2010). A multi-stakeholder approach to the challenges of turtle conservation in the United Republic of Tanzania. *Indian Ocean Turtle Newsletter, 11*, 44–50.

White, F., (1983). *The Vegetation of Africa. A Descriptive Memoir to Accompany the UNESCO/AETFAT/UNSO Vegetation Map of Africa*. Paris, UNESCO.

CHAPTER 12

Biodiversity in Togo

KOUAMI KOKOU,[1] KOMLAN M. AFIADEMANYO,[2]
KOMLA ELIKPLIM ABOTSI,[1] HOUINSODÉ SEGNIAGBETO,[2]
MONDJONNESSO GOMINA,[2] and KOMINA AMÉVOIN[2]

[1]Department of Botany and Plant Ecology, University of Lomé, Togo,
E-mail: kokoukouami@hotmail.com
[2]Department of Zoology, University of Lomé, Togo

12.1   Introduction

The Republic of Togo is located in the Dahomey gap (Figure 12.1), which is an interruption of the West-African tropical forest, where the savanna eco-systems are closer to the coast (White, 1983). Natural forests are limited to the southern Togo Mountains (Adjossou, 2009). Unfortunately, all the natural ecosystems underwent a strong deterioration since 1970, due to the intensification of the industrial agriculture (coffee, cocoa, cotton, and palm

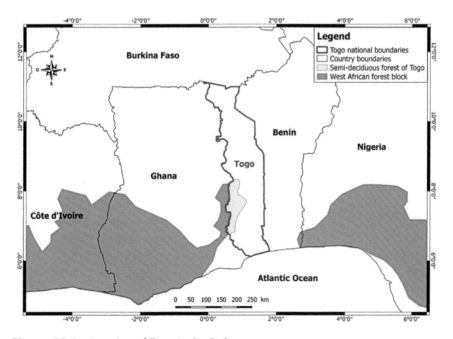

Figure 12.1   Location of Togo in the Dahomey gap.

oil tree), the anarchical logging of the main timbers and the nonexistence of adequate policy.

FAO reported a growing loss of forest cover over the past two decades, with an annual rate of 3.7% over the period 1990–2000, rising to 4.5% in the period 2000–2005, and   to 5.75% during the period 2005–2010 (FAO, 2010). But according to the results of the   National Forest Inventory (NFI, 2015/16), Togo has a forest resource coverage rate of 24.24% (MERF/ GIZ, 2016). These results are based on the   □A this inventory, which is that of the forest code. It stipulates that "the forest is an area covering an area of more than 0.5 ha with trees reaching a height of more than 5 meters and a tree cover of more than 10%."

Because of the weak average forest cover in Togo, the colonial administration had delimited protected areas but the rules imposed on the populations, without taking their interests into account, has led to the invasion and destruction of the existing forests (Tchamiè, 1994). In addition, plantations and reforestations had been undertaken on a large scale in 1900 under the German colonization. It is during this period (1904–1919) that the teak (*Tectona grandis*) became an important timber because of the success of its plantations.

As a result of the natural forests reduction, this study was carried out to update the knowledge on the conservation of the biodiversity in Togo, their economical and cultural importance in order to provide database for sustainable management as indicated by the World Strategic Plan for Biodiversity 2011–2020 and the objectives of Aïchi to which Togo fully adheres (MERF, 2014).

12.2 Major Ecosystems of Togo

Togo is located on the coast of the Gulf of Guinea in West Africa. it is as an area of about 56,600 km^2 and is bordered by Burkina Faso in north, Republic of Benin at east, Ghana at the West and the Atlantic Ocean in south. Located between the 6° and 11° north latitude and between 0° and 2° longitude east, the country extends from north to south over 660 km^2. Its width varies between 50 and 150 km^2. Ern (1979) subdivided Togo into five ecological zones (Figure 12.2, Table 12.1). Many studies have been devoted to plant communities in the different ecological zones (Scholz and Scholz 1983; Brunel et al., 1984; Guelly, 1994; Kokou, 1998; Kokou et al., 2000, 2006, 2008; Kokou and Caballé, 2000; Woegan, 2007; Dourma, 2008; Adjossou, 2009; Adjonou et al., 2013). These studies describe the major vegetation formations and their biodiversity in terrestrial and aquatic ecosystems.

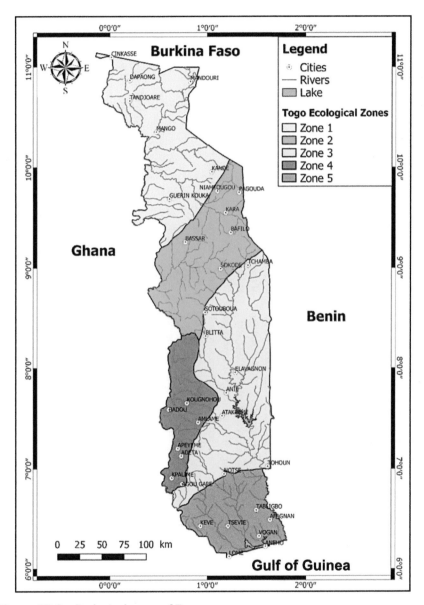

Figure 12.2 Ecological zones of Togo.

12.2.1 *Aquatic Ecosystems: Wetlands*

Togo has about 2210 km² of wetlands, i.e., 4% of the country's area. They consist of marine and coastal areas, streams and ponds, dams and reservoirs,

Table 12.1 Ecological Subdivisions of Togo (Ern, 1979)

Zone	Geomorphology	Ecological features	Ecosystem
1	Plains of the North	Sudanian climate characterized by a rainy season running from June to October and a dry season from November to May, with an average of 6–7 dry months. Total rainfall is between 800 and 1,000 mm. Temperatures vary between 17 and 39°C during the dry season and between 22 and 34°C during the rainy season	Sudanian savannas dominated by Fabaceae Mimosoidae (*Acacia* spp.), Combretaceae (*Terminalia* spp., *Combretum* spp.), dry forests of *Anogeissus*, gallery forests, and grasslands around pools. Several large agroforestry parks of *Vitellaria paradoxa* or *Parkia biglobosa*, *Adansonia digitata* (baobab) or *Borassus* spp.
2	Northern part of Togo Mountains	A rainy season from April to October and a dry season from October to March, characterized by the Harmattan. Yearly rainfall is 1,200–1,300 mm. Average evaporation is very high (2,000–2,100 mm) due to the high insolation (2,451 h per year). The minimal average temperature ranges from 19 to 20°C in January, the maximum reaches 30°C in April. Soils are thin with high proportion of stones or fragments of rocks. Tropical ferruginous soils are also encountered here.	Hilly dry forest and savannah mosaic zones in the north with a soudano-guinean climate as well. It is the domain of the dry dense forest of *Anogeissus leiocarpa* or *Monotes kerstingii* and *Uapaca togoensis* and woodlands of *Isoberlinia doka* and *Isoberlinia tomentosa*. There are also Combretaceae savannas but also agroforestry parks and gallery forests.
3	Plains of the Center	Tropical climate characterized by a rainy season from May till October and a dry season of at least 4 months. Total rainfall is between 1,200 and 1,500 mm per year. The number of rainy days is between 100 and 130 per year. The temperature ranges between 25 and 40°C. Soils are essentially tropical ferruginous.	Guinean savannas are interspersed with islands of semi-deciduous forest in the southern part and with dry forest of *Anogeissus leiocarpa* in the northern parts. The Guinean savannas have a relatively varied A There are also forest galleries whose main species are *Cynometra megalophylla*, *Parinari congensis*, *Pterocarpus santalinoides*, etc.

4	Southern part of Togo Mountains	The climate is a transitional subequatorial climate with a single rainy season, and a low rainfall in August and September. The rainfall amounts from 1,300 to 1,600 mm per year.	Dense semi-deciduous forests. Main species are *Milicia excelsa, Khaya grandifoliola, Erythrophleum suaveolens, Antiaris africana, Terminalia superba, Parinari glabra*. These forests are currently very degraded and transformed into coffee-cocoa agroforests. They are interspersed with Guinean savannas where main species are *Lophira lanceolata, Terminalia glaucescens, Pterocarpus erinaceus, Hymenocardia acida, Crossopteryx febrifuga, Faurea speciosa, Vitex doniana*, etc.
5	Coastal plains	Under a subequatorial climate characterized by a shortage of rainfall in the coastal part (800 mm/year in Lome, the capital). Rainfall increases towards the northern limit where it can reach up to 1200 mm/year. The landscape offers a mosaic of savannahs, agricultural land and prereserved forests (Kokou and Caballe, 2000).	Very degraded plant formations. It is a mosaic of disparate forest islands, with species like *Milicia excelsa, Antiaris africana*, relics of forests galleries with *Cynometra megalophylla, Pterocarpus santalinoides, Cola gigantea*, etc., highly anthropogenic savannas, coastal thickets, halophilic or swampy grasslands, mangroves, fallow land and crops.

seasonally flooded arable lands, canals and drainage ditches, gullies, seasonal or temporary freshwater lakes, permanent freshwater lakes, flood plains, excavations corresponding to mine, gravel and sand mining areas (MERF, 2013). The wetlands of the country can be grouped into five major groups: the Biankouri Wetlands, the Oti Basin, the Mô Plain, the Mono Basin and the Coastal Complex (Figure 12.3).

12.2.1.1 Continental Aquatic Ecosystems

The aquatic ecosystems comprise all natural or artificial continental aquatic environments, lentic or lotic. These include rivers, reservoirs, ponds, lagoons and lakes. The inland waters of Togo occupy an estimated total area of 255,000 ha, which form three main river catchments: the Oti and its tributaries, the Mono and its tributaries, the Zio and Haho and their tributaries (Figure 12.3).

These aquatic environments and wetlands provide the biodiversity of Togo with a   number of plant and animal species. There are about 20 invertebrate phyla including Protozoa, Spongiaries, Cnidarians, Annelids, Echinoderms, Molluscs, Arthropods and a large number of   A
species. Some very rare plant species encountered in the ponds in the south of the Togodo wildlife reserve were not yet been reported in Togo   ☐
Acroceras macrum (Poaceae), *Cyperus iria*, *Pycreus mundtii*, *Torulinium odoratum* Cyperaceae), *Ipomoea muricata* (Convolvulaceae), *Stemondia verticillata* (Scrophulariaceae). In the wetlands of the Oti-Keran Park, some of the listed species belong to new genera (*Aniseia, Cadaba, Courbonia, Iphigenia, Nymphoides*) or new family (Menyanthaceae), which has not previously been reported in Togo.

The main continental lentics environments are the Lake Togo and the
    A
The creation of this   A
bly *Sarotherodon galilaeus, Lates niloticus* and *Clarias gariepinus*. On the other hand, some   ☐A
environment. These include *Brycinus longipinnis, Chromidotilapia guntheri* and *Hemichromis fasciatus*. Similar effects are expected for the construction of the Adjarala dam which is currently underway on the Mono River.

As for the ponds, the Oti River and the Koumongou River, as well as their tributaries, develop large areas of   with many dead branches in the Oti basin. Depressions are often transformed into temporary or permanent pools, especially during the rainy season. Etsè (2012) counted 64 permanent and temporary pools in the Oti basin.

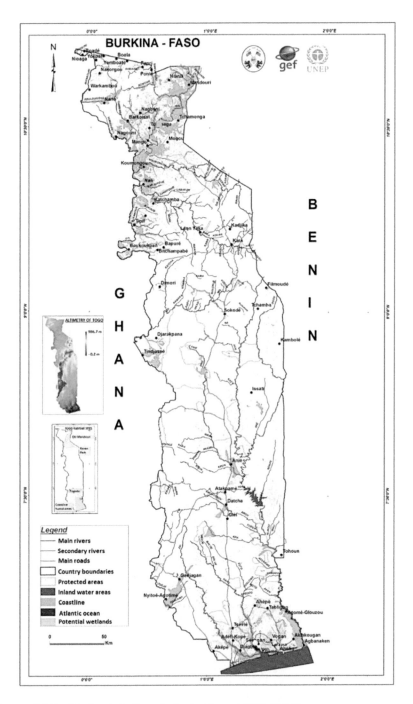

Figure 12.3 Wetlands in Togo (adapted from MERF, 2013).

In the coastal zone, the main pools are those located in the southeast of Togodo National Park. This area, which is now a biosphere reserve, is endowed with important   potential and shelter a rich and fauna. They include hippopotamuses (*Hippopotamus amphibius*), Nile crocodiles (*Crocodylus niloticus*), rare turtle species (*Kinixys* spp, *Trionyx triunguius*), large populations of ducks (*Dendrocygna viduata*), various species of herons (*Ardeola* spp.),   and hunters, waders, shorebirds, terns, etc., a large   A
a highly sought-after   A   *Gymnarchus monarchus*. The Manatee, a rare and protected species, is also found there.

12.2.1.2 Marine Ecosystems

Underwater flora is very poorly known, except algae, on which some beachrock stands have been inventoried. The fauna is very diverse with fish, Sharks, Rays and Batroids, Marine Mammals, Marine Turtles, Birds and Marine Invertebrates, etc.

12.2.1.3 Mangroves

Mangroves in Togo are located in the extreme southeast of the country around the Gbaga channel and its tributary rivers. There are two main species of mangroves: *Rhizophora racemosa* and *Avicennia germinans*, which are associated with *Drepanocarpus lunatus*, *Pterocarpus santalinoides* and the fern *Achrostichum aureum*. Mangroves, subjected to severe human pressure, are now reduced to less than 1,000 ha, whereas they appear to occupy the entire perimeter of the coastal lagoon system in the past. Moreover, the construction of the Nangbéto dam has profoundly disrupted the hydrological functioning endangered mangroves relics. These mangroves are surrounded by vast flooded savannas of *Mitragyna inermis* and *Andropogon gayanus* var. *bisquamulatus*, which are also found in the major bed of rivers and associated depressions.

12.2.2 Terrestrial Ecosystems

12.2.2.1 Semi-Deciduous Forests

They are located in the ecological zone 4 (Ern, 1979) (Figure 12.2). In the northern part of this zone, there are forests of Meliaceae and Sapotaceae, characterized by the following species: *Sterculia tragacantha, Cola gigantea,*

Manilkara obovata, M. multinervis, Aningera altissima. In the southern part, there are several types of forests: (i) on the western slopes of the mountain, there are forests of *Celtis* spp. and forests at *Terminalia superba*; (ii) on the eastern slopes, there are forests at Meliaceae and Moraceae; and (iii) on the high peaks, forests at *Parinari excelsa, P. glabra* and *Polyscias fulva*. The undergrowth of these forests are entirely occupied by cash crops (coffee and cocoa trees).

Other semi-deciduous forest types have been reported, including remnants of coastal forests on old sea dunes or dry semi-deciduous forests. These formations are also in the form of sacred forests containing large feet of *Milicia excelsa,* and *Antiaris africana*, often revered species.

12.2.2.2 Dry Forests

Dry forests are found in ecological zones 1, 2 and 3. The most spectacular formations are characterized by stands of *Anogeissus leiocarpa* in the classified forests of Abdoulaye, Kpessi, Oti-Kéran Park, or with dominant species such as *Monotes kerstingii, Khaya grandifoliola, Aubrevillea kestingii, Parinari glabra, Dialium guineense, Pseudospondias microcarpa, Cola gigantea,* and *Cola millenii* in the Fazao-Malfakassa Park.

12.2.2.3 Forests Galleries and Swamps

Forests galleries or riparian forests floristic composition includes *Pterocarpus santalinoides, Berlinia grandiflora, Uapaca* spp.*, Pentadesma butyracea, Erythrophleum suaveolens, Canarium schweinfurthii*, etc. There are also marshy forests characterized by *Symphonia globulifera, Mitragyna stipulosa, Raphia hookeri, R. sudanica*, etc.

12.2.2.4 Woodlands and Savannas

Woodlands are generally found on the hillsides in ecological zones 1, 2, 3, and 4. They are characterized by dense stands of *Isoberlinia doka* and/or *I. tomentosa, Anogeissus leoicarpa, Uapaca togoensis, Monotes kerstingii*.

The savannas of T

- The Moutain Guinean savannas: They are at the summit of the hills, characterized by *Lophira lanceolata*, forming in some places wooded savannas. At the bottom of the slope where the soils are relatively deep, are localized savannas of *Chasmopodium afzelii* and *Andropogon macrophyllus*. Trees such as *Daniellia oliveri*, *Terminalia glaucesens*, *Pterocarpus erinaceus*, *Lannea kerstingii*, *Vitex doniana*, etc., are found in these savannas;

- The Guinean savannas of the central plain (Zone 3) and the coastal zone (Zone 5): these savannas which extend from the coast to the latitude of Tchamba are shrub savannas, locally wooded. The very rich   is dominated by *Daniellia oliveri*, *Terminalia macroptera*, *Combretum* spp., *Pterocarpus erinaceus*, *Parkia biglobosa*, *Vitellaria paradoxa*, etc. Around the hills of Glitho in the central-eastern region of the country, a particular savanna of the Gymnosperm *Encephalartos barteri* has been developed. Closed to the coast, the Combretaceae are rare, but *Hymenocardia acida*, *Pterocarpus erinaceus*, *Vitellaria paradoxa*, *Acacia campylacantha*, etc. are the most frequent trees. The grass carpet is dominated by *Hypparhenia* spp., *Andropogon* spp., etc. On the coast, there are large tracts of savannas of *Borassus* spp. and *Adansonia digitata*, which are highly disturbed;
- Sudanian savannas: are found in the northern regions of the country, particularly in ecological zones 1 and 2. These are generally shrub savannas dominated by *Terminalia macroptera*, *Combretum* spp., but also spiny trees such as *Acacia gourmaensis*, *A dudgeonii*, *Balanites aegyptiaca*, with a grassy stratum dominated by *Aristida adscensionis*, *Loudetia simplex*, *L. togoensis*, *Andropogon* spp., *Pennisetum pedicellatum*, etc. It is in these areas that the most spectacular agroforestry parks, namely the parks of *Parkia*, *Vitellaria* and *Adansonia* or *Borassus*, are the result of human action on the natural savannas.

12.2.2.5 Maritime Lawns

At the seaside near the beaches, there are very low herbaceous formations. The floristic composition consists essentially of *Sporobolus virginicus*, *Remirea maritima*, *Schizachyrium pulchellum*, *Canavalia rosea*, *Ipomoea brasiliensis* and *Cyperus maritimus*. There are also grasslands of *Imperata cylindrica* and *Sporobolus pyramidalis* on degraded soils and of *Cyperus articulatus*, *Paspalum distichum*, *Leptochloa caerulescens*, *Eleocharis mutata* and *Eleocharis dulcis* in flooded depressions. *Typha australis*, *Echinochloa pyramidalis*, *Oryza longistaminata*, etc., occupy areas with permanent flooding.

12.2.2.6   Forest Plantations

Since the German colonization, large-scale plantations and reforestation have been undertaken in Togo. More than 200 exotic and local species (*Tectona grandis, Erythrophleum suaveolens, Khaya grandifoliola, K. senegalensis, Eucalyptus* spp., *Terminalia superba*) have been planted with the help of international partners. It is now estimated at about 50,000 ha.

12.2.3   *Hotspots Biodiversity in Togo*

12.2.3.1   Protected Areas

Since the colonial period, considerable efforts have been made to conserve biodiversity through the creation of protected areas. Between the year 1939 and 1957, 9% of the national territory was transformed into protected areas. Today, Togo has theoretically 83 protected areas covering national parks, classified forests and wildlife reserves, i.e., 14% of the national area. The most important protected areas for biodiversity conservation are indicated in Table 12.2 and Figure 12.4.

12.2.3.2   Hotspot Out of Protected Areas

The forests of the subhumid mountainous zone of Togo (ecological zone 4) are among the places where plant diversity is particularly high in the sub-Saharan Africa (La Ferla et al., 2001) and where endemism is very pronounced for West Africa (Beentje et al., 1994; Wieringa and Porter, 2004). These wet forests are part of the West African forest block, the most hotspot threatened in the world. Indeed, the work carried out in this zone indicates a floristic richness comparable to that of certain forests considered rich in West Africa. Thus, 928 vascular species, divided into 521 genera, and 115 families, were identified (Adjossou, 2009). This work mentioned that 72 new species were harvested. The results also indicate that 57% of the species can be considered as rare.

There are also other ecosystems throughout the country that are subservient to mountain areas, especially mountains in the northern zone (Ecological zone 2) where biodiversity is high and better conserved. At the level of the granito-gneissic peneplain (zone 3), characterized by a pattern of low-lying hills, there are inselbergs and rows of isolated hills (Meliendo, Haïto and Agou mountains) which appear to be important refuges for biodiversity. On some of the inselbergs and outcrops (e.g., inselberg rocks of Kamina),

Table 12.2 Some Priority Protected Areas for Biodiversity Conservation

Protected area	Area and location	Ecological zone	Date of Designation
Malfakassa-FazaoNational Park	162,000 ha / Between 8°20' and 9°30' latitude North; 0°35' and 1°02' longitude East	2	Act No. 425/51/EF of 15 April 1951
Fosse aux Lions National Park	1,650 ha / Between 10°46' and 10°49' latitude North; 0°11' and 0°14' longitude East	1	Act No. 489–54/EF of 30/05/1954
Oti- Mandouri fauna Reserve	147,840 ha / Between10°18 and 11°00 North; 0°24 and 0°30 East	1	Decree promulgated in 1954
Missahoé □	1,450 ha / Between 6°54 and 7°55 North; 0°34 and 0°38 East	4	Act No. 185–53/EF of 17 Marsh 1953
Galangachi Fauna Reserve	7,650 ha / Between 10°19' and 10°28' latitude North, 0°14" and 0°27' longitude East	1	Act No. 865 of 14–09–1954
Djamdè Fauna Reserve	1,650 ha. / Between 9°28'et 9°33' latitude North, 0°58' et 1°11' longitude East	2	
Alédjo Fauna Reserve	785 ha / Between 9°11' and 9°17' latitude North; 1° and 1°24' longitude East	2	
Abdoulaye Fauna Reserve	30,000 ha / Between 8°22' and 9°05' latitude North; 1°15' and 1°32' longitude East	3	Decree no. 391–51/EF of 7 June 1951
Togodo Fauna Reserve	30,000 ha / Between 6°50' and 7° latitude North, 1°23' and 1°34' longitude East	5, 3	Act no 174/EF of 26 February 1954 Arrete no. 354/EF of 04/07/1952

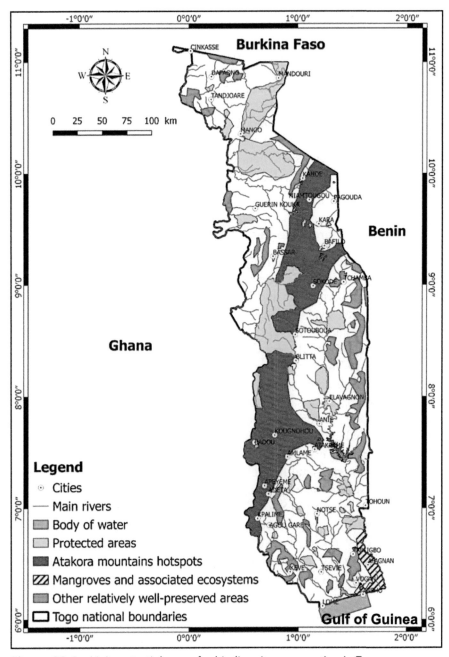

Figure 12.4 Main potential areas for biodiversity conservation in Togo.

various rare Cyperaceae, liana Apocynaceae, *Adansonia digitata, Cyanotis lanata, Chlorophytum* sp. are found.

In southern Togo, the Togodo Wildlife Reserve appears as a regional hotspot of the Dahomey Gap. Indeed, some species   A⊡ endemic in the area, notably *Balanites wilsoniana, Schrebera arborea* and *Strychnos usambarensis*. These species have not yet been reported in any other forest in the country (Adjonou et al., 2010). They are also characterized by a high degree of endangered species whose forest formations in Togodo constitute the last shelters in Togo. Other species listed in the Reserve are also listed on the IUCN Red List as Rare, Critically Endangered or Vulnerable (IUCN, 2001).

12.2.3.3 Biosphere Reserves

The Mono biosphere Reserve (Figure 12.5) is part of a transboundary biosphere reserve currently being established in the lower basin of the Mono River, across southern Benin and Togo. It contains a diversity of ecosystems (forests, savannas, swampy meadows, mangroves, rivers) and animal species that are unique to the region. This diversity is greatly threatened by unsustainable agricultural practices developed by people, as well as poaching and logging. The Mono Transboundary Biosphere Reserve (M-TBR), includes, at the Togolese side, Togodo Protected Areas Complex, Lake Togo, Gbaga Channel, Akissa sacred forest, Hippopotamous ponds Complex of Afito and the sacred forest of Godjé-Godjin. All these sites are of local, national, regional and international interest for the conservation of biological diversity, with a diversity of animal species whose conservation status are subject of international treaties and agreements such as the Convention on Migratory Species of Wild Animals (CMS) or the Convention on International Trade in Endangered Species of Wild Fauna and Flora (CITES). They are also on the IUCN Red List and UNESCO MAB program. Species of particular interest for biodiversity conservation purposes are endemic species, IUCN threatened species, mammals and reptiles (turtles, crocodiles and pythons) currently present on the sites.

12.2.3.4 Ramsar Sites

Togo has currently four sites registered on the List of Wetlands of International Importance (Ramsar Sites), with a total area of 1,210,400 ha:

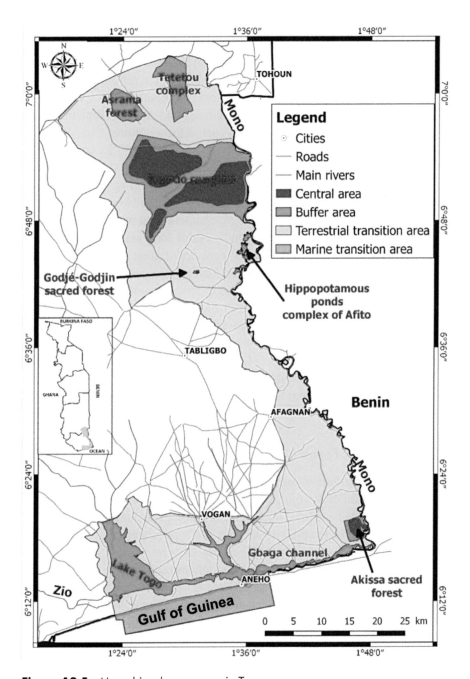

Figure 12.5 Mono biosphere reserve in Togo.

- The Oti-Mandouri watershed (425,000 ha, 10°37'N, 000° 38'E), including the Oti River and its tributaries, marshes, gallery forests, bushes, and savannas. It shelters 27 species of mammals, 37 known species of   as well as crustaceans, molluscs, birds and reptiles. In addition to their crucial conservation role, some vulnerable species, such as hippopotamus (*Hippopotamus amphibius*) and African elephant (*Loxodonta africana*), are an important part of local culture;
- Keran National Park (163,400 ha; 10° 15'N 001° 00'E: contains savanna, gallery forests, wetland habitats of varying sizes, and plant communities, which support local and migratory birds;
- Wetlands of the Coast of Togo: 591,000 ha; 06° 34'N 001° 25'E. Comprising the entire coastline of Togo, characterized by mangroves, rivers, lakes, lagoons, marshes, ponds, and a long sandy beach. These different ecosystems of the littoral zone are of great natural biological, ecological and economic value and host a wide variety of bird, mammal, reptile,   mollusk, crustacean species and many endangered species. This area contributes over 85% of the total annual   A duction in Togo and is also important for transportation of people and goods.
- Reserve of fauna of Togodo (31,000 ha, 06° 50'N 001° 40'E), contains deciduous and semi-deciduous forest studded with ponds and swamps used as a stopover point for migratory birds and provides ideal habitat for waders and other aquatic birds.

12.2.3.5 Sacred Forests

Sacred forests constitute, in areas with high agricultural density, real refuges for the conservation of the biodiversity. Throughout the territory, more than 2,500 sacred forests have been counted in all ecological zones of Togo. They are sometimes extremely small in size, less than one hectare. Their important biological diversity demonstrates the ability of communities to participate in the biodiversity management, conservation and sustainable development through their traditional knowledge. These areas often constitute relics of original Guinean forests or forest islands created by the populations for religious purposes (Voodoo or ancestors cult) (Kokou et al., 1999, 2005).

12.2.3.6 Community Forests

The current legal framework in Togo supports the development of community forests. In this perspective, community forests are increasingly being created as one of the credible alternatives to preserve and restore degraded areas and to conserve biodiversity.

12.3 Floral Diversity in Togo

The species diversity of flora of Togo is not exhaustive, due to the inadequacy of in-depth studies on taxonomic groups, which are of great importance in maintaining and development of ecosystems.

12.3.1 Algae

12.3.1.1 Microalgae

The microalga of Togo belongs to 7 phylums, which are respectively Chromophyta (39%), Chlorophyta (32%), Cyanophyta (17%), Euglenophyta (11%), Rodophyta (1%). Nineteen classes are made up of the whole microalgae in Togo but the most important in terms of number of species are the Bacillariophyceae (26%), Cyanophyceae (17%), Chorophyceae (16%), Conjugatophyceae (12%), Euglenophyceae (11%), Dinophyceae, Fragilariophyceae and Trebouxiophyceae (3%), Xanthophyceae (2%), Chrysophyceae, Cryptophyceae, Synurophyceae and Ulvophyceae contain respectively 1% of the algae flora (Issifou et al., 2014).

Microalgae belong to 134 families of which the most represented are the Desmidiaceae, Euglenaceae, Scenedesmaceae, Naviculaceae, Oscillatoriaceae, Bacillariaceae, Phacaceae, Nostocaceae, Fragilariaceae, Pinnulariaceae, Hydrodictyaceae, Eunotiaceae, Gomphosphariaceae, Chlamydomonadaceae, Oocystaceae, Merismopediaceae, Diplopsliaceae, Selenastraceae, Stephanodiscaceae. These families cover 74% of algae plant community in Togo. Currently, 795 species of microalgae have been recorded in Togo including 83% species totally   only by the genera. The most represented genera (among 282 genera) are *Navivula, Nitzschia, Scenedesmus, Trachelomonas, Closterium, Cosmarium, Oscillatoria, Phacus, Pinnularia, Staurastrum, Strombomonas, Lyngbya* that cover 31% of microalga of Togo (Figure 12.6).

12.3.1.2 Macroalgae Richness in Togo

In total 37 taxa were collected and 27 with them were identified to the genera or to the species (Figure 12.7). Three phylums of macroalgae notably

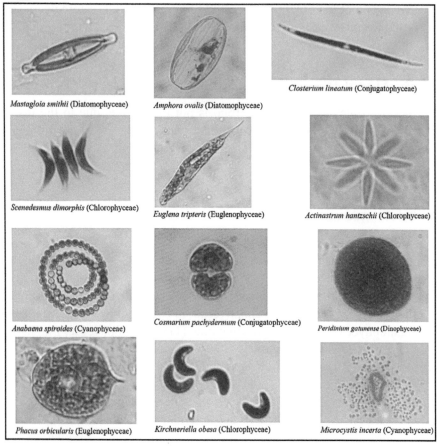

Figure 12.6   Some specimen of microalgae in Togo (Photos Atanley).

the Chlorophyta, Phaeophyta and Rhodophyta are the best represented. The Chlorophyta (green algae) dominates with 9 species.

12.3.2   Fungi

Recent research has made it possible to describe 190 species. Some species have particular habitats, as is the case with Macromycetes, notably Russulaceae (*Russula* spp., *Lactarius* spp.), Boletaceae (*Boletus* spp., *Afroboletus* sp., etc.), Cantharellaceae, Amanitaceae, etc. They are confined to riparian forests or woodlands. Saprophytes are more abundant in the semi-deciduous forest zone where litter is particularly important. Termitomyces are infected with termite mites and are found everywhere.

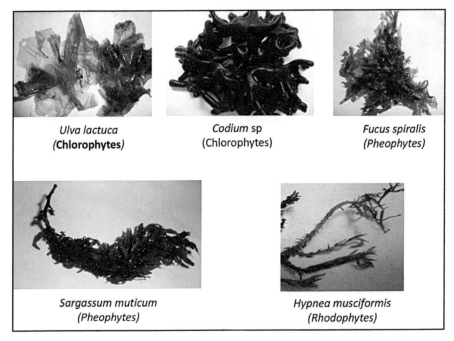

Ulva lactuca
(Chlorophytes)

Codium sp
(Chlorophytes)

Fucus spiralis
(Pheophytes)

Sargassum muticum
(Pheophytes)

Hypnea musciformis
(Rhodophytes)

Figure 12.7 Some specimen of macroalgae of Togo (Photos Lawson).

12.3.3 Bryophytes

The three groups of Bryophytes are all represented in Togo: Mosses with 88 species, Hepatic and Anthocerotes with 44 species. A total of 132 species were identified (Abalo-Loko *in press*).

12.3.4 Pteridophytes

A total of 115 species are known in Togo, 13 of them have been introduced mainly for horticultural purposes (Figures 12.8–12.13). The most represented families are Pteridaceae and Aspleniaceae (19 species each). Next, are the Polypodiaceae and the Selaginellaceae with 13 and 10 species, respectively. In terms of distribution, apart from *Doryopteris nicklesii, Isoetes melanotheca, Ophioglossum gramineum* and *Ophioglossum rubellum*, which are exclusively represented in the Guinean savannas of the ecological zone 3 and *Anemia sessilis* of the rocks of the northern mountains, almost all species belong essentially to the ecological zone 4 (Abotsi et al., 2015).

Figure 12.8 *Asplenium africanum* (Photo Abotsi).

Figure 12.9 *Davallia chaerophylloides* (Photo Abotsi).

Figure 12.10 *Platycerium stemaria* (Photo Abotsi).

Figure 12.11 *Ophioglossum rubellum* (Photo Abotsi).

Figure 12.12 *Adiantum schweinfurthii* (Photo Abotsi).

Figure 12.13 *Pteris tripartita* (Photo Abotsi).

12.3.5 Gymnosperms

There are 13 Gymnosperms. One species is spontaneous, *Encephalartos barteri* in the savannas of east-central Togo (Figure 12.14). The others are introduced for their horticultural qualities.

12.3.6 Angiosperms

12.3.7.1 Systematic Groups of the Flora of Togo

The angiosperm flora of Togo has 163 families distributed as follows (Table 12.3): 125 dicotyledons; 38 monocotyledons; 1,133 genera; and 2,922 taxa.

Figure 12.14 *Encephalartos barteri* (Photo Abotsi).

Table 12.3 Richness of the Flora of Togo

Taxonomic Group	Families	Genera	Species
Dicotyledons	125	872	2,151
Monocotyledons	38	261	771
Total	163	1,133	2,922

12.3.7.2   Life Form and Biogeography of the Species

Plants species in Togo are dominated by two biological forms, which account for about 80%. These include shrubs (43%) and grasses (37%). These two groups are followed by the climbers (15%) and trees (5%). These results clearly show that the flora of Togo is dominated by a high proportion of woody trees, indicating that a significant proportion of plant formations are woodland, wooded and shrub savannas. The low proportion of megaphanerophytes (only 1.30%) is related to the presence of relics of moist and semi-deciduous dense forests in ecological zone 4.

Two major chorological subdivisions dominate the   of Togo and represent about 62.22% of the total species. These species are Guineo-Congolese endemic species (GC, 31.36%), species belonging to the forest massif at the west of the Niger Delta (GCW, 1.14%) and species belonging to the forest massif in eastern Niger Delta (GCE, 0.43%). Transitional species, between the Guineo-Congolese and Sudano-Zambezian regions (GC-SZ) represent 29.29%, and Sudano-Zambezian endemic species (SZ) occupy 21.44% and introduced species are 15.80%, while other categories (Afrotropical, Sudano-Sahelian, Afro-American, Paleotropical) are between 1.14% and 0.003%.

12.4   Diversity in the Usage of the Flora of Togo

12.4.1   Timber

Considering the categorization of timber in Togo, there is no true organized logging activity that can provide information on the value of species. An attempt to classify them into three categories was made, depending on the quality and value of the indicated wood (Akpoto et al., 2015; Adjossou, 2009). The three categories are:

- **Category 1:** Most Popular and Traditionally Exploited Woods including more than 30 species of high value: *Afzelia africana, Detarium senegalense, Distemonanthus benthamianus, Entandrophragma angolense, Erythrophleum suaveolens, Khaya grandifoliola, Mansonia altissimo, Milicia excelsa, Mitragyna stipulosa, Nauclea diderrichii, Piptadeniastrum africanum, Pterocarpus erinaceus, Tectona grandis, Terminalia ivoriensis, Terminalia superba, Triplochiton scleroxylon*, etc.
- **Category 2:** Timber exploited as timber of Category 1 becomes scarce: *Albizia ferruginea, Albizia zygia, Alstonia boonei, Amphimas*

pterocarpoides, Antiaris africana, Berlinia grandiflora, Bombax bu-onopozense, Cola gigantea, Diospyros mespiliformis, Holoptelea grandis, Pterocarpus mildbraedii, Parinari excelsa, Pentadesma bu-tyracea, Prosopis africana, etc.

- **Category 3:** Species exploited by default: *Acacia polyacantha* subsp. *campylacantha, Acacia sieberiana, Anogeissus leiocarpus, Azadirachta indica, Burkea africana, Cassia siamea, Dacryodes klaineana, Detarium microcarpum, Ficus mucuso, Isoberlinia doka, Isoberlinia tomentosa, Lonchocarpus sericeus, Lophira lanceolata, Parkia biglobosa, Terminalia macroptera, Uapaca heudelotii, Terminalia macroptera, Vitex doniana,* etc.

12.4.2 Fuel Wood

The Population of Togo depends mainly on fuelwood energy for various purposes, such as charcoal, firewood, and others. According to Kokou et al. (2009), the species used by fuelwood producers can be classified into two groups:

- **Category 1:** Most Popular and Traditionally Exploited Woods: *Afzelia africana, Anogeissus leiocarpus, Burkea africana, Erythrophleum suaveolens, Lophira lanceolata, Prosopis africana, Terminalia glaucescens, Terminalia laxiflora, Vitellaria paradoxa,* etc.
- **Category 2:** Species exploited by default: *Albizia* spp., *Bridelia ferruginea, Cola gigantea, Daniellia oliveri, Diospyros mespiliformis, Faurea speciosa, Hymenocardia acida, Parkia biglobosa, Parinari curatellifolia, Pericopsis laxiflora, Pseudocedrela kotschyi, Vitex doniana, Tectona grandis,* etc.

12.4.3 Food Plants and Ecosystem Services

Many plant species are used daily for food by Togolese. A total of 101 spontaneous species of flora of Togo have been identified, including many wild or semidomesticated fruit trees such as *Tamarindus indica, Dialium guineense, Saba* spp., *Landolphia* spp., *Adansonia digitata, Monodora myristica, M. tenuifolia,* etc. Apart from some imported spices (*Caryophyllus aromaticus, Pimpinella anisum*), 60% of the species are regularly harvested from local vegetation. Some of these plants are part of the international trade channels, such as *Detarium senegalense, Irvingia gabonensis, Irvingia robur, Parkia biglobosa, Tamarindus indica,* etc. Three spices most used in Togo are seeds of *Monodora myristica*, fruits of *Piper guineense* and of *Xylopia aethiopica*. These spices have very high commercial values in Togo. The seeds of *Vitellaria paradoxa* and *Pentadesma butyracea* are used in the preparation of highly valued oil. In addition, Togolese use some plant organs, notably stems and roots of several species as toothpicks (*Garcinia* spp. especially) and sponges for the oral and bodily cleaning. *Griffonia simplicifolia* is exploited massively like marketable fodder plant. The forests of Togo contain also more than 20 wild yam species that represent an important basis for the improvement of new domestic yams (Hladik and Dounia, 1996). They also offer several other nontimber forest products as various marketable wild fruits (e.g., *Dialium guineense*), comestible mushrooms, honey, meat, fish, shellfish and important tourist potentialities as cascades and pedestrian trails.

12.4.4 Medicinal Plants

Togolese flora is widely used in traditional medicine. Many medicinal plants are traded nationally and internationally for income. Thus the organs (leaves, roots, barks, stems, flowers, fruits, spathes, bulbs, rhizomes and foliar sheaths) of several plants are sold on the stalls in the markets. The diseases or symptoms of diseases for which these plants are often cited are malaria, anemia, headache, agalactia, abscesses, Jaundice, snake bites, eye diseases, oral diseases, hepatobiliary disorders and naso-pharyngeal conditions. The best-selling plants are also those supposed to be aphrodisiacs.

12.5   Conservation Status of the Flora of Togo

12.5.1 IUCN Status

The IUCN threat categories and assessment criteria (2001) were used to determine the conservation status of the species. According to their status, 90% of the species present in Togo have never been evaluated. Data are nonexistent for 2.4% of species. In addition, the results show that 1.2% are vulnerable. These are species that may be in danger in the future if current pressure on ecosystems is maintained. These are species whose populations are continuously decreasing due to various forms of exploitation, mass destruction of habitats or other environmental phenomena such as climate change. They are also species whose populations have been seriously decimated and which currently do not benefit from any protection measures, for example, *Afzelia africana, Cordia platythyrsa, Milicia exelsa, Nauclea diderrichii, Nesogordenia papaverifera, Pterygota macrocarpa.*

The status of some species is alarming and urgent measures should be taken for their full protection. These species account for 0.4% of the are currently in danger and are threatened to disappear ⬜A ritory. Among these species, we can cite *Aldrovanda vesiculosa, Diospyros ferrea, Pavetta lindina, Pavetta mollis, Parinari macrophylla, Placodiscus attenuata, Conocarpus erectus, Strychnos usambarensis* and *Sphenocentrum jollyanum.*

12.5.2 New, Rare, and Threatened Species

More than 50 species were recently collected and could be potentially considered as new species for the flora of Togo. Other taxa are represented by isolated specimens or present in a very reduced ecological area and can be considered also like threatened species. They are *Dacryodes klaineana, Garcinia afzelii, Garcinia kola, Blighia welwitschii, Klainedoxa gabonensis, Ancytrophylum secundiflorum, Cyathea camerouniana,* etc. Currently, some timbers, which are highly exploited could also be classified as threatened: *Afzelia africana, Aningeria altissima, Entandrophragma cylindricum, Aubrevillea kerstingii, Khaya anthotheca, Mansonia altissima, Milicia excelsa, Mitragyna stipulosa, Pterocarpus erinaceaus,* and *P. mildbraedii.* Then 55 species are recognized as exclusive of riparian forests and are potentially threatened because of the current pressure on the riparian forests.

12.5.3 Endemicity

Phyllanthus rouxii, is the only endemic plant reported in Togo. It is described as a hemicryptophyte with a tuberous taproot, what is exceptional for genus *Phyllanthus* L., in West Africa (Brunel, 1987).

12.6 Faunal Resource

Despite its smaller size and being essentially a savanna country, due to its geographical location in the Dahomey Gap, Togo is an area of contact between the western Guinean forest fauna belonging to the block, which stretches from Guinea Savanna to the West of the Volta river and forest elements from Central Africa, stretches from the Democratic Republic of Congo to the East of Togo.

The fauna of Togo is still poorly known although surveys and researches began with German colonization since the nineteenth century. Indeed, several groups of invertebrates are unknown because they have been subjects of very few methodical and specialized surveys. Therefore, except Insects, this synthesis will only focus on the so-called higher taxonomic groups (Fishes, Amphibians, Reptiles and Mammals). In addition to several hundred recorded or unrecorded invertebrates believed to exist in the country, Togo is habitat for at least 98 species of   A⬚
cies of amphibians, 167 species of reptiles, at least 624 species of birds (both passerine and nonpasserine) and about 1,110 species of mammals.

12.6.1 Faunal Diversity

12.6.1.1 Insects

In Togo, the entomofauna is composed of highly diversified aquatic and terrestrial insect species. However, very few representatives of this class are known. To date, 1,420 species of insects belonging to 836 genera, 181 families and 16 orders have been inventoried (Table 12.4). Orders of Coleoptera (351 species), Hymenoptera (324 species), Lepidoptera (209 species), Diptera (167 species), Hemiptera (141 species) and Orthoptera (135 species) are the richest in species, whereas the Neuroptera, Anoploura, Mallophaga, etc., are species, which are poorly represented.

12.6.1.2 Inland Freshwater Fish

Despite the smaller sizes of the watersheds, Togo has a rich fish habitat (Daget, 1950; Paugy and Benech, 1989; Amegavie et al., 1992; Levèque et al., 1992). In total 20 orders, 44 families, 61 genera and 98 species have been inventoried. By way of comparison, a total of 586 species belonging to 192 genera have been identified throughout the West African region (Laleye and Entsua-Mensah, 2009). Among the fishes recorded in the Togolese continental environments, there are 68 strictly freshwater species (Myers, 1949).

The 5 most representative   A⬚
species richness (37 species). These are Mormyridae, Alestidae, Cyprinidae, Clariidae and Cichlidae.   ⬚ A
species in Togo, are numerous; they include Lepidosirenidae, Protopteridae, Megalopidae, Osteoglossidae, Pantodontidae, Gymnarchidae, Hepsetidae, Distichodontidae, Malapteruridae, Poeciliidae, Channidae and Centropomidae, i.e., 27% of the families.

Table 12.4   Number of Species, Genera, and Families of the Different Orders of
Insects Recorded in Togo

	Number of taxa described in Togo		
Order	**Family**	**Genera**	**Species**
Anoplura	1	2	3
Coleoptera	45	165	351
Dermaptera	2	2	2
Dictyoptera	5	9	12
Diptera	25	83	167
Hemiptera	18	91	141
Hymenoptera	31	192	324
Isoptera	2	35	50
Lepidoptera	30	138	209
Mallophaga	2	2	2
Neuroptera	5	7	7
Orthoptera	10	96	135
Psocoptera	1	1	1
Siphonaptera	2	4	7
Thysanoptera	1	8	8
Thysanura	1	1	1
Total = 16 orders	181	836	1420

12.6.1.3   Marine Fish

Bonyfish often caught on the Togolese coasts are mainly anchovies (Engrau-
lidae), moths (Ariidae), sardinella (Clupeidae), groupers (Serranidae), jacks
and horse mackerels (Carangidae), sea carp (Haemulidae) Otoliths (Sciae-
nidae), ethmaloses (Ethmalosa), mules (Mugil), true soles (Soleidae, Cyno-
glossidae), captains (Polynemidae). In the Elasmobranch taxon, sharks with
a pointed snout are very common (Carcharhinidae), Hammerhead sharks
(Sphyrnidae), guitar rays (Rhynchobatidae and Rhinobatidae), stingrays
(Dasyatidae) and torpedoes (Torpedinidae). However, a more careful observation reveals that the catches are very diverse. In fact, nearly 350 species of
marine fish have been identified in 89 families.

In terms of geographical distribution, the composition of the ichthyo-
fauna of the different basins of Togo shows a very low rate of endemism
(2 species out of the 98). Indeed, the Togo Lampeye (*Aplocheilichthys bra-*

cheti) has hitherto been listed only in a few small rivers in the central and coastal regions of Togo while the Keilhack's Lampeye (*Micropanchax keilhacki*) is known only from two localities in Kara and Kante in Northern Togo. Three other species (*Epiplatystogolensis, Foerschichthys flavipinnis* and Aphyosemion (*Chromaphyosemion bitaeniatum*) in the Mono and Zio drainage systems have a restricted range (East Ghana to southeast Nigeria). All other taxa have a wide distribution, including the majority of the Sahelo-Sudanian basins.

12.6.1.4 Amphibians

Sixty species of amphibians are currently known in Togo (Segniagbeto et al., 2007; 2013). The group of Anurans with 59 species is most represented with major families such as Phrynobratrachidae (7 species), Ptychadenidae (7 species) and Hyperoliidae (16 species). Obviously, the inventory of the fauna is not yet finished; there are about 15 Amphibian species present in the West African region whose presence in Togo is likely possible.

12.6.1.5 Reptiles

The number of species currently known is 167 species including 92 species of Ophidians, 43 species of Saurians, 14 species of Chelonians and 3 species Crocodilian species (Segniagbeto et al., 2011, 2014a, 2015a). For the Ophidian group, the presence of other 8 known species in the West African subregion is possible. It is the same for the group of Saurians whose presence of 7 other species is probable. Due to different ecosystems that are spread over the different ecological zones, there are forests, savanna and marine forms, but also arboreal and burrowing forms. Among the snake species there are poisonous and the one that creates the most envenomation is *Echis ocellatus* because of its wide distribution, also its relatively high frequency in almost all the ecological zones of the country and its phenomenon of homocrhomism.

Five of the six internationally known marine turtles are present in Togo (Segniagbeto et al., 2013, 2014a, 2015b, 2016, 2017). Among the   [A cies present, two spawn on beaches in Togo: *Lepidochelys olivacea* and *Dermochelys coriacea*. The other three (*Caretta caretta, Chelonia mydas* and *Eretmochelys imbricata*) are reported only by incidental catches in the nets of inshore   although traces of female green turtle (*C. mydas*) are occasionally recorded on the beaches. An average of 1,000 to 1,200 marine turtles is recorded annually in Togo (Segniagbeto et al., 2016, 2017). The most common species are the olive ridley turtle and the green turtle. Both represent more than 90% of the total number of individuals, followed by the leatherback, loggerhead, and hawksbill turtles.

12.6.1.6 Birds

Some 624 species of birds have been identified in Togo (Fry et al., 2000). They belong to 292 genera, 58 families and 22 orders. Among these species, 408 are known or suspected to be breeding residents, 109 are Palearctic migrants, 80 are intra-African migrants. The most important families are the Accipitridae (46 species), the Sylviidae, the Estrildidae (27 species), the Scolopacidae and the Ploceidae (24 species), the Estrildidae (12 species). Palearctic migrants have been recorded mainly in Anatidae, Apodidae, Ardeidae, Charadriidae, Laridae, and Scolopacidae.

12.6.1.7 Mammals

The nonmarine mammalian fauna of Africa is represented by 1,110 species belonging to 294 genera, 57 families and 16 orders (Kingdom et al., 2013). Among these, 178 species belonging to 108 genera, 38 families and 13 orders are found in Togo. In addition, 40 species of Cetacea are documented accounting to more than one-fourth of the world's marine mammalian fauna. Only elephant shrews or jumping shrews, golden moles and odd-toed ungulate are not recorded in Togo. The main groups in Togo are:

- **Sirenians:** They are represented in Togo by a single species: the manatee of West Africa (*Trichechus senegalensis*). The main zones of observation of the species in Togo are in Lake Togo with two zones of predilection: the Zio river junction with the lake and the Haho river junction with the lake, along the Gbaga channel and in the Mono;
- **Bats:** This group is represented by 52 species belonging to 24 genera and 9 families (Amori et al., 2016*).* The most represented families are Vespertilionidae (15 species), Pteropodidae (10 species), Nycteridae (7 species), Hipposideridae (6 species) and Molossidae (6 species). Among all the species of Bats in Togo, the most characteristic is the yellow fox (*Heidolon helvum*) with colonies distributed in many localities in Togo. Apart from this species, there are also colonies of *Micropteropus pusillus, Epomophorus gambianus, Nanonycteris veldkampii,* etc. which are spread across many parts of the country;

- **Carnivora:** A checklist from Amori et al. (2016) revealed 24 small carnivore species of six taxonomic families in Togo. This includes: 2 species of Canidae (*Canis adustus* and *Lycaon picus*), 5 species of Felidae (*Acinonys jubatus, Caracal aurata, C. caracal, Felis silvestris* and *Leptailurus serval*), 8 species of Herpestidae *(Atilas paludinosus, Crossarchus obscurus, Galerella sanguinea, Herpestes ichneumon, Ichneumia albicauda, Mungos gambianus, M. mungo* and *Crocuta crocuta)*. The others are *Aonys capensis, Hydrictis maculicollis, Ictonyx striatus, Mellivora capensis* (Mustelidae), *Nandinia binotata* (Nandinidae). The rests are genets (*Genetta genetta, G. maculatan, G. thierryi,* and *Civettictis civetta)*. The two large carnivores reported belong to the same genera: *Panthera pardus* and *P. leo*.
- **Ungulates:** There are two groups, or orders, of ungulates, the odd-toed ungulates (Perissodactyla) and the even-toed ungulates (Artiodactyla), but only the latter is native of Togo. An updated systematic checklist of 23 ungulates species under 16 genera and 4 families is done (Amori et al., 2016). They include: pigs (*Hylochoerus meinertzhageni, Phacochoerus africanus* and *Potamochoerus porcus)*, hippopotamuses (*Hippopotamus amphibius*), and antelopes and related forms. Genus in Bovidae are *Alcelaphus* (1 species), *Cephalophus* (4 species), *Damaliscus* (1 species), *Eudorcas* (1 species), *Hippotragus* (1 species), *Kobus* (2 species), *Ourebia* (1 species), *Philantomba* (1 species), *Sylvicapra* (1 species), *Syncerus* (1 species), *Taurotragus* (1 species), and *Tragelaphus* (3 species);
- **Cetaceans:** To date, 16 species of cetaceans have been recorded in Togo (Segniagbeto et al., 2014b, Van Waerebeek et al., 2017). Among the species of marine mammals to frequent the Togolese coast, the most remarkable are: *Megaptera novaeangliae, Balaenoptera bonaerensis, Balaenoptera brydei, Physeter macrocephalus, Globicephala macrorhynchus, Orcinus orca*, etc.
- **Rodentia:** Amori et al. (2016) published the most recent checklist of rodent species of Togo. Data on 47 valid species obtained throughout the country are presented under 28 genera belonging to 7 families, including scaly tailed squirrels (Anomaluridae), African mole rat (Bathyergidae) woodland dormouse and glirids (Gliridae), porcupine (Hystricidae) and trues rats and mices (Muridae). The rests are African pouched rats and northwestern fat mouse (Nesomyidae), squirreld (Sciuridae) and cane rat (Thryonomyidae). The Togo mouse (*Leimacomys buettneri*), also known as Büttner's African forest mouse or the

groove-toothed mouse is a unique muroid rodent known from only two specimens taken from near the type locality of Bismarcburg, near Yege village, Togo, in 1890 (Dieterlen, 1976). Its genus is monotypic. This endemic species is probably extinct.

- **Primates:** To date, 13 species of primates have been recorded in Togo (Amori et al., 2016, Decher et al., 2016). These include *Pan troglodytesverus, Colobus vellerosus, Procolobus verus, Cercopithecus petaurista petaurista, Cercopithecus erythrogaster erythrogaster, Cercopithecus mona, Papio anubis, Erythrocebus patas, Chlorocebus tantalus, Galago senegalensis, Galagoides demidovii, Galagoides thomasi,* and *Perodicticus potto juju.*

12.6.2 Conservation Status of the Fauna of Togo

Figure 12.15 illustrates some species of emblematic insects of the forests of Togo. Unfortunately, the biotope of these emblematic species is eroded from day to day, which threatens their survival. The emblematic species are also hunted mainly in forest areas of Togo and sold in Europe thus constituting a second menace.

Among the Amphibian species of Togo, there are three that are endemic to the forest zone: *Conraua derooï, Hyperolius torrentis* and *Hyperolius baumanni.* The current status of two of these three species is critical (*Conraua derooï* (CR, Figure 12.16), *Hyperolius torrentis* (EN)). Some species of Amphibians are currently traded internationally: *Kassina senegalensis, Kassina cassinoides,* and *Phrynomantis microps.* Since 2015, the population

Figure 12.15 Example of some emblematic insect species hunted in Togo [from left to the right: *Chelorrhina polyphemus* (Coleoptera: Scarabaeidae), *Papilionireus nireus* (Lepidoptera: Papilionidae), *Imbrasiadione* (Lepidoptera: Saturniidae)] (Photos Amévoin).

Figure 12.16 *Conraua derooi,* endemic to semi-deciduous forests area (Photo
Segniagbeto).

of Conraua derooï were exploited for trade but was suspended in 2016 due
to the conservation status of the species.

The status of many rare birds associated with forests degradation in ecological zone 4, riparian forest and woodland is very alarming. Among these
birds, many are threatened (Table 12.5).

Many species of marine animals are threatened such us sharks and rays:
Rhynchobatus lübberti (EN), *Rhinobatos albomaculatus* (VU), *Glaucoste-
gus cemiculus* (EN), *Leptocharias smithii* (NT), *Carcharhinus brevipinna*
(NT), *Scyliorhinus stellaris* (NT), *Carcharodon carcharias* (VU).

There is practically no endemism in reptiles, except for the newly
described species such as *Hemidactylus kyaboboensis* and *Agama parafri-
cana,* whose distribution are very limited. There are two major threats to
populations of reptiles in Togo: habitat degradation and exploitation of populations of different species in international trade. To this end, six species of
turtles are threatened with extinction and are listed on the IUCN Red List.
There are also two of the three crocodile species that are also threatened
(*Mecistops cataphractus* and *Osteolaemus tetraspis*).

Among the species of cetaceans in Togo, the Atlantic humpback dolphin
(*Sousa teuszii*) is the species whose status is currently critical throughout its

Table 12.5 Threatened Birds in Togo

Common name	Scientific name	Status	Habitat
Damara Tern	*Sterna balaenarum*	NT	Migrant to the coast
Cape Gannet	*Morus capensis*	VU	Costal vagrant
Ferruginous Duck	*Aythya nyroca*	NT	Northern savannas
Pallid Harrier	*Circus macrourus*	NT	Savannas
Black-winged Pranticole	*Glareola nordmanni*	NT	Vagrant
Great Snipe	*Gallinago media*	NT	Wetlands in Northern regions
Black Crowned Crane	*Balearica pavonina*	VU	Visitor to the Keran Park
Denham's Bustard	*Neotis denhami*	NT	Northern savannas
African Skimmer	*Rhynchops flavirostris*	NT	Coast and inland rivers
Yellow-casqued Wattled Hornbill	*Ceratogymna elata*	VU	Forest's zone
Rufous-winged Akalat	*Illadopsis rufescens*	NT	Forest's zone
Lagden's Bush-Shrike	*Malaconotus lagdeni*	NT	Forest's in zone IV

range because of its very coastal habitat and interactions with human activi-
☐   ☐   ☐

The current conservation status of most primate species in Togo is very critical (Rome, April 2016 IUCN Status Assessment Workshop on African Primate Species). For the   ☐A
were totally extirpated during the 1970s. Sustainable populations of species such as *Colobus vellerosus* (CR)*, Procolobus verus* (CR), *Cercopithecus petaurista petaurista* (EN) and *Cercopithecus erythrogaster erythrogaster* (CR) are in the protected areas of Fazao Malfakassa, Togodo and the mountain areas of ecological zone 4.

12.7 Main Challenges of Biodiversity Conservation: Causes and Consequences

12.7.1 Challenges Related to Ecosystem Degradation and Species Status

In Togo, the degradation of ecosystems by various anthropogenic pressures (shifting agriculture, logging, livestock and transhumance, bushfires, invasion and colonization of protected areas, chemical pollution by household, industrial and agricultural wastes, organic pollution, poaching of Wildlife trade, overexploitation of fishery resources, etc.) and development projects (dams, mining, etc.) lead to dysfunction and destruction of habitats, and loss of biological diversity. In aquatic ecosystems (lagoons, ponds, sea), various pollutants (water and waste oils, household and industrial wastes, etc.) are constantly being dumped, resulting in degradation and biological diversity.

Mangrove ecosystems are overexploited of their plant and animal resources and the salinity changes induced by the construction of the Nangbéto dam threaten the survival of these ecosystems.

Several plant species are currently threatened with extinction. For example, the only tree of *Diospyros ferrea* (Figure 12.17) was cut while *Parinari*

Figure 12.17 The only tree of *Diospyros ferrea* known to date in Togo (Photo Kokou).

macrophylla has only two representatives on the Togolese coast. *Cyathea camerooniana* is a tree fern in riparian forests of ecological zone 4. This species is currently threatened due to the degradation or destruction of these particular biotopes.

Conservation of a lot of sacred forests is possible because of the religious and protective role of religious priest, even recognized in some villages. But in most of cases, people who play this role belong to an old generation, which is disappearing; the youth do not have the same visions. For example, in a sacred forest in southern part of Togo, the death of the priest led to the sacred forest abandonment by the population, then to its destruction. However, it is in this sacred forest that *Sphenocentrum jollyanum*, a shrubby Menispermaceae never signaled in the   of Togo has been collected some years before (Kokou, 1998). Another forest visited later in which *Pancovia sessiliflora* has been recorded for the   time in Togo, turned into a farm.

Regarding wildlife, the questioning of the boundaries of the protected areas by the population in the early 1990s, following the sociopolitical disturbances that Togo went through, has resulted in massive culling of wildlife in parks and reserves. Thus, poaching has led to the disappearance or scarcity of many species in Togo. Indeed, Togo's fauna has experienced a sharp reduction of several species, especially in large mammals during the last 20 years. Several vertebrate species once common and very abundant in Togo have become very rare or are extirpated because of their overexploitation. Population of elephants has completely disappeared from the Fosses aux Lions National Park. Some species of large predators (*Panthera leo*, *P. pardus*, etc.) have completely disappeared from Togo's protected areas.

The main threat to marine turtles in Togo is the incidental catches at sea, which represents more than 60% of the individuals surveyed annually. These catches provoke the death of the individuals by choking and many corpses of turtles are found on the beaches. Approximately 20–25% of the individuals surveyed are cadavers and this phenomenon concerns mainly the two most frequent species. What is important to note for the case of the green turtle is that individuals of different age classes are regularly recorded during follow-up work both male and female individuals.

The species of herbivorous mammals belonging to the genera *Cephalophus*, *Alcelaphus*, *Kobus*, *Syncerus*, *Hippotragus*, *Tragelaphus*, and bongo (*Tragelaphus eurycerus* drastically reduced their populations or have completely disappeared).

12.7.2   Challenges Associated with the Overexploitation of Plant Resources

The abusive and uncontrolled exploitation of plant resources to satisfy the requirements of woodfuel, timber and service wood is also a factor of erosion of biodiversity. Degradation of plant formations has accelerated in the ecological zone 2, 3 and 4, formerly better conserved than zones 1 and 5. Firewood and charcoal, the main source of domestic energy for 80% of population is now heavier for these areas. Also, the irrational exploitation of timber has worsened in recent years as a result of the timber trade with Asia. The use of the chainsaw has rapidly supplanted the manual saws used until recently and has contributed to the scarcity of valuable forest species such as *Milicia excelsa, Khaya grandifoliola, Khaya senegalensis, Triplochiton scleroxylon.* The technique of artisanal logging and processing using the mobile saws give very low average yields, about 11% (Akpoto et al., 2015). Today, timber from the forest zone (ecological zone 4) has become very scarce and its supply from neighboring countries is becoming more important day by day. These practices are the primary causes of the destruction of wildlife habitats and the erosion of animal and plant diversity.

12.7.3   Challenges Related to Overhunting, Illegal Poaching and International Pet Trade

In Togo, one of the major cases of poaching is for bushmeat, or meat consumed from wildlife such as reptiles, birds and mammals. In rural areas, many people rely on natural resources for their survival and generate cash income through the sale of bushmeat.

A

there is a big international market (mostly illegal) of wild animals and their parts mainly for their esthetic and medicinal value. This is unfortunately the case of rare lizards, tortoises and birds: the hinge-back tortoises (*Kinisyx* spp.), the monitor lizards (*Varanus* sp.), the gray parrot (*Psittacus erithacus*), the red-headed lovebird (*Agapornis pullaria*).

Togo is not spared from the scourge of international ivory   A ☐
in the port of Lomé, Togo, a Vietnamese and two Togolese were arrested for ivory   after the seizure of 1.6 tons of ivory, ready to be shipped to Vietnam.

12.7.4 Challenges of Non-Compliance with Environmental Commitments and Obligations

In Togo, the Decree No° 2006–058/PR establishes the list of works, activities and planning documents submitted to the Environmental Impact Assessment (EIA) and the main rules of this study. Unfortunately the implementation of this decree suffers from shortcomings such as (i) management plan not often implemented, (ii) no environmental monitoring for several projects that have received the certificate of conformity, (iii) failure of environmental ethics, including lack of rigor in the conduct of studies (often no specialists in teams and plagiarism), (iv) evaluation of noncompliant reports, (v) poor environmental governance in the EIA process, etc. Therefore, instead of contributing to the conservation of biodiversity, EIAs are contributing more to the loss of biodiversity (e.g., *Mammea africana*, is a plant species with only one tree identified. It has been eliminated because a high voltage power line cross it habitat); *Conocarpus erectus* (Figure 12.18), is a species, which habitat occupies a narrow space on the beach of Lomé. It may disappear in Togo after the extension of the harbor if the company doesn't make effort to find another similar biotope).

12.7.4 Challenges Related to the Proliferation of Invasive Alien Species

In most of ecosystems, there is a proliferation of invasive alien species, and the most frequently encountered are exotic terrestrial or aquatic species. As mainland species, *Chromolaena odorata* (Asteraceae), a species of fallows especially of the Guinean climate, *Azadirachta indica* (Meliaceae) even become the woody vegetation of certain localities throughout the country, *Leucaena leucocephala* (Mimosaceae), *Titonia diversifolia* (Asteraceae) and *Mimosa invisa* (Fabaceae), etc. In the ecological zone 4 described as one of the hotspot of Togo, the intensification of the teak plantation (*Tectona grandis*) nowadays dangerously threatens the local flora). These species seriously disrupt the development and maintenance of natural ecosystems.

In aquatic environments, the main aquatic species that invade ponds and lagoons in Togo are *Pistia stratioites* (Araceae) and *Eichhornia crassipes* (Ponteriaceae) (Figure 12.19) but also the fern species Azollaceae (*Azolla africana* Desv.) and Salviniaceae (*Salvinia auriculata* Aubl.), resulting in the eutrophication and asphyxiation of all the biological diversity of these ecosystems.

Figure 12.18 *Conocarpus erectus*, threatened species after the extension of Lomé harbor (Photo Kokou).

Figure 12.19 Lagoon invaded by *Eichhornia crassipes* (Photo Kokou).

12.8   Strategies to Safeguard Biodiversity in Togo

Togo being part of the Convention on Biological Diversity is committed to contributing to a significant reduction in the rate of loss of biological diversity in order to reduce poverty and improve the well-being of populations. Although Togo has not developed national indicators for the conservation and sustainable use of biodiversity, the formulation of the National Biodiversity Strategy and Action Plan (NBSAP) based on the 2011–2020 Global Strategic Plan and its Aichi goals, is a great step.

In the NBSAP 2011–2020 of Togo, measures are taken to ensure the conservation of biodiversity are *in situ* and *ex situ* conservation, sustainable use of the components of biological diversity, education and public awareness, impact assessments and information exchange. In addition, with the Belgian Partnership under the Convention on Biological Diversity, Togo has an information exchange center (CHM) to disseminate useful information on Biodiversity.

Other activities under national CBD strategies and action plans have focused on the implementation of REDD+ and the protection of protected areas. The main activities undertaken include the consensual of preexisting protected areas, creation of Biosphere Reserve, national forest inventory, the baseline studies for the implementation of a national system of monitoring, reporting and   sustainable development of agricultural and forestry value chains, in particular the fuelwood subsector.

Togo is member of the Global Biodiversity Information Facility (GBIF), and   from numerous advantages, including support from the GBIF Secretariat and partners such as JSTOR, JRS Biodiversity International for the Mobilization, publication and use of data available on the GBIF website. In addition, Togo is actively participating in the project of the French government on biodiversity "Sud Expert Plante," which enabled the National Herbarium to be renovated, to the number of available herbarium, to share and to introduce several forest species that are threatened with extinction in an arboretum.

12.9   Conclusion

This review demonstrates the richness of the biodiversity of Togo despite the small size of the country. But there is still work to be done to bring this biodiversity up to date. Many ecological zones and ecosystems or habitats are still less prospected.

The exploitation of biodiversity of Togo presents enormous opportunities that several socioprofessional groups use for various purposes; it occupies an important place in the national economy.

Several taxa are currently at high risk for a variety of reasons, including habitat degradation due to agriculture, logging and mining, poaching, invasion of exotic species.

To reverse the situation, it is imperative to bring communication to all levels in order to disseminate information and raise awareness among the general public about the conservation and sustainable use of biological diversity. All the means are useful to reinforce the knowledge and to raise awareness of Togolese towards the   and the fauna. National expertise needs to be developed, which requires a clear improvement in all areas of biodiversity science in order to have a more precise knowledge of biological resources and ensure proper planning of their use. This involves training national senior level biodiversity frameworks and capacity building on negotiations on biological resources.

Finally, it is important to develop regional and international cooperation to address the various challenges posed by the management and use of biodiversity for the well-being of local populations.

Keywords

- conservation status
- challenges
- faunal resource
- floral diversity

References

Abalo-Loko, G. A., Check list of liverworts and hornworts from Togo.

Abotsi, K. E., Radji, R., Dubuisson, J. Y., & Kokou, K., (2015). The Pteridaceae family diversity in Togo, *Biodiversity Data Journal*, *3*, e5078.doi: 10.3897/BDJ.3.e5078.

Adjonou, K., Djiwa, O., Kombaté, Y., Kokutse, A. D., & Kokou, K., (2010). Etude de la dynamique spatiale et structure des forêts denses sèches reliques du Togo: implications pour une gestion durable des aires protégées. *Int. J. Biol. Chem. Sci.*, *4*(1), 168–183.

Adjonou, K., Kokutse, A. D., Kokou, K., (2013). Dynamique spatiale et diversité floristiquede la Réserve de Faune de Togodoau Sud Est du Togo (Afrique de l'Ouest). *Scripta Botanica Belgica*, *50*, 63–72, 201.

Adjossou, K., (2009). Diversité, structure et dynamisme de la végétation dans les fragments de forêt humide du Togo, les enjeux pour la conservation de la biodiversité. *Th Doc Univ Lomé*.

Akpoto, A., Kokutse, A. D., Radji, R., Adjonou, A., & Kokou, K., (2015). Impact of small-scale logging in semi deciduous forest of Togo (West Africa). *J. Biodivers Manage Forestry*, *4*, 1 http://dx.doi.org/10.4172/2327–4417.1000138.

Amegavie, K., Kuwadah, A., & Etou, K., (1992). La faune aquatique: Poissons. In: Salami-Cadoux, M. L., *Evaluation et suivi des répercussions de l'aménagement hydro-électrique de Nangbeto (Togo) sur les ressources halieutiques et sur les maladies parasitaires des populations humaines et animales.* Rapports de fin de projet- Campus-Nangbeto.

Amori, G., Segniagbeto, G. H., Assou, D., Decher, J., Spartaco, G., & Luiselli, L., (2016). Non-marine mammals of Togo (West Africa): An commented annotated checklist. *Zoosystema., 38*(2), 201–244.

Beentje, H. J., Adams, B., & Davis, S. D., (1994). Regional overview. In: Davis, S. D., Heywood, H. A. C., (eds.). *Centres of Plant Diversity.* Cambridge: IUCN Publications Unit., vol. 1, pp. 101–264.

Brunel, J. F., (1987). Sur le genre *Phyllanthus* L. et genres voisins de la Tribu des Phyllanthoidae Dumort. (Euphorbiaceae) en Afrique intertropicale et Madagascar. Thèse ès-Sci. Univ. Strasbourg.

Brunel, J. F., Scholz, H., & Hiepko, P., (1984). *Flore analytique du TOGO.* Phanérogames. GTZ, Eschorn.

Daget, J., (1950). Poissons d'eau douce de la région côtière du Togo et du Dahomey. *Notes AP., 46,* 57–59.

Decher, J., Segniagbeto, G. H., Luiselli, L., (2016). Comparing the mammalian fauna of German colonial Togo at the time of Paul Matschie (1861–1926) working at the Zoological Museum Berlin, with a mammal checklist of the current Togolese Republic. *Mammalian Biology - Zeitschrift für Säugetierkunde, 81,* 3–18.

Dieterlen, F., (1976). Bemerkungen über *Leimacomys buettneri* Matschie 1893.(Dendromurinae, Cricetidae, Rodentia). *Säugetierkundliche Mitteilungen, 24,* 224–228.

Dourma, M., (2008). Les forêts claires à *Isoberlinia doka* Craib & Stapf et *Isoberlinia tomentosa* (Harms) Craib & Stapf (Papilionaceae) en zone soudanienne du Togo: écologie, régénération naturelle et activité humaine. Th. Doct. Sc. Nat., Univ. Lomé, Togo.

Ern, H., (1979). Die Vegetation Togos. Gliederung, Gefährdung, Erhaltung. *Willdenowia, 9,* 151–155.

Etsè, K. E., (2015). Contribution de la télédection et sig à l'étude de la biodiversité et des potentialités écotouristiques des mares de la plaine de l'oti au Togo. Master en Sciences et Technologies de l'Espace, CRASTE-LF, Maroc.

Food and Agriculture Organization (FAO), (2010). *Evaluation des resources forestières mondiales* Rome.

Fry, C. H., Keith, S., & Urban, E. K., (2000). *The Birds of Africa.* London, Academic Press.

Guelly, K. A., (1994). *Les savanes de la zone forestière subhumide du Togo.* Thèse de Doctorat. Univ. Pierre et Marie CURIE, ParisVI.

Hladik, A., & Dounias, E., (1996). Les ignames spontanées des forêts denses africaines. Plante à tubercule comestible», In: Hladik, C. M., Hladik, A., Pagezy, H., Linare, O. L., Koppert, G. J. A., & Froment, A., (eds.). *L'alimentation en forêts tropicales. Interaction bioculturelle et perspective de développement.* Unesco, Paris. pp. 275–294.

Issifou, L., Atanlé, K., Radji, R., Lawson, H. L., Adjonou, K., Edorh, M. T., Kokutsé, A. D., Mensah-Atto, A., & Kokou, K., (2014). Checklist of tropical algae of Togo in the Guinean Gulf of West Africa. *Sientific Resaerch and Essays, 9*(22), 932–958.

IUCN, (2001). *IUCN Red List Categories: Version 3. 1.* Prepared by the IUCN Species Survival Commission. IUCN, Gland, Switzerland and Cambridge, UK.

Kingdom, J., Happold, D., Bytinski, T., Hoffman, M., Happold, M., & Kalina, J., (2013). *Mammals of Africa*, Bloomsbury, London, vol. 1–6.

Kokou, K., & Caballé, G., (2000). Les îlots forestiers de la plaine côtière togolaise. *Bois et Forêts des Tropiques, 263*, 39–51.

Kokou, K., (1998). *Les mosaïques forestières au sud du Togo: Biodiversité, dynamique et activités humaines.* Thèse de Doctorat, Université de Montpellier II, France, pp. 140.

Kokou, K., Adjossou, K., & Hamberger, K., (2005). Les forêts sacrées de l'aire Ouatchi au sud-est du Togo et les contraintes actuelles des modes de gestion locale des resources forestières. *Revue électronique Vertigo, 6*, 3.

Kokou, K., Atato, A., Bellefontaine, R., Kokutse, A. D., & Caballe, G., (2006). Diversité des forêts denses sèches du Togo (Afrique de l'Ouest). *Revue d'Ecologie (Terre et Vie), 361*, 225–246.

Kokou, K., Batawila, K., Akoègninou, A., & Akpagana, K., (2000). Analyse morpho-structurale et diversité floristique des îlots de forêt protégés dans la plaine côtière du sud du Togo. *Etude Veget. Burkina Faso*, 33–48.

Kokou, K., Caballé, G., & Akpagana, K., (1999). Analyse floristique des îlots forestiers du sud du Togo. *Acta bot. Gallica 146*(2), 139–144.

Kokou, K., Nuto, Y., & Atsri, H., (2009). Impact of charcoal production on woody plant speciesin West Africa: A case study in Togo. *Sci. Res. Essay, 4*(8), 881–893.

La Ferla, B., Taplin, J., Ockwell, D., & Lovetti, J., (2001). Continental scale patterns of biodiversity: Can higher taxa accurately predict African plant distributions? *Bot. J. Linn. Soc., 138*, 225–235.

Laleye, P., & Entsua-Mensah, M., (2009). Freshwater fishes of western Africa. In: Smith, K. G., Diop, M. D., Niane, M., & Darwall, W. R. T., (eds.). *The Status and Distribution of Freshwater Biodiversity in Africa.* Gland, Switzerland and Cambridge, UK, IUCN.

Lévêque, C., Paugy, D., & Teugels, G. G., (2003). *Poissons d'eaux douces et saumâtres de l'Afrique de l'Ouest*, Tome 2, IRD Editions. Publications Scientifiques du Muséum, MRAC.

MERF/GIZ, (2016). *Inventaire Forestier National. ProREDD*, Lomé.

Ministère de l'Environnement et des Ressources Forestières (MERF), (2013). *Cartographie des zones humides du Togo.* Lomé.

Myers, G. S., (1949). *On the Usage of Anadromous, Catadromous and Allied Terms for Migratory Fishes.* Copeia.

Paugy, D., & Benech, V., (1989). Les poissons d'eau douce des bassins côtiers du Togo (Afrique de l'Ouest). *Revue d'Hydrobiologie Tropicale, 22*(4), 295–316.

Scholz, H., & Scholz, U., (1983). Flore descriptive des Cypéracées et Graminées du Togo. *Phanerog. Monogr., 15*, pp. 360.

Segniagbeto, G. H., Bour, R., Ohler, A., Dubois, A., Roedel M. O., Trape, J. F., Fretey, J., Petrozzi, F., Aïdam, A., & Luiselli, L. A., (2014a). Turtles and tortoises of Togo: Historical data, distribution, ecology and conservation. *Chelonian Conservation and Biol., 13*(2), 152–165.

Segniagbeto, G. H., Bowessidjaou, J. E., Dubois, A., & Ohler, A., (2007). Les Amphibiens du Togo: état actuel des connaissances. *Alytes, 24*(1–4), 72–90.

Segniagbeto, G. H., Okangny, D., & Luiselli, L., (2013). The Endemic *Conraua derooi* in immediate conservation need in Togo. *Frog Log, 108*, 23–24.

Segniagbeto, G. H., Trape, J. F., Afiademanyo, K., Roedel, M. O., Ohler, A., Dubois, A., et al., (2015a). Checklist of the lizards of Togo, (West Africa), with comments on systematics, distribution, ecology, and conservation. *Zoosystema, 37*(2), 381–402.

Segniagbeto, G. H., Trape, J. F., David, P., Ohler, A. M., Dubois, A., & Glitho, I. A., (2011). The snake fauna of Togo: Systematics, distribution, and biogeography, with remarks on selected taxonomic problems. *Zoosystema, 33*(3), 325–360.

Segniagbeto, G. H., Van Waerebeek, K., Bowessidjaou, E. J., Ketoh, G. K., Kpatcha K. T., Okoumassou, K., & Ahoedo, K., (2014b). An annotated checklist of the cetaceans of Togo, with a first specimen record of Antarctic minke whale *Balaenoptera bonaerensis* Burmeister 1867 in the Gulf of Guinea. *Integretive Zool., 9*, 1–13.

Segniagbeto, H. G., Okangny, D., & Fretey, J., (2015b). First observation of a loggerhead, *Caretta caretta*, in Togo, West Africa. *African Sea Turtles Newsletter, 3*, 15–16.

Segniagbeto, H. G., Okangny, D., Afiademagno, K., Kpotor, K., Denadi, D., Fretey, J., & Luiselli, L., (2016). Spatio-temporal patterns in occurrence and niche partitioning of marine turtles along the coast of Togo (West Africa). *Herpetozoa, 29*(1/2), 15–26.

Segniagbeto, H. G., Okangny, K., Denadi, D., Fretey, J., & Luiselli, L., (2017). Body size and stability in the spatio-temporal distribution pattern of sea turtles along the coasts of Togo: Implications for conservation and ecotourism. *Herpetozoa, 30*(1/2).

Tchamie, K. T. T., (1994). Enseignements à tirer de l'hostilite des populations locales à l'égard des aires protégées au Togo. *Unasylva, 176*, 22–27.

Van Waerebeek, K., Uwagbae, M., Segniagbeto, G., Bamy, I. L., & Ayissi, I., (2017). New Records of Atlantic Humpback Dolphin (*Sousa teuszii*) in Guinea, Nigeria, Cameroon and Togo underscore pressure from fisheries and marine bushmeat demand. *Revue d'écologie. 72*(2).

White, F., (1983). *The Vegetation of Africa*. UNESCO, Paris.

Wieringa, J. J., & Poorter, L., (2004). Biodiversity hotspots in West Africa, patterns and causes. In: Poorter, L., Bongers, F., N'Kouamé, F., Wawthorne, W. D., (eds.). *Biodiversity of West African Forests, An Ecological Atlas of Woody Plant Species*. CABI Publishing. pp. 61–72.

Woegan, Y. A., (2007). *Diversité des formations végétales ligneuses du parc national de Fazao-Malfacassa et de la reserve de faune d'Alédjo (Togo)*. Th. Doct. Sc. Nat., Univ. Lomé, Togo.

CHAPTER 13

Biodiversity in Zambia

STANFORD MUDENDA SIACHOONO

Department of Zoology and Aquatic Sciences, School of Natural Resources, Copperbelt University, Kitwe, Zambia, E-mail: stanford.siachoono@cbu.ac.zm

13.1 Introduction

The biodiversity of Zambia is best understood from an evolutionary perspective. Figure 13.1 gives the geological time scale as seen from the geologist's point of view. It was in fact made for the author by the University of Zambia Geology Department, for the purpose of understanding the geological time scale for Zambia in another past assignment.

Gaston and Spicer (2004)   □A
examples. However the standing global   □A
the Convention on Biodiversity (UNEP 2003), this is a land mark   □
which   biodiversity as, *"the variability among living organisms from all sources including terrestrial, marine and other ecosystems and ecological complexes of which they are part; this includes diversity within species, between species and of ecosystems"*. Gaston and Spicer (2004) however broaden their   □A
tute the building blocks of biodiversity, namely genetic diversity, organismal diversity and ecological diversity and in general how to measure biodiversity. They further amplify these elements as: (1) ecological diversity includes, biomes, bioregions, landscapes, ecosystems, habitats, niches, and populations; (2) genetic diversity includes populations, individuals, chromosomes, genes, and nucleotides; and (3) organismal diversity includes domains or kingdoms, phyla, families, genera, species, sub species, populations, and individuals.

Fossil and molecular evidence support the theory that organisms evolved from a single common ancestor. Biodiversity has increased since that time of inception about 3.5–4 billion years ago and the present time. The appearance of multi-cellular organisms 1.4 billion years ago seems to have increased the earth's biodiversity in respective areas (Figure 13.1). This change is witnessed in the Paleozoic Era with the sudden appearance of the   meta-zoans with hard parts. The   of organisms in respective time zones is also attributed to the invasion of new habitats, or they followed major extinctions. For example Gaston and Spicer (2004) note that among land plants, the dominance of primitive vascular plants gave way to pterido-phytes, the ferns, and lycopsids, the club mosses, gave way to the predomi-nance of the gymnosperms that were   overtaken by angiosperms. The vertebrate story is similar, the early amphibians and reptiles stocks gave way to a number of successful amphibians and reptiles, later birds and mammals.

From the Geological Time Scale in Figure 13.1, it is clear that some spe-cies have been lost along the way. Dinosaurs are a good example and there are fossil records in Zambia to support their existence in the Jurassic Period. Extinction is a common occurrence in the rise of biodiversity globally. The level of biodiversity will rise if there is an increase in the level of speciation than the rate of extinction. There is only stability if the rate of extinction is the same as that of speciation. The geological time scale shows that at each Period there is a loss of biodiversity mainly due to climatic shifts (Gas-ton and Spicer, 2004). They suggest   A has become extinct in large taxonomic groups that include (1) canvassing experts, who have studied particular groups of species over long periods and have gained an understanding of the numbers that are unknown to sci-ence; (2) extrapolating into the future the growth in cumulative numbers of the taxonomically described species through time by species description; (3) the proportion of undescribed species should involve estimating over-all numbers of species from the ratio of previously unknown to previously known species in large samples of specimens and then extrapolating from the overall numbers of known species; (4) well studied areas provide overall numbers of species globally and have been estimated by extrapolating from those few areas for which numbers of species are reasonably well known; (5) well studied groups approach involves estimating overall numbers of species based on global numbers in well-known groups and estimates of the ratio of the number of species to others in those few regions where the latter are well known.

Zambia is a landlocked country that lies at the end of the Great Rift Val-ley between latitudes 8° and 18° S and longitudes 22° and 33° E. It shares the political borders with eight neighbors, namely Angola, Democratic Republic of Congo, Tanzania, Malawi, Mozambique, Botswana, Namibia, and Zimbabwe. The political boundaries have a number of shared biodiver-sity resources because of the common habitats. The country expands with an area of 752,618 km^2 (Dowsett et al., 2008). The country has an undulating land features that range from 900 to 1,400 m above sea level. It has neverthe-less some high peaks notably the   Mountains and the Nyika plateau which have peaks of more than 2,000 m above sea level. Rainfall averages

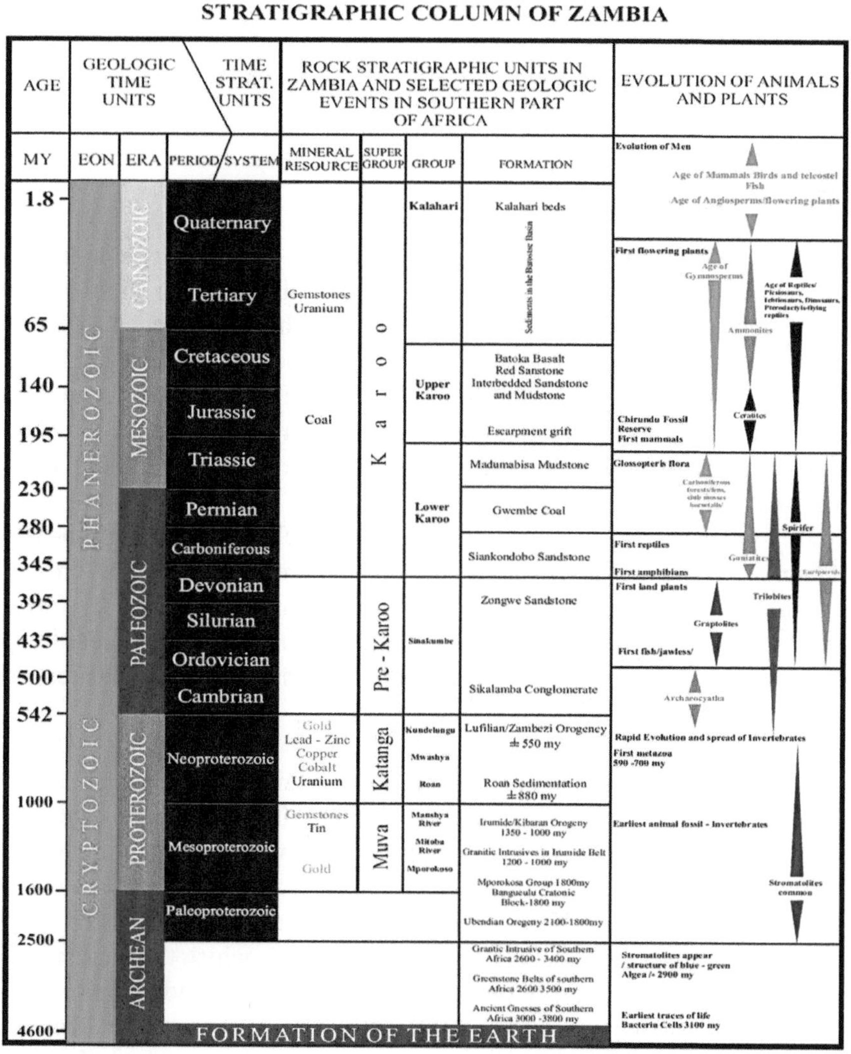

STRATIGRAPHIC COLUMN OF ZAMBIA

AGE	GEOLOGIC TIME UNITS			TIME STRAT. UNITS	ROCK STRATIGRAPHIC UNITS IN ZAMBIA AND SELECTED GEOLOGIC EVENTS IN SOUTHERN PART OF AFRICA				EVOLUTION OF ANIMALS AND PLANTS
MY	EON	ERA	PERIOD/SYSTEM		MINERAL RESOURCE	SUPER GROUP	GROUP	FORMATION	
1.8		CAINOZOIC	Quaternary				Kalahari	Kalahari beds	Evolution of Men / Age of Mammals Birds and telcostel Fish / Age of Angiosperms/flowering plants
65	PHANEROZOIC		Tertiary		Gemstones Uranium				First flowering plants / Age of Gymnosperms / Age of Reptiles/ Pleistmans, Ichthomars, Dinosaurs, Pterodactyls/dying reptiles / Ammonites
140		MESOZOIC	Cretaceous	Karoo		Roo	Upper Karoo	Batoka Basalt Red Sanstone Interbedded Sandstone and Mudstone	Ammonites
195			Jurassic		Coal			Escarpment grift	Chirundu Fossil Reserve First mammals / Ceratites
230			Triassic					Madumabisa Mudstone	Glossopteris flora / Carboniferous forests/fens, club mosses horsetails
280			Permian				Lower Karoo	Gwembe Coal	Spirifer
345		PALEOZOIC	Carboniferous					Siankondobo Sandstone	First reptiles / Gontaites First amphibians / Euryperids
395			Devonian			Sinakumbe		Zongwe Sandstone	First land plants / Trilobites
435			Silurian	Pre - Karoo					Graptolites
500			Ordovician					Sikalamba Conglomerate	First fish/jawless/
542			Cambrian						Archaeocyatha
1000	CRYPTOZOIC	PROTEROZOIC	Neoproterozoic	Katanga	Gold Lead - Zinc Copper Cobalt Uranium	Kundelungu	Kundelungu Mwashya Roan	Lufilian/Zambezi Orogency ± 550 my / Roan Sedimentation ± 880 my	Rapid Evolution and spread of Invertebrates / First metazoa 590 -700 my / Earliest animal fossil - Invertebrates
1600			Mesoproterozoic	Muva	Gemstones Tin / Gold	Manshya River Mitoba River Mporokoso		Irumide/Kibaran Orogeny 1350 - 1000 my / Granitic Intrusives in Irumide Belt 1200 - 1000 my / Mporokoso Group 1800my Bangaeulu Cratonic Block-1800 my / Ubendian Orogeny 2100-1800my	Stromatolites common
2500			Paleoproterozoic						Stromatolites appear / structure of blue - green Algea /> 2900 my
4600		ARCHEAN						Granitic Intrusive of Southern Africa 2600 - 3400 my / Greenstone Belts of southern Africa 2600 3500 my / Ancient Gnesses of Southern Africa 3000 -3800 my	Earliest traces of life Bacteria Cells 3100 my

FORMATION OF THE EARTH

Figure 13.1 Zambian geological timescale.

between 800 to 1,400 mm per annum. The country has notably three seasons namely, the cool dry period lasting from May to August/September, followed by the hot and dry period from September to late October and there after the hot and wet period up to March/April.

The biodiversity distribution is greatly ☐ A

The resulting precipitation drains into four major rivers, namely, the Kafue,

Luangwa, Chambeshi and the Zambezi. The Zambezi basin creates the largest catchment zone that is shared by six of neighbors in form of drainage. Three of the major rivers have wetlands habitats that are of great importance to the biodiversity of Zambia.

13.2 Habitats

Zambia falls into the woodlands and savannas. The phyto regions include the Zambezian region transitional zone between the Kalahari Highveld and the Zambezian region (Chidumayo et al., 2011). The country does not have a coastal habitat. The country has however a large portion of the water bodies in southern Africa that is demonstrated by the relief maps showing the water resources in the sub-region. It is comprised of rivers, natural lakes and dams. A number of fish species are found in these water bodies in addition to the host of invertebrates and aquatic plants.

Dowsett et al. (2008) have discussed in detail eleven types and sub-types of habitats in Zambia as they affect the ecology and distribution of bird species in Zambia. These are the Zambezian woodland, the deciduous forests and thickets, the evergreen forests, evergreen scrublands, grasslands, open waters swamps, rivers, cliffs and boulders, and man-made habitats that are as result of settlements or agricultural activities.

They   ☐A
grassland, grassland, thicket and the shrub land for the ease of the reader as
working   for their habitat descriptions. The stratigraphic geological time scale gives the general distribution of the various and common soils in Zambia.

The Zambezian woodland is the major woodland type in Zambia and is divided into three types, namely Miombo woodlands which the predominant terrestrial vegetation cover in the country. The Mopane woodland, the undifferentiated woodland or wooded grassland. The Miombo woodland cover is widespread in the Zambezian region that ranges from Angola to Tanzania and accounts for close to 80% of the woodland cover in Zambia. This woodland type is dominated mainly by tree species of *Brachystegia, Isoberlinia* and *Julbernadia*. This habitat is common for agricultural cultivation especially for grain crops such as maize.

The deciduous forests and thicket is the second vegetation type and is divided into the *Baikiaea* deciduous forest or and other dry thicket or forests. The *Baikiaea* forests are dominant in the low rainfall areas of the south western Zambia that is predominantly Kalahari sands. This forest type is dominated by the *Baikiaea plurijuga* tree species hence the term given to the habitat type. Dry thickets forests occur mainly in the Karoo sands and are interspaced with termite mounds.

The evergreen forests may be divided into dry, moist and riparian forests. The difference between the dry and moist evergreen forests is not very clear and distinct as they seem to overlap. The riparian forests are however the dominant cover and consist of mainly narrow strips of trees that are ever-green growing on banks of rivers and water courses.

The grasslands habitats may be divided into three habitats, namely the montane, watershed,   plains, and dambos. These are predominantly found in the Kalahari sands especially in the western and the north-western part of the country.

Swamps are a common habitat mainly along river channels and form a   A □
swamps that are home to unique species such as the Black and the Kafue
□   □   □

Open waters have been mentioned in the opening part of this part as a major habitat together with rivers. Zambia has both man-made and natural open water. The open natural water bodies are to the north and are favored by higher rainfall patterns. They also form the Congo basin water systems and the drainage system.

13.3 Fungi

Fungi are generally not grouped together with plants because it does not have the chlorophyll that it can use to make food using sunlight. Fungi are found in diverse habitats and survive by breaking down organic materials from dead or even living organism. This makes it be treated as good or bad depending on the activity that it affects or causes. The bad side is that the spoilage of goods such as food, textiles, and timber mainly by molds and yeasts. They also cause known plant and animal diseases.

The good side of fungi is in the soil fungi that are important in breaking down the cellulose and lignin in plant remains and is responsible for main-taining soil fertility. Some fungi are used in the food industry such as bread and cheese making, brewing of alcoholic beverages and the medical   for drugs such as penicillin.

Evolutionary, early fossil records of primitive fungi date back to the pre-Cambrian period some 1 billion years ago (Harkonen et al., 2015). Other authors argue that during the Paleozoic Era much of the fungi was aquatic and it is presumed that fungi colonized land during the Cambrian period long before the plants emerged (Brundrett, 2002).

Most accounts of fungi species are on the visible fruiting bodies that represent different groups of mainly higher fungi. These spore bearing fruiting bodies of fungi are considered higher mushrooms as opposed to the lower fungi that may be only unicellular except for yeast. Pearce (1981) gives an introductory account of the edible of wild mushrooms in Zambia and how to use them. Harkonen et al. (2015) further   this account by giving nutrition and ecological accounts of the various species of Zambian mushrooms. These two accounts however miss the other   ☐A
        ☐A
al., 2016).

Harkonen et al. (2015) classify the Zambian fungi into three major groups, namely, the Glomeromycota, Ascomycota and Basidiomycota. The   group is comprised of about 200 species that are important in ecological processes and live as mycelium in the soil and have no visible fruit bodies. The Ascomycota are the largest group of Zambian fungi and number about 64,000 species while the Basidiomycota is one group with about 30,000 species. They develop spores that arise from cells called basidia hence their name.

13.3.1 Lichens

Lichens seem to be the evolutionary bridge between plants and fungi because it is an association of fungi and algae. The two are intertwined to form a single thallus. The fungal part being the mycobiont while the algae part is the phycobiont. Lichens are common in Zambia especially on tree trunks where they grow. Studies on lichens are also very poor.

13.4 Plants

Plants emerge in the late Paleozoic Era during the Permian period (Figure 13.2) as horse tails, club mosses and carboniferous forests in the lower Karoo formation and generally referred to as the *Glossopteris flora*. There is very little work done on lower plants in Zambia.

Generally plants globally and in Zambia may be divided into non-vascular plants, the bryophytes mainly the mosses, liverworts, and hornworts. They have evolved from algae some 400 million years and invaded land. The vascular plants are composed of gymnosperms and angiosperms. That is also a host of both lower and higher vascular plants. The lower vascular

Figure 13.2 Truffles from Zambia.

plants or the pteridophytes, the ferns are good example use their rhizome as their stem. A number of such plants are abundant with some being a source of food in their wild form. Phiri and Ochyra (1988) gave a preliminary account of mosses of Zambia.

Gymnosperms appear   after the lower plants. In Zambia their numbers as forests have diminished, and only the fossil remains are seen at the Chirundu fossil forests (Figure 13.1). This is the only evidence that they once formed a large part of the landscape in the country in the course of evolution and land formation.

13.4.1   Lower Plants

Lower plants include water based pyhto-planktons and other single celled plants that are abundant in fresh water ponds and dams. They are the principal food sources for the fish species that live there. Algae, especially *Spirogyra* is abundant in all stationary and moving water bodies. They appear as slime aggregates of filaments made of cell structures with chlorophyll. Mosses and ferns grow mainly on land in moist places in wooded areas near water bodies. In addition to these there is a host of herbaceous macrophytes that occur in abundance in most streams that are also poorly studied. Current there is some academic information on them because of the water purification process that they contribute. They are also poorly known.

13.4.2 Higher Plants

Higher plants are well studied especially species of economic importance and clusters of trees that mark areas of interest for global warming, water shed management and areas that provide for agricultural crop land. Economic tree species include the hard wood timber plants that are harvested for this exploitation. Storrs (1995) gives an account of the various hard wood and soft wood tree species in Zambia.

There is also a host of lower vascular plants. The lower vascular plants or the pterodophytes, the ferns are good example, use their rhizome as their stem. A number such plants are abundant with some being a source of food in their wild form as tubers.

Gymnosperms appear   after the lower plants. In Zambia their numbers as forests have diminished, and only the fossil remains are seen at the Chirundu fossil forests (Figure 13.1). This is the only evidence that they once formed a large part of the landscape in the country in the course of evolution and land formation. Phiri (1996) and Fanshawe (1973) have documented the taxonomic accounts of the plants in Zambia. Phiri (2005) divides the plant kingdom in Zambia into four groups, namely, pteridophytes, gymnosperms, angiosperms as monocotyledons and dicotyledons. Gymnosperms, apart from the fossil forest remains in Chirundu, Phiri (2005) lists only two families with two species existing in the country, namely the *Podocarpus latifolius*, and *Encephalartos schmitzii*. Both of these are found almost in the central Zambia in Mpika area. The largest group is that of the angiosperms, the   plants that have 189 families and more than 6,000 species, the pterodophytes have only 30 families and 166 species.

Plants are a vital component of the biodiversity spectrum in Zambia. Fanshawe (1973), Phiri (1996, 2005) have documented the taxonomic accounts of the vascular plants in Zambia. Storrs (1995) also gives an account of Zambian Trees and their distribution in the various habitats. Palgrave (1988) gives a more comprehensive account of the trees of the southern Africa sub-region to an academic audience.

The Zambezian region in which the Zambia falls is characterized by a number of tree species that are the key in the habitat names. This region is characterized by the genera of tree species that include, *Acacia, Brachystegia, Colosphospermum, Combretum, Terminalia, Hyparrhenia,* and *Richardsiella.* The vegetation physiognomy therefore appears as forests, woodlands, thickets and grasslands. These units of vegetation that arise out of this are; Miombo forests, Kalahari woodlands, Mopane woodland, *Acacia-Combretum* woodland, thickets, and grasslands.

The plateau region is mostly covered by Miombo tree species that are dominated by *Brachystegia, Isoberlinia* and *Julbernadia*. Other dominant tree species of the Miombo include are *Anisophyllea* sp., *Pornifera* sp., *Erythrophleum africanum, Marquesia macroura, Parinari curatellifolia,* and *Percopsis angolensis*. The wet Miombo woodlands variants occur in the high rainfall areas and include, *Brachystegia floribunda, B. puberula, B. wangermeana, Cryptosepalum exfoliatum, Isoberlinia angolensis, Julbernadia paniculata, Magnistipula butayei, Maranthes floribunda,* and *Parinari excelsa*. These taxa are usually absent in the dry Miombo that are dominant in the southern part of the country. Notably, in the escarpments the Miombo *Brachystegia* tree species include *B. boehmii, B. bussei, B. manga, B. microphylla,* and *B. taxfolia*.

The Kalahari woodlands are mainly supported by Kalahari sands ground structures. The tree species are mainly those of hard woods, namely, *Burkea africana,* and *Giubourtia coleosperma* in association with elements of the miombo such as *Brachystegia longifolia, B.speciformis, B. bakeriana, Julbernardia globiflora,* and *J. paniculata*.

Grasslands and   plains grasslands are common in Zambia. Grass species are mainly those of *Hyparrhenia* sp. and *Sectaria* sp.

13.5 Animals

13.5.1 Invertebrates

Invertebrates' fauna assemblage in Zambia has a diverse background as they are sub phyla. These life forms are the earliest to rise in form of the single celled Protozoa. The evolutionary trend is that these appear as early life forms so all invertebrates and modern vertebrates trace their evolutionary tree to the early protozoans. Invertebrates appear in the Cambrian period during the Paleozoic Era some 590 million years ago. The information on invertebrates in general is very poor. Only that which is used for academic work and information on problem invertebrates that cause problems for mankind is available.

13.5.1.1 Bacteria and the Early Prokaryotic Cells

The early life forms are said to have taken the form of single cells that existed as prokaryotic cells. This included Archaea and Bacteria. A number of viruses also exist as part of the biodiversity. Most of these are associated with diseases, such as the Acquired Immune Deficiency Syndrome (AIDS) in humans. The common cold, and other viruses that attack crop plants.

The eukaryote cell evolved from the prokaryote cell some 1.5–2 billion years ago. This was the rise of multicellular organisms leading   to invertebrates.

The Protozoa, are eukaryotic and studies on them have been limited to the species that cause diseases both to humans and other animals especially livestock. In Zambia these have been recorded as malaria species of the genus *Plasmodium*, namely, *P. ovale, P. malaria, P. falciparum,* and *P. vivax.* The common one being *P. falciparum* that is responsible for most of the malaria attacks on human beings through their established cycle with the mosquito insect.

The *Trypanosome* protozoans have a life cycle that is associated with the tsetse   in Zambia. They cause a blood disease called *nagana* in livestock and sleeping sickness in human beings, but strangely they do not affect wildlife that they live association with. This seems to be an evolutionary trend. The *schistosomes* have a life cycle that is associated with a snail as an intermediate host and they affect human beings with a condition known as *schistosomiasis.*

Zambia being land locked does not have the predominant marine invertebrates such as jelly   sponges and others that evolved earlier as life forms. The common phyla in Zambia include the Platyhelminthes, Annelids, Nematodes, Arthropods and Molluscs. However attention is paid mainly to the disease causing or intermediate hosts to organisms that poses a problem to either human beings or livestock. The account of the invertebrates is
                     ☐   ☐   ☐
The early phylum in the Zambian invertebrates is the Platyhelminthes, the   Ⱥ
occupies a wide range of habitats. The common ones in Zambia are the parasitic species, namely, the tape worm and the   They are a major concern in livestock farming and they may also infect humans if there is carelessness in the preparation of meat products especially pork and beef.

The Annelids phylum is composed of worms that are ringed. These include earth worms and leeches. Earthworms are found in moist places and are key to breaking down organic matter and thereby improving the soil structures. Leeches on the other hand are blood sucking worms that dwell mainly in fresh water bodies. The Nematodes are the round worms that also occupy a wide range of habitats. Some are free living but a majority of them are parasitic and are found in guts of animals especially vertebrates. They also affect some plants especially the crops.

The Molluscs group is found in both marine and fresh water environments and several classes. In Zambia the two classes stand out mainly the Gastropods and Bivalves. Most are free living and others are intermediate hosts for diseases such as *schistosomiasis*.

The Arthropod group is probably the largest group of the invertebrates. It is divided into   A mainly marine forms. Other are the Arachnids, the spiders, mites and scorpions, the Myriapods comprising of the millipedes, and centipedes, the Crustaceans, these are mainly aquatic and include, the crabs. The Hexapods that include the class insect are the most successful invertebrates. In Zambia attention is paid to the pests and disease carrying species.

13.5.2 *The Fish*

Romer (1971) gives an account of the global evolutionary rise of all the major vertebrates in the world. In addition to this their physiological attributes that makes them unique organisms. The fish group is said to be the evolutionary ancestral stock for all higher vertebrates. Fish dominate both marine and fresh water and making them one of the most diverse groups of vertebrates. The Devonian period may be called the age of fish because of the array fossils found in this period. Fish may be grouped into four classes, namely *Agnatha, Placodermi, Chondrichythyes,* and *Osteichthyes* Romer (1971). The Osteichthyes are the higher bony fishes. The other primitive fish groups, namely the Agnatha, Placoderms and the Chondrichythyes have remained marine forms. Zambia has a number of fresh water bodies that include rivers, lakes and man-made water bodies. All these water bodies have unique fish species. Utsugi and Mazingaliwa (2002) give a checklist account and distribution records of Zambian fish species. This account is not exhaustive but gives one an overview of what is obtaining in the different habitats and water bodies.

There are some orders that have well developed abundance or breed successfully to create the   industry. Lake Tanganyika has commercial   because of the Order Clupeiformes with the family Clupeidae that has the sardine group of   A this the lake has   A   *Lates stappersii* species that are abundant and form the backbone of the   A the lake not to mention the Family *Cichlidae* that provide for the aquarium trade source. In addition to this the family has a number of breams that are

the common species one   in the market for sale. The breams as they are
called are many but the common one are the ones belonging to the genus
Tilapia. The Order Siluriformes, the   group is not only diverse but it is
found in a number of water bodies around the country. It is also a common

☐   ☐ ☐   ☐ ☐   ☐ ☐

13.5.3 *Amphibians and Reptiles*

Amphibians and Reptiles are usually grouped together in the study called
Herpetology. Amphibians were the first to invade land. The turning point
was the development of legs and the shift from the use of gills to lungs for
breathing. The Class Amphibia is the basic group of land dwelling vertebrates with three Orders, namely Anura for toads and frogs, the Urodella for
newts and salamanders and the Apoda for the worm like burrowers. Channing (2001) gives an account of a broad overview of amphibians in the Central and Southern Africa. Part of the 200 amphibian species known in the
sub region that includes Angola, Botswana, Lesotho, Malawi, Mozambique,
Namibia, South Africa, Swaziland, Zambia and Zimbabwe, the Anuras
not only occupy a variety of habitats but they are the most abundant. Fossil records of the late Paleozoic Era linked to the Carboniferous and early
Permian suggest that there were numerous and varied amphibians of a more
primitive nature. They flourished in the late Paleozoic Era (Romer, 1971).

Reptiles descended from the ancient amphibians. They solved the reproductive system. Amphibians as the name suggests need water to lay their eggs
and hatch them until they develop into mature adults. Reptiles overcame this
dependence on water by developing the amniotic egg. The leathery shelled
egg is laid on land unlike the amphibian egg that is laid in water. Modern
reptiles consist of crocodiles, lizards, snakes and the tortoises. These are
abundant in the tropics because they are cold-blooded. They are remnants of
great array of reptiles that radiated in the late Paleozoic Era into a variety of
forms that made the Mesozoic Era to be known as the age of reptiles.

Lizards and snakes, Order Squamata are the most successful group of
reptiles. Snakes are said to be derived from lizards with reduced limbs.
A number of tropical forms have developed poison glands with a variety
of poisons. Broadley (1971) gives a checklist account and distribution of
amphibians and reptiles in Zambia. This is the only comprehensive work
done on herpetology   ☐A
give an account of the snakes of Zambia as an atlas and a   guide that
is very comprehensive. Simbotwe and Garber (1979), Simbotwe and Pat-

terson (1983) and Simbotwe (1980) give ecological and reproductive study accounts of mainly lizard species in Zambia.

13.5.4 Birds

Birds or Aves as they are called sometimes are often termed as glorified reptiles. Their ancestral bird, the *archaeopteryx* from the late Jurassic fossil deposits is often considered as a lucky find, as bird bones are very brittle hence the fossil finds are rare. The *archaeopteryx* had teeth and a long tail. Teeth in birds have been lost since the Mesozoic period (Romer, 1971). Birds have, evolutionary, developed the ability to fly except the few flightless species, using modified pectoral limbs and also a bipedal adaptation. Flying is the highest modification and reinforced with ability to maintain a constant body temperature. Birds have also evolved an improved blood circulation system. Benson et al. (1971) give the early accounts of the birds of Zambia in as far as the classification and the distributions of bird in Zambia. Dowsett et al. (2008) gave a detailed and updated account of the birds of Zambia checklist, habitat preferences, breeding cycles and host of knowledge on birds of Zambia. The checklist accounts for 750 species of birds in Zambia. The Zambian bishop is the only endemic species known to occur mainly around central Zambia.

There other accounts of birds of southern Africa (Newman, 1991; MacLean, 1985; Sinclair and Davidson, 2006) that add value to the knowledge on bird of Zambia and are also good   guides to a bird watcher who needs a backup on identifying birds in the   A☐ account of the habitat preferences by identifying Important Bird Areas in Zambia. These have been   A ☐   - tected areas status. In both accounts authors acknowledge the bird behavior is such that some birds are migrants. They prefer to migrate to Europe during the summer months and   back here when winter comes in Europe. This opportunity has often been exploited by bird lovers who would like to plot the migration paths for these birds by ringing them when chance allows.

The general   ⌐A

in 20 Orders. Zambia does not have records of natural occurring   A birds, the Paleognathous or the ratites. The Ostrich is the closest in Africa in this group. Its distribution however skips Zambia from East Africa and is found in the southern region Namibia, Botswana and South Africa. The only species found in Zambia are those found in game ranches.

13.5.5 Mammals

Mammals are the highest evolved vertebrates (Romer, 1971). They are called mammals because of their mammary glands that they use for feeding their off springs. Romer (1971) classifies mammals into three groups, namely the monotremes (*prototheria*), marsupials (*metatheria)* and the placental (*eutheria)* or the true mammals. The mammal fauna in Zambia is primarily that of the *eutheria* group or the placental mammals. Ansell (1978) gives a checklist and systematic classification of more than 237 species of Zambian mammal fauna, in 13 orders, their habitats and distribution. Skinner and Smithers (1990) account for the comprehensive systematic and classification of mammals in the Southern African sub-region.

Ansell (1978) starts his systematic   □A group that radiates from the ancestral *eutheria* group (Romer 1971) and forms the founding stock of the *eutheria* group. This group is composed of mainly insectivorous shrews from which others trace their evolutionary origins. These include the bats, primates, lagomorphs, rodents, carnivores, primitive ungulates pangolins, and aardvarks. The ungulates (hooved animals) are further divided into sub ungulates that are represented by the elephants and ungulates. The ungulates are composed of the artiodactyls (even-toed) and the Perrisodactyls (odd-toed) that are the major groups of both medium and large herbivores that are successful in Zambia.

Tourism in Zambia is based on large mammals that are often used for photographic and sport hunting. The BIG FIVE, the lion, elephant, buffalo, leopard, and rhino are often used as the   □A tion (Figure 13.3 and 13.4). Rodents are equally very successful in Zambia and are often responsible for the post-harvest invasion of grain harvests on farmers.

13.6 Management and Conservation of Biodiversity in Zambia

The management and conservation of biodiversity in Zambia is achieved both at land use level and at legislative and policy levels. At land use level the country has set aside 20 National Parks (6% of the total land area) as protected areas. This implies that no other land use can take place in these areas apart from the protection of the biodiversity that is there. National parks in return offer tourism as the only industry that operates in these areas. It is non extractive and aims at offering recreational facilities in the respective areas. These areas are managed at an institutional level under the National Parks

Figure 13.3 Buffalo, a member of the BIG FIVE.

Figure 13.4 The Lion a member of the BIG FIVE.

and wildlife department that provided for competent leadership and management services on behalf of the Zambian Government and the population at large.

Apart from land use, other biodiversity management strategies in Zambia include the respective legal and policy frameworks that have been put in place to protect and manage such resources. Wildlife, for example has the protection by land use and also by legislation and policy instruments. Others like   A

nated capture   areas. Some biodiversity is protected in this way by

☐ ☐   ☐   ☐   ☐

13.6.1 Global Connections

Biodiversity management in Zambia would not probably receive the attention it does at the moment if there were no global influences. In Zambia these have a long background. In 1980 (IUCN), for example, the World Conservation Union produced the *World Conservation Strategy* document that had influence on the Zambian government who followed suit, like many other countries in the sub-region, and produced the *National Conservation Strategy for Zambia* (1985) with help from the World Conservation Union. This document acknowledged the conservation needs for the biodiversity in the country and created a platform for the added legislative frameworks. The environmental impact assessment became a reality for any investor who aspired to disturb any habitat.

The management and conservation has continued to be strengthened with being signatories to protocols such as the Convention on Biodiversity (CBD) (2003), Convention on Trade in Endangered Species (CITES) (1975), the Ramsar convention on conservation of wetlands and partnering institutions like the World Wide Fund for Nature (WWF) in projects that promote sustainable use of natural resources especially in wetlands.

13.7 Conclusion

The account of biodiversity would be very bulk if all the details of individual species were given in this chapter. However one may be able to follow the groups of special interests given the introduction to the various groups in this chapter. This account also holds a myriad of subjects implying that there are different study interests ranging from evolutionary biology, to respective disciplines of both zoology and botany. Respective scholars will therefore start their subject interest with this introduction of the biodiversity of Zambia brief that sets the landscape for the understanding of the subject matter.

Acknowledgments

My Colleague Gillian Kabwe (PhD) was kind enough to read the manuscript and made a lot of improvements to its flow. I'm truly grateful for her help. Professor Tembo from the School of Mines at the University of Zambia provided the Geological time scale chart which has been helpful in the explaining the evolutionary pathway for the Zambian Biodiversity. Chris MacBride provided the pictures of lions in their natural environment from the Bush camp in the Kafue National Park. Many thanks for the permission to use the pictures for this publication.

Keywords

- animals
- fungi
- habitats
- plants

References

Ansell, W. F. H., (1978). *Mammals of Zambia*. National Parks and Wildlife Services. Chilanga, Zambia.

Benson, C. W., Brooke, R. K., Dowsett, R. J., & Irwin, M. P. S., (1971). *The Birds of Zambia*. London, Collins.

Broadley, D. G., (1971). *A Checklist of Amphibians and Reptiles of Zambia*. PUKU 6. National Parks and Wildlife Services. Chilanga, Zambia.

Broadley, D. G., Doria, C. T., & Wigger, J., (2003). *Snakes of Zambia, An Atlas and field Guide*. Chimaira Publishers.

Brundrett, M. C., (2002). Coevolution of roots and mycorrhizae of land plants. *New Phytologist, 154*(2), 275–300.

Channing, A., (2001). *Amphibians of Central and South Africa*, Constock Publishing Associates.

Chidumayo, E., Okali, D., Kowero, G., & Lawrence, M., (2011). *Climate Change and African Forests and Wildlife Resources*, The African Forest Forum, Nairobi (eds.), Kenya.

Convention on Biodiversity (CBD) Handbook, (2003). UNEP.

Convention on Trade in Endangered Species (CITES), (1975). Washington DC, Nairobi, Kenya.

Dowsett, R. J., Aspinal, D. R., & Dowsett-Lemaire, F., (2008). *The Birds of Zambia, An Atlas and Handbook,* Turaco press and Aves a.s.b.l., Liege, Belgium.

Fanshawe, D. B., (1973). *Checklist of the Woody Plants of Zambia Showing Their Distribution*. Forest Research Bulletin no 22. Government Printers. Lusaka.

Gaston, K. J., & Spicer, J. I., (2004). *Biodiversity: An Introduction, 2nd edn*. Blackwell Publishers, Malden, USA.

Government of the Republic of Zambia / International Union for the Conservation of Nature and Natural Resources (IUCN) (1985). *The National Conservation Strategy for Zambia.* Gland, Switzerland.

Harkonen, M., Niemela, T., Mbindo, K., Kotiranta, H., & Piearce, G., (2015). *Zambian Mushrooms and Mycology.* Finish Museum of Natural History, Helsinki, Finland.

International Union for the Conservation of Nature and Natural Resources (IUCN), (1980). *World Conservation Strategy.* IUCN, WWF and UNEP, Gland, Switzerland.

Leonard, P., (2005). *Important Bird Areas in Zambia.* Bird life International, Cambridge, U.K.

MacLean, G. L., (1985). *Robert's Birds of Southern Africa, 4th edn.* CTP Book Publishers, Isando, South Africa.

Newman, (1991). *Newman's Birds of Southern Africa.* Southern Book Publishers, Gauteng, South Africa.

Palgrave, K. C., (1988). *Trees of Southern Africa, 3rd edn.* Struik Publishers, Cape Town, South Africa.

Pearce, D. G., (1981). *An Introduction to Zambia's Wild Mushrooms and How to Use Them.* Forest Department of Zambia.

Phiri, P. S., & Ochyra, R., (1988). A preliminary account of mosses of Zambia. *J. Bryol., 15*, 177–197.

Phiri, P. S., (1996). The floristic status of grasses of the south Luangwa National Park and the Lupande area. In: *The Biodiversity of African Plants. Proceedings of the XIVth AETFAT Congress*, Wageningan. Kluwer Academic Publishers.

Phiri, P. S., (2005). *A Checklist of Zambian Vascular Plants.* Southern African Botanical Diversity Network Report No32 SABONET, Pretoria.

Romer, A. S., (1971). *The Vertebrate Body, 4th edn.* W.B. Saunders and Company Publishers, Philadelphia, USA.

Siachoono, S. M., Shakachite, O., Muyenga, A., & Bwalya, J., (2016). Underground treasure: A preliminary inquiry into the ecology and distribution of Zambian truffles. *Intern. J. Biol., 8*(2), 1–8.

Simbotwe, M. P., & Garber, S. D., (1979). Feeding habits of Lizards in the genera, *Mabuya, Agama, Ichnotropis*, and *Lygodactylus* in Zambia. *Kansas Academy of Sciences, 82*(1), 55–59.

Simbotwe, M. P., & Patterson, J. W., (1983). *Ecological Notes and Provisional Checklist of Amphibians and Reptiles Collected From Lochnivar National Park Zambia.* Black Lechwe Scientific Publications.

Simbotwe, M. P., (1980). Reproductive biology of the Skinks, *Mabuya striata* and *Mabuya quinquetaeniata. Herpatologica, 36*(1), 99–104.

Sinclair, I., & Davidson, I., (2006). *Southern African Birds. A Photographic Guide.* Struik Publishers, Cape Town, South Africa.

Skinner, J. P., & Smithers, R. H. N., (1990). *Mammals of the Southern African Sub-Region.* University of Pretoria.

Storrs, A. E. G., (1995). *Know Your Trees. Some Common Trees Found in Zambia.* Forest Department. Zambia.

Utusugi, K., & Mazangaliwa, K., (2002). *Field Guide to Zambian Fishes and Planktons and Aquaculture.* Fisheries Department: Zambia.

Index

A

Abelmoschus esculentus, 193, 218
Abies maroccana Trabut, 141
Abiotic
  factors, 176, 177
  parameters, 146, 176
Abrus melanospermus Hassk, 26
Abutilon
  macropodum, 184
  pannosum, 41
Acacia
  commiphora, 298
  ehrenbergiana, 12, 27
  erythrocalyx, 9
  flava, 121
  gerrardii, 5, 28
  hockii De Wild, 26
  laeta, 10
  macrostachya, 23, 180
  nilotica, 44, 182, 279, 282
  raddiana, 13, 52, 55, 142
  senegal, 180, 190
  senegali, 121
  seyal, 7, 10, 13–15, 180, 278, 282
  sieberiana, 13, 365
  tortilis, 9, 10, 14, 42, 46, 47, 53, 278
Acalypha senegalensis, 184
Acanthodactylus
  boskianus, 45
  scutellatus, 40, 42, 51
Achillea santolina, 39, 40, 54
Achyranthes argentea, 4
Acraea punctimarginea, 319
Adansonia digitata, 23, 86, 178, 185, 186,
  193, 278, 282, 344, 350, 354, 365
Adenorhinos, 320
Adonis dentata, 39
Aerva
  javanica, 41, 50
  lanata, 48
Aethiomastac embeluspraensis, 89
Aframomum inversiflorum, 65
African region, 202, 231, 257, 368, 370

Afrixalus uluguruensis, 320
Afro-montane forest, 298
Afrotrilepis jaegeri, 217, 230
Afzelia africana, 4, 23, 27, 86, 185, 191,
  364–367
Agama spinosa, 48
Agapornis
  lilianae, 319
  pullaria, 378
Agathophora alopecuroides, 41
Agave, 57
Agricultural
  activities, 123, 124, 126, 228, 263, 322,
    390
  biodiversity, 150, 216, 231, 236, 237
  ecosystems/agrosystems, 227, 139
  expansion, 104, 237, 249, 257, 284
  land, 4, 58, 205, 207, 213, 240, 303–345
  Plant Genetic Resources and Research
    Centre, 280
  practices, 167, 175, 354
  production, 156, 262, 266, 277, 304
  Research Centre (ARC), 101
  sector, 153, 156, 175, 277, 290, 300
  security and productivity, 290
Agrobiodiversity, 134, 235, 263
Agrochemicals, 238, 325
Agro-climatic zones, 300
Agro-ecological zones, 82, 83, 141, 290
Agroforestry, 182, 183, 197, 335, 344, 350
Agroforestry systems, 290
Agro-pastoralists, 281
Aizoaceae, 116, 252, 255
Aizoon canariense, 50
Alafia
  multiflora, 221
  scandens, 2
Albizia ferruginea, 27, 28, 364
Albula vulpes, 234
Alcedo atthis, 53
Alcelaphus buselaphus major, 194
Alchornea cordifolia, 181, 221
Alectra basserei, 184
Alextesb longipinnus, 233

Printed and bound by CPI Group (UK) Ltd, Croydon, CR0 4YY

23/10/2024

01777704-0013